Sampling and Analysis
of
Airborne Pollutants

Sampling and Analysis
of
Airborne Pollutants

Eric D. Winegar
Lawrence H. Keith

LEWIS PUBLISHERS
Boca Raton   Ann Arbor   London   Tokyo

Library of Congress Cataloging-in-Publication Data

Sampling and analysis of airborne pollutants / edited by Eric D.
  Winegar and Lawrence H. Keith.
        p.   cm.
    Includes bibliographical references and index.
    ISBN 0-87371-606-X
    1. Air--Pollution--Measurement.   2. Air--Analysis.   I. Winegar,
Eric D.   II. Keith, Lawrence H., 1938–
    TD890.S25   1993
    628.5'3'0287--dc20

                                                      92-32315
                                                      CIP

PRINTED IN THE UNITED STATES OF AMERICA
1 2 3 4 5 6 7 8 9 0

Printed on acid-free paper

To Robin

Contents

DATA INTERPRETATION TECHNIQUES

Preface

The purpose of this book is to present some of the best and latest work on air sampling and analysis methods and data interpretation. It is an outgrowth of the American Chemical Society's Division of Environmental Chemistry Symposium held in Washington, DC, in August, 1990, entitled Measurements of Airborne Compounds: Sampling, Analysis, and Data Interpretation. The purpose of that symposium was to gather a wide array of leaders in the air sampling and analysis field to present their work in an interdisciplinary atmosphere. This gathering fostered cross-fertilization between technical disciplines. Hence the diversity of topics presented in this book. Workers conducting ambient air programs, for example, can benefit from the exposure to statistical methods for interpreting the data they gather.

Continuing concern about our air environment and recent developments in the regulatory arena such as the Clean Air Act Amendments of 1990 have created an expanded need for research into sampling and analysis methods. Part of the Amendments is an extensive list of new chemicals to be measured. The chemical measurement community will be facing the challenges of developing new methods for the accurate measurement of these compounds. It is hoped that this book will assist this effort in some small way.

Demands for data on unusual compounds for risk assessment needs have also fueled this development. Of course, the public's awareness of environmental issues has certainly been one of the stronger primary driving forces in the environmental arena.

The book is organized into five parts: Sampling Methods and Approaches, Volatile Organic Compound Analysis, Aerosol Sampling and Analysis Developments, Optical Remote Sensing, and Data Interpretation Techniques. These five parts correspond to some of the major areas of current air pollution research.

Our hope in compiling this work was to inform other workers of some of the practical aspects of these topics and to serve as a reference source for users of air sampling methods. Our ultimate desire is to understand what impact air pollution has on human and animal health and how it affects our biosphere. If this book can serve as a small aid in understanding the complexity of airborne pollutants, then we will deem the effort a success.

Eric D. Winegar is currently Director of Research and Technical Services at Air Toxics Limited, an environmental laboratory and services firm in Rancho Cordova, California. He was previously Senior Scientist and Group Leader of the Environmental Chemistry Group at Radian Corporation in Sacramento, California. He received his PhD in chemistry from the University of California where he conducted physical chemistry research and participated in work on the organic aspects of impacts of carbonaceous aerosols on visibility. Dr. Winegar has been involved in a wide array of air toxics programs at Radian Corporation and at Air Toxics Limited, including taking active part in method development efforts. Past activities have included work in source testing, quality assurance, and groundwater monitoring. In addition to his interest in air sampling and analysis methods, Dr. Winegar follows developments in chemometrics and other aspects of environmental measurements.

Dr. Winegar is author or coauthor of more than 30 technical papers and presentations, and is active in local and national American Chemical Society functions as a member of the Division of Environmental Chemistry. He has also served as volunteer, vice-chair, editor of the proceedings, and chair of the annual regional Air and Waste Management Association conference on "Current Issues in Air Toxics."

Lawrence H. Keith is a Principal Scientist at Radian Corporation in Austin, Texas. A pioneer in environmental sampling and analysis, method development, and handling of hazardous compounds, Dr. Keith has published many books and presented and published more than one hundred technical articles involving these subjects. Recent publications have involved electronic book formats and expert systems. Dr. Keith serves on numerous government, academic, publishing, and environmental committees and is a past chairman of the ACS Division of Environmental Chemistry. Prior to joining Radian Corporation in 1977, he was a research scientist with the U.S. Environmental Protection Agency.

Contributors

M. Ahuja
California Air Resources Board
Sacramento, CA 95812

Lowell Ashbaugh
California Air Resources Board
Sacramento, CA 95812

W. R. Betz
Supelco Incorporated
Supelco Park
Bellefonte, PA 16823

John L. Bowen
Desert Research Institute
Post Office Box 60220
Reno, NV 89506

David A. Brymer
Radian Corporation
8501 MoPac Boulevard
Austin, TX 78720

Joan T. Bursey
Radian Corporation
3200 E. Chapel Hill Rd./Nelson Hwy.
Research Triangle Park, NC 27709

Ruth L. Carlson
Radian Corporation
8501 MoPac Boulevard
Austin, TX 78720

K. R. Carney
Institute for Environmental Studies
Louisiana State University
Baton Rouge, LA 70803

Glen R. Cass
Environmental Engineering Science
  Department
California Institute of Technology
Pasadena, CA 91125

D. P. Y. Chang
Civil Engineering Department
University of California
Davis, CA 95616

Judith C. Chow
Desert Research Institute
Post Office Box 60220
Reno, NV 89506

John Craig
University of California
San Francisco, CA 94143

Jihong Dai
Byrd Polar Research Center
The Ohio State University
Columbus, OH 43210

Raul Dominguez, Jr.
South Coast Air Quality Management
  District
21865 E. Copley Drive
Diamond Bar, CA 91765

P. Fellin
Concord Environmental Corporation
2 Tippett Road
Downsview, Ontario M3H 2V2

Clifton A. Frazier
Desert Research Institute
Post Office Box 60220
Reno, NV 89506

Kochy K. Fung
Atmospheric Assessment Associates
Woodland Hills, CA

Alan W. Gertler
Desert Research Institute
Post Office Box 60220
Reno, NV 89506

Larry Gruenke
University of California
San Francisco, CA 94143

S. Katharine Hammond
Environmental Health Division
Department of Family & Community
  Medicine
University of Massachusetts Medical
  School
Worcester, MA 01655

S. A. Hazard
Supelco, Inc.
Supelco Park
Bellefonte, PA 16823

Lynn M. Hildemann
Department of Civil Engineering
Stanford University
Stanford, CA 94305

K. S. Ho
Rohm and Haas Company
727 Norristown Road
Spring House, PA 19477

Steven D. Hoyt
Environmental Analytical Service
170-C Granada
San Luis Obispo, CA 93401

James J. Huntzicker
Dept. of Environmental Science and
  Engineering
Oregon Graduate Inst. of Science &
  Technology
19600 N.W. Von Neumann Drive
Beaverton, OR 97006

M. A. Jackisch
Institute for Environmental Studies
Louisiana State University
Baton Rouge, LA 70803

B. M. Jenkins
Agricultural Engineering Department
University of California
Davis, CA 95616

Paul L. Kebabian
Center for Chemical and
  Environmental Physics
Aerodyne Research, Inc.
45 Manning Road
Billerica, MA 01821

Charles E. Kolb
Center for Chemical and
  Environmental Physics
Aerodyne Research, Inc.
45 Manning Road
Billerica, MA 01821

S. J. Lambiase
Supelco, Inc.
Supelco Park
Bellefonte, PA 16823

Dwight Landis
ATEC, Incorporated
Post Office Box 8062
Calabasas, CA 91302

Margaret R. Leo
Blasland, Bouck & Lee
8 South River Road
Cranbury, NJ 08512

Vivian Longacre
Environmental Analytical Service
170-C Granada
San Luis Obispo, CA 93401

Howard I. Maibach
University of California
San Francisco, CA 94143

William C. Malm
CIRA-Foothills Campus
Colorado State University
Fort Collins, CO 80523

Monica A. Mazurek
Environmental Chemistry Division
Department of Applied Science
Brookhaven National Laboratory
Upton, NY 11973

Robert A. McAllister
Consultant
1229 Nottingham Drive
Cary, NC 27511

Stephen R. McDow
Dept. of Environmental Sciences and
  Engineering
University of North Carolina
Chapel Hill, NC 27514

Timothy R. Minnich
Blasland, Bouck & Lee
8 South River Road
Cranbury, NJ 08512

Leonard Newman
Environmental Chemistry Division
Department of Applied Science
Brookhaven National Laboratory
Upton, NY 11973

Larry D. Ogle
Radian Corporation
8501 MoPac Boulevard
Austin, TX 78720

R. Otson
Health and Welfare Canada, EHC
Tunney's Pasture
Ottawa, Ontario K1A 0L2

E. B. Overton
Institute for Environmental Studies
Louisiana State University
Baton Rouge, LA 70803

J. Paskind
California Air Resources Board
Sacramento, CA 95812

O. G. Raabe
Director, Institute for Environmental
  Health Research
University of California
Davis, CA 95616

Joann Rice
Radian Corporation
3200 E. Chapel Hill Rd./Nelson Hwy.
Research Triangle Park, NC 27709

James M. Roberts
CIRES
University of Colorado
NOAA
Boulder, CO 80309

Wolfgang F. Rogge
Environmental Engineering Science
  Department
California Institute of Technology
Pasadena, CA 91125

C. E. Schmidt
Environmental Consultant
19200 Live Oak Road
Red Bluff, CA 96080

Robert L. Scotto
Blasland, Bouck & Lee
8 South River Road
Cranbury, NJ 08512

Bernd R. T. Simoneit
College of Oceanography
Oregon State University
Corvallis, OR 97331

James F. Sisler
CIRA-Foothills Campus
Colorado State University
Fort Collins, CO 80523

Philip J. Solinski
Blasland, Bouck & Lee
8 South River Road
Cranbury, NJ 08512

C. F. Steele
Institute for Environmental Studies
Louisiana State University
Baton Rouge, LA 70803

Michael Stroupe
Nutech, Incorporated
Research Triangle Park, NC 27711

Y. Z. Tang
Concord Environmental Corporation
2 Tippett Road
Downsview, Ontario M3H 2V2

Ellen Mosley-Thompson
Byrd Polar Research Center
The Ohio State University
Columbus, OH 43210

Lonnie G. Thompson
Byrd Polar Research Center
The Ohio State University
Columbus, OH 43210

S. Q. Turn
Agricultural Engineering Department
University of California
Davis, CA 95616

Appathurai Vairavamurthy
Department of Applied Science
Brookhaven National Laboratory
Upton, NY 11973

Margil W. Wadley
South Coast Air Quality Management
 District
21865 E. Copley Drive
Diamond Bar, CA 91765

Denny E. Wagoner
Radian Corporation
3200 E. Chapel Hill Rd./Nelson Hwy.
Research Triangle Park, NC 27709

John G. Watson
Desert Research Institute
Post Office Box 60220
Reno, NV 89506

Ronald C. Wester
University of California
San Francisco, CA 94143

R. B. Williams
Agricultural Engineering Department
University of California
Davis, CA 95616

R. L. Wong
Institute for Environmental Studies
Louisiana State University
Baton Rouge, LA 70803

Acknowledgments

First and foremost to be acknowledged in the creation of this book are the many distinguished authors who offered their hard-won successes as contributions to the field. Their dedication to excellent science is proven in the following pages.

In addition, we acknowledge and thank the many reviewers of each chapter. Their insights added substantially to the quality of each contribution. We thank Bruce Appel, Air Industrial Hygiene Laboratory, California Department of Health; Lynn Barrie, Atmospheric Environment Service; Patricia Boone, United States Environmental Protection Agency; Thomas C. Curran, United States Environmental Protection Agency; Bart M. Eklund, Radian Corporation; Robert R. Freeman, Air Toxics Limited; W.B. Grant, NASA Langley Research Center; M.W. Hemphill, Texas Air Control Board; R.K.M. Jayanty, Research Triangle Institute; Brian Leaderer, John B. Pierce Laboratory; Robert G. Lewis, EPA Atmospheric Research and Exposure Assessment Laboratory; Scott Mgebroff, Texas Air Control Board; Charles D. Miller, Radian Corporation; Mark Pitchford, United States Environmental Protection Agency, Environmental Monitoring and Systems Laboratory; C.E. Schmidt; Robert K. Stevens, United States Environmental Protection Agency; Roger L. Tanner, Desert Research Institute; John Trijonis, Santa Fe Research Corp.; William M. Vaughan, Environmental Solutions, Inc.; R.S. Viswananth, Tulsa City County Health Department; Hal H. Westberg, Laboratory for Atmospheric Research, University of Washington; C. Herndon Williams, Radian Corporation; and David T. Williams, Environmental Health Centre, Environment Canada. We also thank Kathi Kuhlman for her expert document processing skills in producing the manuscript.

Sampling Methods and Approaches

CHAPTER 1

Standard Technical Approach for the Analysis of Ambient Air Samples

Denny E. Wagoner, Joann Rice, and Joan T. Bursey

TABLE OF CONTENTS

0-87371-606-0/93/$0.00+$.50
© 1993 by Lewis Publishers

I. DEVELOPMENT OF AIR TOXICS GAS CHROMATOGRAPHY/MASS SPECTROMETRY (GC/MS) ANALYTICAL METHODOLOGY

In the Research Triangle Park Laboratory of Radian Corporation, air analysis by capillary gas chromatography/mass spectrometry began with the introduction of capillary analysis to the volatile organic sampling train (VOST) methodology.[1] The desorption of Tenax®/Tenax®* charcoal tubes through a purge and trap unit for volatile components of stack gas was first performed[1] using a wide-bore capillary column (0.32-mm i.d.) which was directly coupled to the ion source of the mass spectrometer.

Combustion samples are frequently high concentration samples that can saturate both the chromatographic system and the analytical system. When Megabore®** columns (0.53-mm i.d.) became commercially available, a Megabore® column was used in order to enhance the capacity of the analytical system for combustion samples. In order to preserve the direct coupling to the ion source of the mass spectrometer, the flow of helium through the Megabore® column was decreased to approximately 2 mL/min in order to allow the vacuum system of the mass spectrometer to accommodate the flow of carrier gas. This flow rate of 2 mL/min is far below the 15 mL/min recommended by most column manufacturers for optimum chromatography, but use of such a high flow rate was not possible with a directly coupled column.

In order to optimize column flow rate and hence chromatography, the direct connection to the ion source was eliminated and the column was coupled to the ion source through a glass jet separator. Glass jet separators were designed for operation with packed chromatographic columns and operate most effectively at a flow of approximately 30 mL/min. In order to optimize the enrichment and efficiency of the jet separator, use of makeup gas at a flow of approximately 15 mL/min (total flow of 30mL/min with column and makeup combined) was initiated, and the capillary technology with the enhanced chromatographic capacity of the Megabore® column was made available for the analysis of gaseous combustion samples in the VOST analysis.

When GC/MS analysis of ambient air samples was required, the approach of Megabore® capillary chromatography coupled to the ion source of the mass spectrometer through a glass jet separator with makeup gas was used. This approach has been uniformly successful, allowing ambient air sample analyses without requiring the drying of samples (Figure 1). This approach has also been applied to the purge and trap analysis of water, waste, and soil samples, as well as thermal desorption analysis of sorbent tubes in the TO-1 and TO-2 methodology.[2]

---

*   Registered trademark of Enka N. V.
**   Registered trademark of J & W Scientific.

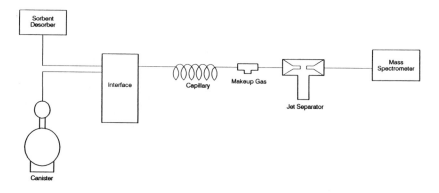

FIGURE 1.   Schematic diagram of MS analytical system.

A jet separator specially designed for Megabore® capillary columns to couple directly to the ion source and obviate the use of makeup gas is presently commercially available for a limited number of analytical systems, but we have not evaluated this technology.

II. DEVELOPMENT OF GAS CHROMATOGRAPHY/MULTIDETECTOR (GC/MD) METHODOLOGY

At the time of the development of Megabore® VOST methodology, the determination of the nonmethane organic compound content of ambient air samples taken at various locations throughout the United States was being performed. A question arose in this program as to whether the organic compounds being observed could possibly be halogenated hydrocarbons. A gas chromatograph with an electron capture detector was used to analyze a number of the ambient air samples to identify halogenated organic compounds and estimate a quantity. Since the preliminary assessment of the gas chromatographic analytical methodology was successful, the next sampling cycle of the nonmethane organic compound program included the development of a gas chromatographic analytical system to perform specific compound characterization and quantitative analysis.

At the same time as the mass spectrometer technology was being developed and evaluated, a parallel program for design and construction of a multidetector gas chromatographic system for analysis of air toxics in ambient air samples was being conducted.[3] Gas chromatography was used as the primary analytical method in this program because of the sensitivity which the technique offered. A flame ionization detector (FID) was used as a universal detector, with a nondestructive photoionization detector (PID) in series to offer enhanced sensitivity for aromatic compounds. A new design for the photoionization detector allowed direct coupling of the PID to the FID. Earlier versions of the PID were susceptible to leaks and required the column effluent to be split. An electron capture detector (ECD)

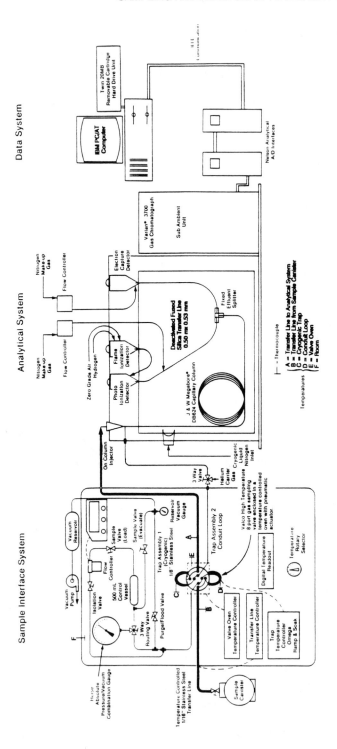

FIGURE 2.   Gas chromatograph multiple detector system.

was added in parallel with the PID/FID using a postcolumn fused silica splitter in order to offer enhanced sensitivity for halogenated compounds, which were of primary interest in a program concerned with analysis of air toxics (Figure 2). An alternative approach which has been employed uses a Hall electrolytic conductivity detector instead of an electron capture detector. The Hall detector is very specific for halogenated organic compounds, but is intolerant of water. The gas chromatograph with multiple detectors proved to be reliable (downtime of approximately 5%) and highly accurate for performing qualitative and quantitative analysis of ambient air samples.

A. Canister Cleaning Procedures

The ability to clean canisters for repeated use is essential to the success of any ambient air program. However, in order to clean canisters it is necessary to design, construct, and evaluate the equipment required, since no canister cleaning apparatus is commercially available for cleaning large numbers of canisters simultaneously. No single process has been proven conclusively superior for producing cleaned canisters. Canisters are cleaned with the use of vacuum/pressurization cycles, using heavy vacuum and humidified ultrahigh purity air (Figure 3). At the end of the multiple evacuation/pressurization cycles, the canister is filled with humid zero air. After a period of equilibration (at least overnight), a sample of this humid zero air is withdrawn and analyzed. If the canister meets the project-specific cleanliness criteria, the canister is considered clean and is evacuated prior to shipment to the field. An alternative approach to the cleaning of canisters uses heat in the evacuation/pressurization process, with the same project-specific criterion applied for cleanliness. Canisters are cleaned successfully using both approaches.

However, a phenomenon known as offgassing is frequently encountered with cleaned canisters, no matter which cleaning process is used. If a cleaned canister, which has successfully passed cleanliness criteria, is filled with humidified zero air and allowed to sit for a period of days, trace levels of various organic compounds may be observed in the cleaned canister. The occurrence of this phenomenon is why canisters which are used to collect samples that contain high levels of organic compounds should not be used subsequently for the collection of ambient air samples. Whichever system is used for cleaning canisters, clean humidified air for pressurization and an effective vacuum system are essential components of the process.

B. Compound Stability in Canisters

The ability to analyze organic compounds in canisters depends upon the stability of those compounds in the canisters when a field sample has been taken. A stability study was therefore designed and executed to determine whether the concentration of selected compounds in humidified air would change while these compounds were being stored in 15-L SUMMA®-treated stainless steel canisters

---

Registered trademark of Molectrics Corporation.

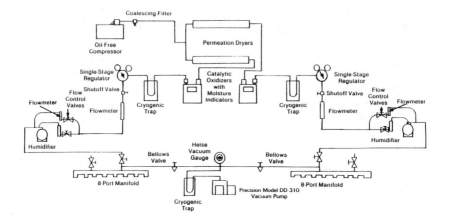

FIGURE 3.   Canister cleanup apparatus.

for a period of 79 days. Test mixtures were prepared from a certified gas standard from a high pressure cylinder having an accuracy certified to within ±5%. Analyses were performed at approximately 2-week intervals. Test canisters were humidified to simulate ambient air samples.

The stability was determined to be acceptable for the following compounds:

methylene chloride	*cis*-1,3-dichloropropene
1,1,2-trichloroethane	tetrachloroethene
dibromochloromethane	chlorobenzene
o-xylene	m-xylene
p-xylene	bromoform
bromofluorobenzene	1,1,2,2-tetrachloroethane
m-dichlorobenzene	p-dichlorobenzene
o-dichlorobenzene	*trans*-1,2-dichloroethene
1,1-dichloroethane	chloroprene
perfluorobenzene	bromochloromethane
chloroform	1,1,1-trichloroethane
carbon tetrachloride	benzene
1,2-dichloroethane	perfluorotoluene
trichloroethene	1,2-dichloropropane
bromodichloromethane	trans-1,3-dichloropropene
toluene	n-octane

The sample canisters required more than 8 hr after filling to reach equilibrium, as demonstrated by reproducibility of analytical results immediately after filling. For all compounds studied, overall precision was 11.2% and overall accuracy (bias) averaged 9.8% for 79 days.

C. Preparation of Analytical Standards

A crucial component of accurate quantitative analysis is the preparation of accurate and reproducible standards. The ability of the analytical laboratory to perform an accurate quantitative analysis is dependent upon the ability of the laboratory to generate accurate working standards. The initial approach to the preparation of standards for the U. S. Environmental Protection Agency (EPA) Urban Air Toxics Monitoring Program involved the preparation of a stock canister standard using gravimetric injection of neat liquid standards into a canister. The highest success level using this approach was approximately ±50%. This level is adequate for semiquantitative estimates but, where strict criteria are imposed for the quantitative analysis, better accuracy and reproducibility are essential. The approach using liquid standards was not sufficiently reproducible for preparation of standards on a routine basis, and alternative methodology for the preparation of standards was developed.

The approach presently used for the reproducible preparation of standards involves dynamic flow dilution (Figure 4). This apparatus was designed to take certified gaseous standards supplied in cylinders and dilute these standards with zero-grade air that has been passed through a catalytic oxidizer to destroy hydrocarbons. The diluent is then humidified to approximately 70% relative humidity. The gases are blended in a SUMMA®-polished mixing sphere and bled into cleaned evacuated canisters for subsequent use as accurate analytical standards for any of the analytical systems. The flow dilution apparatus allows the use of three or more certified cylinders in the preparation of a single multicomponent analytical standard. The flow dilution system has the following advantages:

- dilutes 1:2 to 1:1000 with approximately 10% precision
- does not contaminate the gas mixture
- may use dry air, dry nitrogen, or any other gas as a diluent
- may use three or more certified gas cylinders to generate diluted standards
- may independently regulate the amount each cylinder is diluted
- has completely heat-traced lines to prevent condensation of any organic compounds or moisture
- makes efficient use of expensive certified cylinders of gaseous standards

Calibration standards are prepared in humidified zero air to simulate field samples. Average relative humidity in the United States is approximately 70%, thus this value is a reasonable level to use in calibration samples.

A dynamic flow dilution system allows the use of certified cylinders containing multiple components that can be blended and diluted accurately with humidified zero air. Since the introduction of the dynamic flow dilution system, preparation of accurate and reproducible calibration standards is routinely attained (±30% accuracy down to 2 ppbv based on NTIS gas standards). The presence of water vapor in canisters is essential in maintaining sample integrity. Various observations at several laboratories have indicated that the presence of water vapor

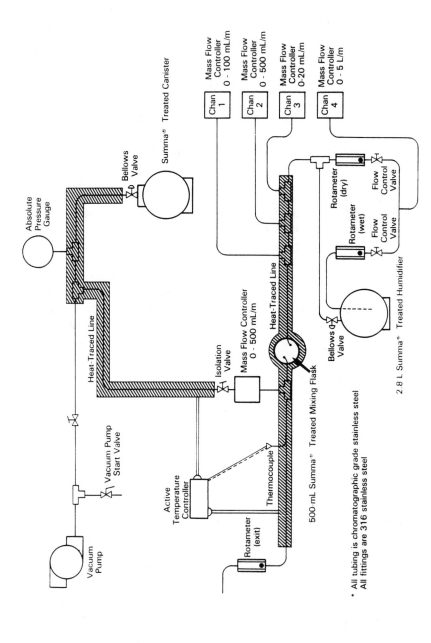

FIGURE 4.   Dynamic flow dilution appartus.

neutralizes active sites in the sampling/canister system.[4] In spite of the SUMMA® deactivation process, compound losses may still occur in a canister which contains compounds present in dry air.

D. Sorbent Calibration

Calibration for sorbent tubes is a much more straightforward procedure. Clean sorbent tubes are also essential to making accurate qualitative and quantitative determinations. There is no universally accepted procedure for cleaning sorbents, although most sorbent cleaning processes include extraction with organic solvent, possibly even with a sequence of organic solvents, prior to vacuum drying and/ or heated vacuum drying. The final step in the preparation of sorbent is heating the prepared sorbent tube with inert gas flow. The cleaned tube is thermally desorbed and analyzed as a quality check to establish cleanliness prior to shipment to the field. The sorbent tube which has been demonstrated to be cleaned is then sealed into a glass, airtight container and put into another airtight container for shipment to the field. Extensive precautions and care in handling are required to prevent the contamination of sorbent in the field.

When sorbent analyses are performed, calibration solutions prepared in methanol may be used to calibrate the tubes by the process of flash evaporation. A solution of compounds of interest in methanol is used, because when the solution is vaporized and swept onto the sorbent tubes in a stream of inert gas, the methanol is readily swept from the tube, leaving the compounds of interest. The same flash evaporation process can be used to put internal standards and surrogate compounds onto the sorbent tube prior to analysis to provide a continuing check on the operation of the instrumentation and the possibility of matrix effects.

E. Analytical Interface for Canisters and Sorbents

Sorbent and canister samples are introduced to the analytical system in a similar way: the sample is either thermally desorbed from the sorbent and cryogenically focused or extracted from the canister and cryogenically focused with a custom designed and constructed sample interface (Figure 5).

The sample interface delivers reproducible volumes of ambient air samples under pressure or under vacuum to any analytical device without drying the sample. Sampling canisters are shipped to the field under vacuum and the sample is taken to a point just below atmospheric pressure to allow significant leakage to be detected. Field samples must therefore be transferred from a canister which is under 0.5–14.0 in. of mercury vacuum. The system interface was designed and constructed to minimize contamination and memory effects and to maximize repeatability. The potential for contamination of the ambient air samples was minimized by the design and selection of materials for construction. The connecting tubing of the interface is 1/8- and 1/16-in. o.d. chromatographic-grade stainless steel. Each routing valve, shutoff valve, and fitting is constructed of 316 stainless steel. The preconcentration trap assemblies are constructed of chromatographic-grade stainless steel and filled with 60/80 mesh glass beads to facilitate

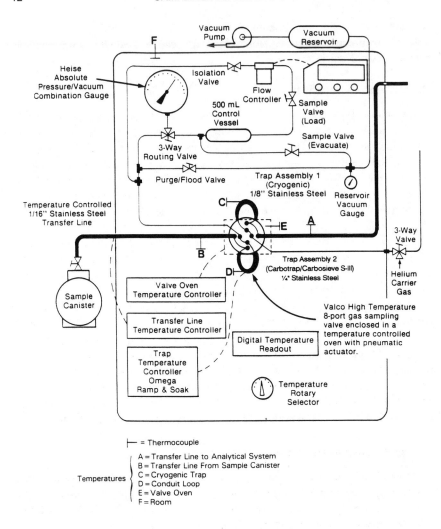

FIGURE 5.   Sample interface system.

condensation of organic compounds and to minimize aerosol formation. Daily baseline checks are performed to monitor potential contamination of the interface. These checks consist of sampling and analyzing cleaned humidified air immediately after calibration of the analytical system. These zero-air analyses have demonstrated that the interface remains essentially contamination free (<0.2 ppbv per target compound). The elimination of compound memory from successive samples is achieved by the use of heat traced temperature controlled components. Each part of the interface which contacts the sample both before and after the analytical trap is temperature controlled. The eight-port gas sampling valve is enclosed in an oven and maintained at 160°C with an active temperature controller. Transfer lines are maintained at the same temperature with a separate active temperature controller. The analytical sample trap is heated to over 200°C during the thermal desorption cycle to ensure the removal of any residual organic compounds.

Strict temperature control of the thermal desorption apparatus for solid sorbents is essential: desorption temperatures that are too high will cause decomposition of some of the polymeric sorbents that are used in ambient air sampling. Repeatability of canister sample injection is accomplished by the use of a high resolution pressure/vacuum gauge to measure sample loading pressure accurately and reproducibly. Organic compounds are trapped in a liquid argon bath at −185°C. Thermal desorption is accomplished using helium carrier gas and a programmable temperature controller powering a 1000-W heater embedded in a brass block containing the sample trap. Repeated measurements of calibration standards show that the samples are delivered to the analytical system at a constant volume with a deviation of less than 3.7%. The accuracy and reproducibility of the calibration process are demonstrated by the analysis of canister audit samples supplied by U.S. EPA on a regular basis to audit the analytical operations of the Urban Air Toxics Monitoring Program (Table 1).

This interface will also allow introduction of internal standards directly with the canister analytical samples. The internal standards can either be introduced into the cryotrap by means of a fixed volume sample loop from a certified cylinder or by a gaseous injection from a mixture in a static dilution bulb. Internal standards are introduced directly onto sorbent tubes using the flash evaporation technique: a solution of standards in methanol is vaporized and carried through a sorbent tube in a flow of inert gas. The direct analogy for canisters would require introduction of internal standard compounds into the canister as it is returned from the field and would introduce 'an unacceptably high risk of contaminating the sample, so introduction of standards into the cryotrap as organic compounds from the canister sample are being condensed has been chosen as a better approach. Approaches for introduction of surrogate compounds into canisters such as introduction of surrogate compounds into the canisters prior to shipment to the field are being investigated, but no procedures have yet been established.

F. Samplers and Certification

Representative analysis of volatile organic compounds either from canisters or from sorbents begins with the field collection. The best analysis in the world cannot rescue a sample that is contaminated or that has not been correctly collected in the field. As in most areas of air toxics sampling and analysis, a variety of sampling systems are in operation. One of the simplest systems is used in canister sampling. Since the canister is shipped to the field under vacuum, the sampling operation requires only that air in the field be allowed to flow into the canister at a qualified rate.

One very common sampling device is a critical orifice through which field air is introduced into the canister. The critical orifice will allow the collection of unattended integrated air samples, sampling over periods ranging from 2 to 24 hr.

Table 1. UATMP Audit Results

Compound	True concentration (ppbv)[a]	Average % bias[a]	
		GC/MD	GC/MS
Vinyl chloride	3.3	5.3	17.9
Chloroform	3.6	9.1	4.2
Carbon tetrachloride	3.4	7.3	3.6
Methylene chloride	3.4	15.2	31.4
1,2-Dichloroethane	3.7	4.6	-2.0
Trichloroethylene	4.2	0.6	-1.9
Benzene	3.2	0.0	10.1
Tetrachloroethylene	3.7	3.1	3.1
Bromomethane	3.6	9.8	6.4
1,1,1-Trichloroethane	3.6	9.7	2.8
1,2-Dichloropropane	3.7	1.1	-0.9
Toluene	3.6	6.4	12.5
Chlorobenzene	3.7	6.0	1.0
Ethylbenzene	3.5	-17.2	-12.4
o-Xylene	3.7	-5.6	21.2
Overall average	3.6	3.7	6.5

[a] Average of three external audits.

Many other field samplers are available, either through various commercial suppliers or from equipment designed and constructed by various laboratories. The cleaning of the sampling devices is critical, since the sampling device can readily cross-contaminate samples if the cleaning procedures are not adequate.

Samplers are certified by determining the recovery of spiked challenge compounds. The sampler is then purged with humid zero air until none of the challenge compounds are collected. The ability of the system to transfer compounds from the atmosphere in the field to the interior of the canister without the introduction of contamination is verified. A standard operating procedure has been established for the certification and validation of samplers. When the sampling system has been certified, it is ready for field use. In conjunction with U.S. EPA on the Urban Air Toxics Monitoring Program,[5] the laboratory has designed and built field sampling units (Figure 6) that are installed in various states and then operated by the state personnel, with telephone consultation and troubleshooting by laboratory personnel. The sampler is designed to allow simultaneous sampling for aldehydes.

G. Screening and Dilution of Samples

Analytical good practice calls for any given analyte to be at a concentration within the calibration range to obtain a valid analytical determination. With sorbents, the skill in performing the analysis lies in the selection of an appropriate calibration range for a given series of analyses. If an analyte desorbed from a sampled tube saturates the analytical system, no valid data can be obtained for that analyte. If an analyte desorbed from the sorbent tube is not saturated, but occurs at a concentration that must be extrapolated beyond the calibration range, this

15

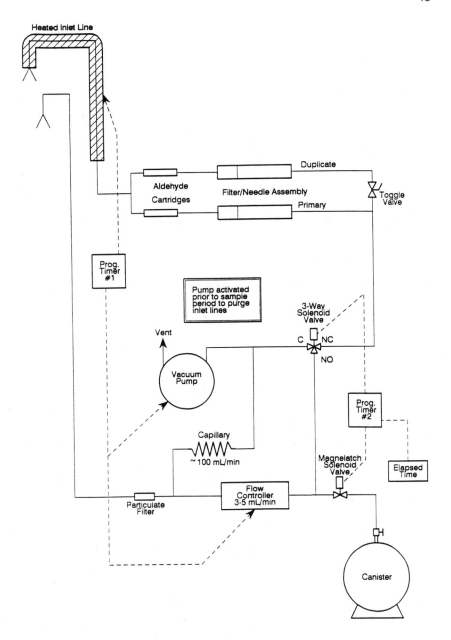

FIGURE 6.   Sampling assembly for the UATMP.

concentration can be challenged. Common analytical practice is then to add a calibration point at a value above the analyte concentration. However, there is always the potential problem with sorbents that saturation may be encountered, and there is no way to resolve the problem of analytical system saturation.

In the case of canisters, there are a number of approaches that may resolve the problem of saturation or exceeding the calibration range. A canister contains enough sample to perform several analyses. If an analysis of a canister sample shows an analyte exceeding the calibration range, the contents of the canister can be diluted with clean air or nitrogen to an appropriate level, and the analysis can be repeated. Dilution of a canister sample is straightforward: the canister is pressurized with the diluent gas and is equilibrated for at least 18 hr for static mixing. Then the pressure is bled off to a known ambient level for analysis. A second equilibration period is allowed to occur prior to analysis.

Another approach that can prevent repeating analyses is to screen samples prior to analysis. A screening analysis can be performed rapidly, and an appropriate dilution level can be determined as the result of the screening analysis. The screening can be performed using preconcentrated direct flame ionization detection (PFPID), a nonspeciated analysis which produces a value in terms of total carbon, or a rapid GC/MD analysis can be performed to determine whether the problem with concentration involves one of the analytes of interest or other compounds. Screening and dilution, as appropriate, of canister samples can ensure that all analytes fall within the calibration range.

III. ANALYSIS

Canister analysis using GC/MD is performed using a sample size of approximately 250–800 mL. The GC/MS analysis uses approximately 500 mL of sample. Since the analytical systems showed no problem in accommodating the water introduced with the air samples, no effort was made to select or develop procedures to dry canister samples. There are numerous procedures available to dry air samples, but drying procedures also tend to remove more polar water-soluble molecules that may be present in the air sample, as well as water. Since the analysis is performed on undried samples, polar compounds may be observed and quantified in either GC/MD analysis or GC/MS. Among the compounds routinely analyzed by GC/MD are methanol, ethanol, isopropanol, and diethyl ether.

Sensitive MS analysis was performed for 1,4-dioxane using selected ion monitoring (where the MS focuses on particular masses rather than scanning a mass range): detection limits for dioxane in this mode were estimated to be <0.1 ppbv and a linear calibration was obtained (Figure 7). The ability to analyze whole air samples without drying is necessary to be able to determine polar molecules. Acrylonitrile is another example of a polar molecule of general interest which cannot be analyzed successfully if the sample is dried, but that readily produces a linear calibration in whole air standards (Figure 8).

The strength of GC/MS in the analytical procedure is the ability to characterize compounds for which no calibration has been performed. Even with a system of multiple detectors, the gas chromatographic analysis can identify with confidence only those compounds for which a standard is available. If compounds coelute, the mass spectrometer as detector can deconvolute the compounds unless they have

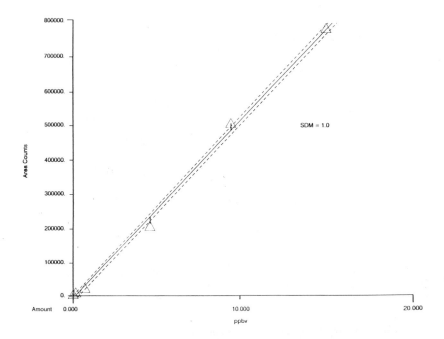

FIGURE 7. GC/MS calibration for 1,4-dioxane.

common ions, i.e., they are isomers. In many instances, the gas chromatograph is incapable of deconvolution of coeluting compounds. In general, the gas chromatographic analysis tends to be less costly than the mass spectrometric analysis. The mass spectrometer allows accurate calibration for some specified number of compounds, with the ability to characterize additional compounds and supply an estimated concentration. Both techniques have advantages and disadvantages.

The strengths of the mass spectrometric approach are recognized by the Contract Laboratory Draft Statement of Work for Ambient Air, where GC/MS is the analytical methodology mandated.[6] When a long-term monitoring program is conducted with a few analytes of special interest, the most cost-effective approach which will yield data of appropriate quality for the compounds of interest is usually GC/MD. At a Superfund site where the presence of some compounds is expected, but other unanticipated compounds may be observed, the ability to characterize these additional compounds requires the use of MS as the analytical methodology.

IV. CONCLUSIONS

Ambient air analysis is a rapidly changing field where new technology is constantly being developed and applied. However, the following basic precepts of analytical chemistry are still applicable:

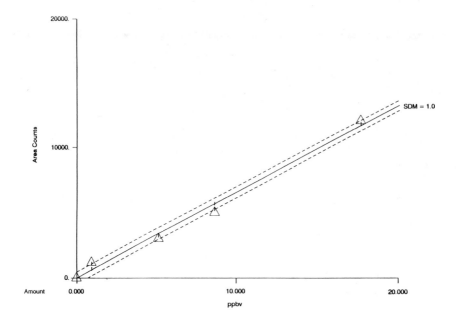

FIGURE 8.   GC/MS calibration for acrylonitrile.

- Accurate standards are required to perform accurate analyses.
- Calibration should be performed in the same way as analysis.
- Alter a sample as little as possible before analysis.
- Cleanliness and control of both sampling and analytical parameters are essential.
- Quality control and quality assurance are critical for documenting the quality of analytical data: performance evaluation samples and other tests of analysis quality demonstrate competent performance and document the quality of the data generated.
- Contamination must be avoided in order to generate valid field data: certification of the cleanliness of field samplers documents that field equipment was free from contamination upon leaving the laboratory.

ACKNOWLEDGMENTS

The authors wish to thank the U.S. EPA, Office of Air Quality Planning and Standards TSD, and the Quality Assurance Division of the Atmospheric Research and Exposure Assessment Laboratory for their support in the Urban Air Toxics Monitoring Program. The authors also wish to thank Raymond G. Merrill, Jr., Dave-Paul Dayton, Robert F. Jongleux, Robert A. McAllister, Larry D. Ogle, and Walt Crow for their contributions to the work of Radian Corporation in the area of air toxics method development and analysis.

REFERENCES

1. Buedel, T., R. Porch, J. Bursey, J. Homolya, R. Fuerst, T. Logan, and R. Midgett. "Application of Megabore Capillary Gas Chromatography/Mass Spectrometry in the Analysis of Samples Generated by the Volatile Organic Sampling Train," *Biomed. Environ. Mass Spectrom.* 18:138–144 (1989).
2. Winberry, W.T., Jr., N.T. Murphy, and R. M. Riggan, "Compendium of Methods for the Determination of Toxic Organic Compounds in Ambient Air," EPA-600/4-84-041, Second Supplement (June 1988).
3. Dayton, D-P., and J. Rice, "Development and Evaluation of a Prototype Analytical System for Measuring Air Toxics," Final Report to the U. S. Environmental Protection Agency, Environmental Monitoring Systems Laboratory, EPA Contract No. 68-02-3889, WA No. 120, Radian Corporation, Research Triangle Park, NC (1987).
4. Fitz-Simons, T., T.A. Lumpkin, and W.A. McClenny, "Report of the Air Monitoring in the Kanawha Valley, West Virginia," Unpublished U.S. EPA Report available from the Atmospheric Research and Exposure Assessment Laboratory, Research Triangle Park, NC (January 1987).
5. McAllister, R. A., P. L. O'Hara, W. H. Moore, D-P. Dayton, J. Rice, R. F. Jongleux, R.G. Merrill, Jr., and J. T. Bursey. 1988 Nonmethane Organic Compound and Urban Air Toxics Monitoring Programs, Final Report, EPA Contract No. 68D80014 (July 1988).
6. U. S. Environmental Protection Agency Contract Laboratory Program. Draft Statement of Work for Analysis of Ambient Air, U. S. Environmental Protection Agency, Office of Emergency and Remedial Response, Hazardous Site Evaluation Division, Analytical Operations Branch, Washington, DC (August 1990.)

CHAPTER 2

USE OF A MICROCHIP GAS CHROMATOGRAPH FOR AMBIENT AIR ANALYSIS

K. R. Carney, R. L. Wong, E. B. Overton,
M. A. Jackisch, and C. F. Steele

TABLE OF CONTENTS

0-87371-606-0/93/$0.00 + $.50
© 1993 by Lewis Publishers

I. INTRODUCTION

Inhalation is a major and relatively unavoidable route of exposure to toxic materials. Because of this, volatile toxics present a unique cause for concern in the environment. The value of monitoring volatile toxics in the workplace and out-door environments has been well established, and exposure in indoor residential environments is rapidly becoming an important issue. Additionally, regulations deriving from recent amendments to the Clean Air Act will soon require much more emphasis on monitoring for volatile air toxics. Now, more than ever, a need is apparent for reliable, low-cost field deployable methodology for field screening of volatile organic compounds, particularly in air.

A. Field Deployable Techniques

Field techniques can provide several advantages over laboratory-based tech-niques, namely, more rapid return of information to the user, large reduction in sample handling and storage requirements, and lower direct costs. Laboratory-based techniques in many, if not most, instances provide higher sensitivities and better precision than corresponding field techniques. When sample quality is considered, however, rather than instrumental output alone, field techniques may yield more accurate results than laboratory techniques because of the difficulty in assuring that the sample analyzed in the laboratory is representative of the location sampled in the field. Considerable effort is often required to maintain sample quality during the lag time prior to laboratory analysis. Even when sample integrity is assured for laboratory analyses, the ability to receive immediate feedback from field analyses allows prompt adjustments in sampling strategies to ensure that the site is described accurately by the analytical results for samples that are sent to the laboratory.

Furthermore, the cost for an analysis in the field is a fraction of that for a similar analysis in the laboratory, especially when the costs of sample handling, storage, and transport to the laboratory are included. Thus a cost-effective sampling strategy might involve using more field analyses of, perhaps, lower quality and fewer laboratory analyses. Such a strategy could ultimately improve the value returned for each analysis performed in the laboratory. For some applications, field analyses alone may provide the desired data quality; for other applications field techniques may augment the laboratory analysis of field samples. Only rarely would an environmental field investigation fail to benefit from the availability of an applicable field technique. The utility of field tech-niques not only applies to analytical investigations, but also to processes such as production activities or waste removal activities. In these cases, the immediate feedback from field analyses could easily outweigh a slight loss in data quality.

Certain criteria are important for techniques and methodologies intended for use in the field. Mobility is, by definition, an important factor; the ability to transport the analyzer to the sample is the crux of field methodology. The term mobility immediately brings to mind factors such as weight and size that physically determine how easily one may move an instrument to the field. Mobility, in practice, also implies low use of consumable resources such as electricity, compressed gases, and solvents. Clearly mobility is a relative term and for some circumstances, a laboratory GC with its data system, carrier gas, electrical generator, and other ancillary equipment may be placed in a truck and be considered mobile. Generally, however, equipment considered highly mobile can be transported without specialized transportation accommodations. Simply put, the equipment can be thrown in the trunk of a car, driven to a sampling site, and put to use quickly.

B. Instrumentation

The current state of the art for general analysis of volatile organics in the field is represented by a variety of portable gas chromatographs. These instruments present a range of detection limits, resolution, speed, and data systems. The current and foreseeable paragon of environmental analysis is the mass spectrometer in the guise of gas chromatography/mass spectrometry (GC/MS), MS/MS, or liquid chromatography (LC/MS). The advantage of GC/MS for sample component identification is seemingly indisputable although earlier workers have indicated that dual column chromatography can provide equivalent information.[1] The current forerunner in MS technology for field deployable instrumentation is the ion-trap detector (ITD) or ion-trap mass spectrometer (ITMS). Even these most portable of mass spectrometers are quite demanding in terms of capital outlay, electricity, and weight. The development of field portable GC/ITD or other GC/MS technology represents the high end of field instrumentation development.

The low end of field instrumentation development is represented by portable gas chromatographs, having more or less selective detectors, which can be deployed in larger numbers, moved about more easily, and operated by personnel with less training. A unique development in portable gas chromatography has been the development of so-called microchip GC technology. This technology uses photolithographic machining techniques to produce injection and detection systems on silicon microchips. The resulting components are ideally suited for use with microbore high resolution capillary columns. The resulting gas chromatographs are small and lightweight, less than 100 in.[3] and 8 oz, excluding carrier gas supply and electronics. Two advantages derive from the microchip GC. First, the use of short very high resolution columns provide an unparalleled resolution/speed combination. Very fast, well-resolved chromatograms are easily obtained. Typically chromatograms, for which an example is shown in Figure 1, are about

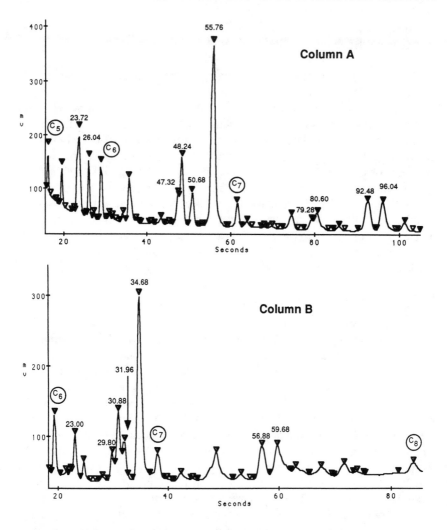

FIGURE 1.   Gas chromatograms of gasoline vapor at 60°C using M200 chromatograph.

3 min long and have peak capacities approaching 100. A second advantage,
provided by the small size, is that two separate GCs may be packaged in a single
unit.

II. MICROCHIP GAS CHROMATOGRAPH

The Microsensor model P200 is such a dual microchip gas chromatograph. The
entire unit (including carrier gas, power supply, control and data acquisition

electronics, and two isothermal microchip gas chromatographs) occupies less than 1000 in.^{3} and weighs less than 25 lbs. Power requirements are in the 5- to 15-W range, and the use of 0.100-mm columns limits gas consumption to approximately 2 ml/min; thus the P200 fits well the definition of highly mobile. This instrument has several desirable characteristics: dual GC columns that provide fast high resolution separations, a universal detector, precise temperature control, and reproducible retention behavior. The instrument also has some important characteristics which limit its utility for analysis of toxic materials in ambient air. The most important of these are a fairly high detection limit, limited volatility range for analytes in a single analysis, and the limited column capacity characteristic of high-resolution microbore columns. Additionally, while the dual column nature provides more qualitative information than a single-column GC, the retrieval of that information is nontrivial.

A. Dual Column Capability

With two parallel columns, a single sample mixture can be analyzed simultaneously on two different stationary phases. This adds greatly to the reliability of retention-based identifications. Ideally for two columns yielding totally uncorrelated retention times and having the same peak capacity, the resolving power would be increased by a power of two over that for a single column.[2] For example, given a peak capacity of 100 for a typical chromatogram, two completely uncorrelated chromatograms would have a peak capacity of 10,000. Of course, any pair of GC columns will be correlated to some extent as the vapor pressures of the compounds in a mixture play some role in any GC separation. For the two liquid phases most commonly found in the P200, OV-73, and OV-1701, the increase over a single column may be as low as a power of 1.2. Thus, if a single-column chromatogram has a peak capacity of 100, the dual column analog will have an equivalent peak capacity of 250. The more dissimilar the two phases, the higher the exponential power of the increase. These increases in resolving power can be seen in Figure 2, which is a two-dimensional representation of the retention times for approximately 60 compounds on the two P200 columns. The ellipses correspond to approximately the peak half-height width (FWHM) for chromatographic peaks from the P200 columns. Projection of the ellipses onto either of the axes shows the analogous single-column resolution. That the retention data are somewhat correlated can be seen by the distribution of retention times about a line of unit slope. In spite of a fairly high degree of correlation, the use of dual columns substantially increases the number of compounds that can be resolved into a two-dimensional space. Further improvements should be expected with more disparate stationary phases.

B. Detector

Each GC module has a solid-state microthermal conductivity detector; data acquisition electronics provide 100 data points per second for each column. The

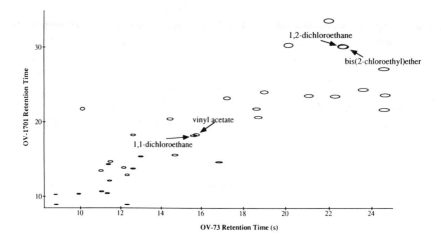

FIGURE 2.   Partial 60°C library with 1% acceptance regions.

incorporation of a thermal conductivity-type detector has certain advantages for a general-purpose field deployable volatile organics analyzer. Because the response of the thermal conductivity detector is fairly constant over a wide range of compounds,[3] one may often use pseudo standards for semiquantitative screening analysis. The weight response of a thermal conductivity detector to most compounds is within 30% of its response to 2,2,4-trimethylpentane (isooctane). Exceptions to this rule are low molecular weight compounds (<35 amu), halogenated compounds, and heavy metal containing compounds, e.g., $Pb(C_2H_5)_4$, which exhibit lower than expected response factors. Consequently, if the operator has no standard for a given compound or the identity of the compound is unknown, then the response factor of a similarly eluting compound may be used to obtain a fair quantitative estimate adequate for screening purposes. An often overlooked fact is that the thermal conductivity detector is a low impedance device compared to, for example, the flame ionization detector. The difficulty and cost associated with producing a low-noise, high-gain electrometer capable of acquiring 100 data points per second are considerably less for a low impedance device.

C. Retention Precision

The combination of micromachined injection valves, microprocessor controlled injection, and ±0.1°C temperature control provides very reproducible retention behavior. Standard deviations (SD) for retention times are typically about 0.2% for nonpolar and moderately polar compounds. For very polar compounds, such as alcohols, the figure is in the 2–4% range. The capability of producing such highly repeatable retention times further enhances the reliability of qualitative information provided by a dual microchip GC. Figure 3 shows retention times for a series of chromatograms run over the course of approxi-

OV-73 column at 60°C

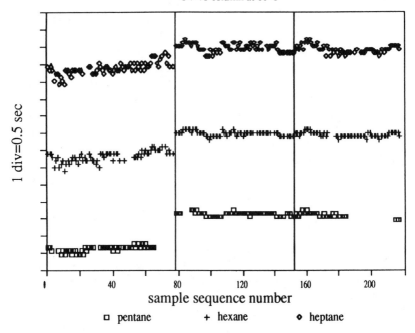

sample sequence number

□ pentane   + hexane   ◇ heptane

FIGURE 3.   M200 retention time variation over 4 weeks.

mately 4 weeks. The vertical rules represent 5-day idle periods during which no samples were run. During the first 5-day idle period, the instrument was shut down for 2 days and the carrier gas was changed; otherwise the instrument was powered continuously. Between the idle periods, samples were run over the course of 4 days. The retention time precision is very good over the continuous operating period, repeatable to within 50 msec, with no drift over the entire period; but when the instrument was shut down and restarted, the retention times shifted significantly.

D. Detection Limit

Detection limits for the current microchip GC with thermal conductivity detectors generally are between 0.2 and 1 ppmv.[4,6] These usually are regarded as a limitation of the microthermal conductivity detector. In fact, a 1-ppmv detection limit corresponds to a detection of approximately 10^{-12} g of material which is competitive with flame ionization detection. Injection volumes must be small (<1 μL) with a 0.100-mm column if the high resolution advantage of the column is to be preserved. Thus the high detection limits are due to the well-known tradeoff between resolution and sensitivity rather than an inherent limitation of the detector.

A portable sample concentrator developed at Louisiana State University (LSU) uses fairly straightforward sample preconcentration to increase analyte loading onto the column without increasing injection volumes and lower detection limits by 2 to 3 orders of magnitude.[4] The LSU "toolbox" concentrator is a modular unit that can be used with any instrument capable of analyzing gas-phase samples. A volume, for example, of 1000 mL of air is passed over a multibed sorbent trap and the concentrated analytes desorbed into a small volume on the order of 1 – 2 mL. The resulting 2-mL sample is then analyzed in the same manner as an unconcentrated gas-phase sample. The toolbox concentrator has previously been used with GC/ MS for ambient air analysis.[5] Lowering the effective detection limits of the P200 into the low parts per billion volume range by using the toolbox concentrator begins to make the advantages of the microchip GC available for ambient air analysis.

E. Volatility Range

In the general pattern of portable gas chromatographs, the P200 has an un-heated sample inlet system and analyzes samples in the gas phase only. The main limitation for the P200 is the volatility of analytes; sample preprocessors such as the LSU portable sample concentrator[4,7] allow volatile analytes to be taken from a solid or liquid matrix into a gas-phase sample suitable for analysis by a portable GC. Nevertheless, the unheated injection system imposes a lower limit on the volatility of compounds that can be analyzed with the current system.

A more limiting difficulty is the absence of temperature programming capability in the current microchip GC. The volatility range of components that can be analyzed in a single run is fairly small. A clear disadvantage of the thermal conductivity detector (TCD) is that it responds to water and air. Any early eluting compounds must be resolved from the air and water background if they are to be detected. The low temperatures required for detection of the most volatile components degrades the chromatography of later eluting compounds in the sample both by diffusion controlled band broadening and potentially by column overloading. Column overloading is not currently a problem because the analyte loadings are small for ambient air even with sample preconcentration. Higher degrees of sample preconcentration could potentially create overloading problems for compounds having low vapor pressures.

A single isothermal run with a 4-m P200 column is capable of analyzing compounds over a range of about 400 Kovat's retention index units. For example, compounds eluting between butane and octane or, at a higher temperature, between pentane and nonane may be analyzed in a given run. Because of the short run times one may conveniently change the temperatures between chromatograms and make one run at a low temperature and make a second run at a higher temperature. Alternatively one may configure the instrument with identical columns and use different temperatures on each. At best these methods are inconvenient and at worst, greatly reduce the general applicability of the instrument with regard to selectivity and qualitative information content.

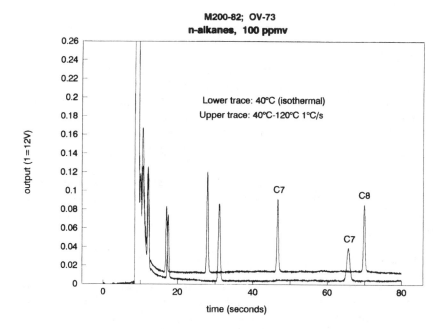

FIGURE 4. Comparison of isothermal and temperature programming.

Preferable is the alternative of temperature programming a single run. The currently available microchip GC with only a thermal conductivity detector available suffers considerably from baseline drift with temperature programming. Programming rates useful for these short run times must be in the neighborhood of 0.5°C/sec. Under such conditions, the baseline drift can be as much as several volts. For various instruments baseline drift as low as 100 mV and as high as 4 V have been seen for a temperature change from 40 to 120°C at 1°C/sec. The observed baseline drift is fairly reproducible for a given instrument but nevertheless makes peak detection difficult. We have developed temperature programming capability for the microchip GC and while the details will be discussed at a future time, examples are shown in Figure 4 to illustrate the potential advantage of temperature programming.

III. M2001 SOFTWARE

The M2001 software package was developed at LSU-Institute for Environmental Studies (IES) under the auspices of the National Oceanic and Atmospheric Administration (NOAA). Based on the Apple Macintosh®* line of personal computers, M2001 is an instrument control and data acquisition program for use in the field with the Microsensor M200 and P200 microchip gas chromatographs. It was originally designed to facilitate use of the microchip GC by personnel

* Registered trademark of Apple Computer Incorporated.

responding to chemical emergencies or acutely hazardous situations. Design criteria were simplicity of use, general applicability, and minimal user intervention during data analysis. To these ends, the software uses the Macintosh® "point and click" interface to adjust instrument parameters such as sample injection time, run time, column temperature, etc. An automatic peak integrator requiring no user intervention finds and integrates the chromatographic peaks from each of the two columns. The "smart" integrator uses fuzzy logic algorithms to choose tangent skimming or peak merging when calculating areas of partially resolved peaks. User definition of slope sensitivities or timed events is unnecessary. Because of the high level of automation in the data treatment, changes in operating conditions pose no difficulty with regard to timed events. Finally, an identification algorithm makes sample component identifications using a permanent retention index library. Instrument calibration at run time comprises two parts; a quantitative calibration and a qualitative calibration to make minor changes in the retention index library.

A. Integrator/Quantitation

The automatic integrator tracks the baseline trend and uses deviations from this trend to define the beginning of peaks. Peak apexes are determined using a simple first derivative test. Peak stops are found by comparing the trace with the extrapolation of the baseline trend from the peak start. The method is sufficiently robust that even though the decision criteria are hard coded, the peak finder and integrator accommodate well changes in operating conditions.

The system supports single-point or multipoint calibration for up to 12 calibrants. For sample components not included in the original calibration, semiquantitative estimates are based on the nearest eluting calibrant. This should give estimates within less than ±50% for the majority of analytes.

B. Identifier/Qualitation

The third major feature of M2001 is the automatic identifier which uses the substantial qualitative information in the dual column chromatograms to identify sample components. As discussed above, the high retention time precision of the microchip GC makes it a high resolution identifier. The qualitative calibration maps the retention index library onto the current retention time axes using the same calibrants as in the quantitative calibration. The use of a retention index library system with this run time recalibration further improves the reliability of identifications and can obviate the need for transporting to the field authentic standards of all potential compounds. Thus, a simple mixture of alkanes in nitrogen or air serves as reference for all compounds in the library.

After the chromatogram has been integrated, the peak lists from the two columns are fed into an identification routine that uses a game-playing algorithm[8] to determine the combination of compounds from the library that produces chromatograms most similar to those obtained from the sample. The identifier uses

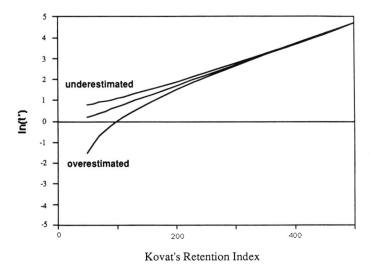

Kovat's Retention Index

FIGURE 5.   Effect of incorrect column dead times estimates.

retention times on both columns, as well as relative peak areas to account for as many peaks as possible. It accounts for the possibility that some peaks correspond to coeluting analytes and that some peaks may be due to compounds that are not in the library.

C. Retention Index Library

The retention index library is based on Kovat's system using normal alkanes as retention references. The current retention index library contains 107 compounds and was created using a single instrument over the course of several months. The 107 target compounds consisting of aliphatic hydrocarbons, chlorinated hydrocarbons, aromatics, ketones, and alcohols were individually run in mixtures of n-alkanes in nitrogen. Concentrations were kept low enough (below 2000 ppmv) that column overloading did not alter observed retention times.

The resulting retention data for the 107 compounds and n-alkanes were then used to create a linear mapping of retention times, into retention indices. Because of the extraordinary precision in retention times the average retention times for the alkanes provided reference retention times with very low standard errors, a factor of 5 less than the time resolution of the data acquisition. The procedure used to calculate retention indices for the library is very similar to the calibration procedure used when analyzing samples. The mapping of retention time onto retention index is a three-parameter curve fit describing the linear relationship between the logarithm of the adjusted retention time and retention index. In addition to the slope and intercept that define the line, a third parameter — column dead time — is incorporated into the adjusted retention time. Figure 5 illustrates the importance of this third parameter. The overestimate and underestimate curves correspond to an error of about 2% in the dead time estimate. The retention index scale in Figure 5 corresponds to a temperature of 40°C; for higher temperatures the deviation from linearity becomes more significant at higher retention indices. Traditionally, the retention time of an unretained peak has been used to estimate column dead time. The retention of air to the extent of 0.5 ± 0.2 sec results in a major overestimate of the column dead time. Therefore M2001 uses an iterative approach, similar to that of Wronski et al.,[9] to estimate column dead time. Repeated least squares fits are obtained by linear regression while adjusting column dead time estimates to maximize the correlation coefficient. The resulting line is then used as the mapping function to convert between retention times and retention indices. The main difference between the calibration and library computation is that the very low uncertainty in the reference retention times and column dead times provides a much more accurate mapping between retention index and retention time. Replicate determinations of library retention indices over the course of several months further ensure that the accuracy of the analysis time calibration, not uncertainties in the library, ultimately determines the calibration accuracy. Retention indices calculated for the library entries showed no change over the course of 6 months. Standard deviations were approximately 0.2 index units for most compounds.

Though Kovat's indices are somewhat independent of temperature and pressure, the dependence was large enough to require that the library be created under a fixed set of conditions in order to conserve the maximum amount of available qualitative information. The library was created in three temperature segments to allow a choice of temperatures without diminishing the accuracy of the library. A 40°C segment contained information on approximately 30 compounds having retention indices between 350 and 800. A 60°C segment contained information on approximately 90 compounds having retention indices between 400 and 1000. Finally, a 100°C segment contained information on approximately 80 compounds with retention indices between 600 and 1100. The substantial amount of overlap between the segments helps ensure that the library contains the desired compounds at the temperature required for a particular separation.

Flow rates were chosen to achieve gas velocities between 100 and 200% of the optimum determined by Golay plots for each column. A second factor in the choice of flow rates was the fact that retention is generally higher on the more polar column. Because of this, some compounds that elute on the OV-73 column will elute outside of the run time window on the more polar OV-1701 column if identical flows are used. To reduce the number of times that compounds are observed only on the OV-73 column the more polar column was operated at a slightly higher carrier gas velocity. The use of this restricted set of operating conditions assures the maximum retention of qualitative information in the library, yet it still allows some flexibility in choosing conditions that provide optimum separation.

Fundamental to the general usefulness of the M2001 identification scheme is the stability of the retention index library. The library must accurately transport

Table 1. Retention Index Variation Over 4-Week Period

Name	Period 1	Period 2	Period 3	SD
Vinyl acetate	586.6	585.6	585.3	0.75
Hexane	600.9	600.0	599.8	0.67
Methylethyl ketone	604.7	604.7	604.4	0.19
Ethyl acetate	620.6	620.1	620.2	0.30
Chloroform	623.7	623.7	623.7	0.03
1,1,1-trichloroethane	652.8	652.9	652.8	0.10
Benzene	669.6	669.3	669.4	0.14
Cyclohexane	669.4	669.8	669.7	0.21
Carbon tetrachloride	669.8	669.8	670.1	0.19

the relative retentions of the included compounds from the laboratory to the field. For a large library of compounds to be feasible the library must be stable with time as well; frequent recalculation of large numbers of retention indices quickly becomes too cumbersome to be practical. Likewise, if the library does not accurately represent this information for different instruments, the creation of separate libraries for each instrument is an unwelcome proposition. The two primary questions then are

1. How stable are library entries over time?
2. How well does the library transport to other microchip GCs?

The results summarized in Table 1 show the stability of the library entries with time for a single instrument. These results correspond to the same 4-week period as in Figure 1 and are typical of the 107 compounds in the library. Particularly significant is that although retention times show a significant shift between the first and second periods, the retention indices show no such shift. Furthermore, after 6 months the retention indices showed no change from the earlier values shown in Table 1.

The answer to the second question is central to the ultimate value of the M2001 identifier/library system. The well-known variability of retention indices between different batches of the same column material made the transportability of the library between instruments suspect. Several microchip GCs were used to test the transportability of the library. A group of 14 compounds, representing the range of functionality found in the library, were analyzed on three additional microchip GC units. At this point, the library was 6 months old and (as mentioned above) retention indices obtained on the unit used to create the library showed no change from the original values. The transportability of the system was indicated by the ability of the M2001 software to accurately predict retention times of library compounds based on the qualitative calibration using a series of n-alkanes. Each instrument was evaluated over the course of 5 days. At the end of each day, the power and carrier gas flow to all instruments were turned off. This required restarting and recalibrating each instrument at the beginning of each day followed by analysis of each of the 14 compounds. Generally the system performed very

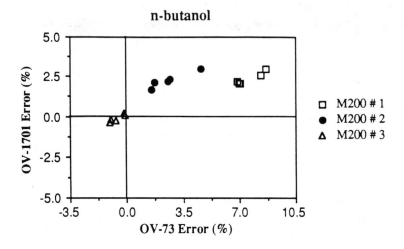

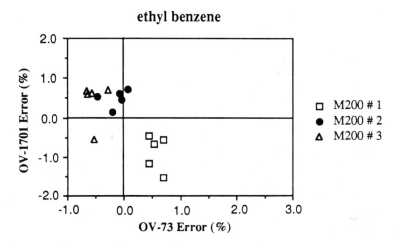

FIGURE 6.   Precision of predicted retention times.

well and predicted retention times to within 1% of the observed values. For
nonpolar or slightly polar compounds such as ethylbenzene, predictions were
within 1% for all three instruments (Figure 6a). For more polar compounds or
compounds that tend to adsorb, such as n-butanol or dibromochloromethane, the
predictions were less accurate and varied more from instrument to instrument
(Figure 6b). In the worst case, for n-butanol the predicted retention time was 8%
too high, though 4% was more typical. The typical deviations between observed
and predicted retention times for the 14 compounds are shown graphically in
Figure 7. Note that the prediction for n-butanol, a polar hydrogen-bonding

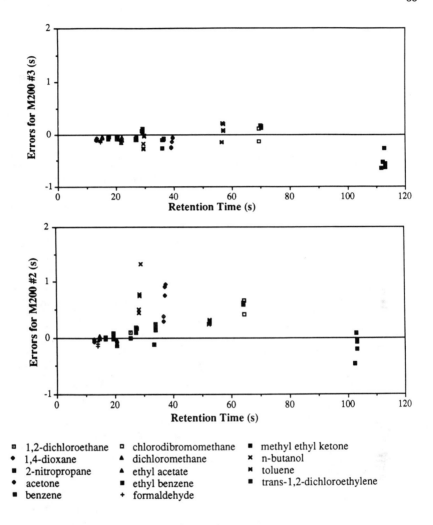

FIGURE 7.   Accuracy of predicted retention times (OV-73).

compound, is uniquely poor. For most of the other test compounds, including some of quite high polarity, the prediction errors are 2% or less.

IV. CONCLUSIONS

The current microchip gas chromatograph provides a unique combination of portability and qualitative information content. Short microbore capillary columns provide reasonably good component resolution in a fraction of the time required for more conventional chromatographs. Injectors and detectors fabricated from silicon microchips minimize extracolumn losses of the resolution advantage provided by small bore columns. The transportability of retention information from one instrument to another further extends the general applicability of the system. Advantages of the microthermal conductivity detector may offset many of its shortcomings. For example, the universal nature of the detector, having a fairly constant response factor over a range of compounds, makes the instrument useful for semiquantitative analysis when authentic standards of particular analytes are not available.

As a stand-alone instrument for ambient air analysis, however, the microchip GC suffers principally from inadequate sensitivity. The use of a portable sample preconcentrator lowers detection limits to the point that the instrument begins to be useful for ambient air monitoring. The lack of a heated injection system is not a serious problem for analysis of volatile compounds. A more important concern is the lack of temperature programming capability. This limits the range of compounds that can be analyzed in a single chromatographic run and degrades detection limits for later eluting compounds. Furthermore, the lack of temperature programming may require that the instrument be operated at lower temperatures than is desirable for late eluting compounds in order to resolve early eluting compounds. This can aggravate the problem of column overloading that could occur if some components in the samples are concentrated to levels above 2000 ppmv.

Overall, the positive attributes of the microchip GC make it a potentially valuable tool for use in situations that require mobile equipment. Use with a portable sample concentrator can lower detection limits to the low parts per billion volume range and make the microchip GC a viable option for some ambient air monitoring applications.

REFERENCES

1. Huber, J. F. K., E. Kenndler, and G. Reich. "Quantitation of the Information Content of Multidimensional Gas Chromatography and Low-Resolution Mass Spectrometry in the Identification of Doping Drugs," *J. Chromatogr.,* 172, 15 (1979).
2. Giddings, J. C. "Multidimensional Chromatography," *Anal. Chem.,* 56, 1238A (1984).
3. Grob, R. L. *Modern Practice of Gas Chromatography* (New York: John Wiley & Sons, Inc., 1985), p. 235.
4. Backhouse, T. H., MS Thesis, Louisiana State University, Baton Rouge, LA (1990).
5. LSU Institute for Environmental Studies Instrument Development Group, "Air Monitoring in the Area of St. Gabriel, Louisiana," Report to the Geismar Area Industrial Technical Group, St. Gabriel, LA (1988).
6. Chang, J. S. C., S. M. Gordon, and R. E. Berkley. "Laboratory Evaluation of Microsensor Technology Gas Analyzer," Proceedings 1990 EPA/A&WMA International Symposium on Measurement of Toxic and Related Air Pollutants, Raleigh, NC, (May 1990), 830.
7. LSU Institute for Environmental Studies, Final Report, National Oceanic and Atmospheric Administration, U.S. Department of Commerce Contract #50-ABNC-7-0100 (1990).

8. Steele, C. F., J. J. Stout, K. R. Carney, and E. B. Overton. "Computer Software for Chromatography in the Field," *Proceedings 1990 EPA/A&WMA International Symposium on Measurement of Toxic and Related Air Pollutants,* Raleigh, NC (May 1990), 855.
9. Wronski, B., L. M. Szcepaniak, and Z. Witkiewicz. *J. Chromatogr.* 364, 53–61 (1986).

CHAPTER 3

Theory and Application of the U.S. EPA Recommended Surface Emission Isolation Flux Chamber for Measuring Emission Rates of Volatile and Semivolatile Species

C. E. Schmidt

TABLE OF CONTENTS

0-87371-606-0/93/$0.00 + $.50

© 1993 by Lewis Publishers

I. INTRODUCTION

Measuring the emission rate of organic and inorganic compounds from area sources presents a variety of challenges for atmospheric scientists. Emission rate data are preferred over downwind air concentration data (mass/volume) for most applications because emissions rate data can be used for many purposes such as input to dispersion models to estimate downwind concentrations over various time periods and for various meteorological conditions. Emission rate data are thus useful for assessing long-term air impacts and for health risk assessments. Emission rate data can also be used to satisfy the air pathway analysis requirements for national priority list (NPL) waste site investigation and to satisfy permit requirements at controlled waste management facilities or treatment processes.

There are many acceptable approaches for assessing emissions from area sources. These approaches will be described in limited detail herein. However, the focus of this chapter is to present, in detail, information regarding the theory and application of the EPA recommended direct emissions assessment technology. Discussions and references are provided for direct emissions measurement of land surfaces; nonaerated, nonmixed liquid surfaces; mixed and aerated liquid surfaces; and fugitive emissions from process control measures or equipment. Use of emissions rate data for air pathway analysis and health risk assessment is also presented.

II. FOUR APPROACHES FOR ASSESSING EMISSIONS

Recently, the EPA has formalized the approach for conducting air pathway analysis (APA) at hazardous sites in support of site restoration activities and exposure assessment for undisturbed and disturbed (i.e., during remediation) waste sites.[1-4] This approach is described in a four-volume series of guidance documents, one of which is focused on the baseline assessment of emission rate from area sources such as hazardous waste landfills and lagoons. Volume 2 of this series describes four approaches for estimating emission rates of gaseous and particulate matter including: direct emission measurement, indirect emission rate assessment, fenceline monitoring and modeling, and predictive modeling.[2] These four basic approaches for measuring or estimating emission rate, described in and paraphrased from APA Volume 2, are discussed below. The reason for including this material in this chapter is to offer measurement options other than direct measurement and to illustrate applicability of the flux chamber technology.

A. Direct Assessment Approach

Direct emission measurement technologies are often the best approach for assessing emission rates from area sources. The emission rates that are typically generated can be used as input for dispersion models to predict ambient concentrations at various locations under varying meteorologic conditions. The techniques generally consist of isolating or covering a small surface using a chamber or enclosure. The concentration of gas species from the emitting surface is measured within the chamber or from an outlet line. These concentration measurements, along with other technique-specific parameters, are then used to calculate an emission flux or relative concentration value. By combining data from multiple measurement locations, the emission flux (rate per area per time) can generally be related to an emission rate for the entire source. Most direct measurement techniques are best suited for the measurement of volatile species.

The direct emission assessment approach is generally preferable to other approaches because it has been proven to be relatively cost effective for obtaining emission rate, and it avoids modeling or estimation techniques. Direct emission measurement technologies and equipment are generally relatively simple and straightforward. The cost of the direct emission measurement technologies varies with the number and type of analyses to be performed. Most of the direct techniques, however, are cost-effective relative to other approaches since several measurements can usually be made in a given day. Real-time instruments can be used with all the direct technologies to provide immediate data for decision making during the sampling program, and for the relative ranking of the emission rate at locations across the site. This procedure can be used to reduce the number of samples requiring laboratory analysis by screening for those samples with significant concentrations.

Direct emission rate assessment has several advantages when compared to the other approaches. The single, most significant advantage of this approach is that all data needed for the emissions assessment can be measured and an emission rate can be calculated from the measured data. No modeling or extrapolation is required, and the accuracy and precision of the emissions data from the direct assessment approach generally is of improved data quality and utility over emissions estimates from other approaches. The approach is also free from upwind contamination effects and, to a large extent, meteorologic effects.

The preferred direct emissions assessment technology, as described in APA Volume 2, is the EPA recommended surface emission isolation flux chamber.[2,5] This direct emissions assessment technology has been used for a variety of applications successfully and has recently been reported by EPA as the most useful and commonly used measurement/estimation technology.[6]

B. Indirect Assessment Approach

Indirect emission estimation technologies generally consist of measuring the ambient concentration of the emitted species from an array of samples and then applying these data to an equation (air dispersion model) to determine the emission rate. Many of the equations were developed to determine downwind concentrations resulting from stack emissions. For area emission sources, the source is treated as a virtual point source or line source. The most commonly used indirect technology is the transect approach.[7] The transect approach uses a specific air dispersion model and concentration data collected from an array of point samples configured to transect the plume. Remote sensing and line-source vs point-source sampling is also a viable approach.[8]

Indirect technologies are very similar and most involve clusters of ambient air samplers (point samplers). The concentration profile technique involves a vertical array of samplers directly over the source.[9] The boundary layer technology is a simplified version of the transect technology and involves several downwind samplers each at a different height.[10] Because of their cost and complexity, the indirect technologies are usually used for measuring emissions when direct measurement techniques prove to be unsuitable.

A disadvantage of indirect emission measurement technologies is that the results are highly dependent upon meteorologic conditions. They require meteorologic monitoring to properly align the sampling systems and to analyze the data following sample analysis. Changing meteorologic conditions significantly affect the likelihood of collecting useful data. Unacceptable meteorologic conditions may invalidate much of these data collected, requiring an additional sampling effort. These technologies also may produce false negative results if the emitted species are present in low concentrations which are below the sampling and analysis detection limits, or if upwind sources cannot be fully explained. Also, it may not be feasible at some sites where the source area is excessively large, or where insufficient space exists downwind of the source to set up the sampling array without disturbance of the air flow pattern by obstructions (e.g., buildings, tanks).

The types of volatile and particulate species that can be measured by these technologies are essentially unrestricted and are dependent on the sampling media selected and analysis technique rather than on the emission measurement technology. Either volatile or emissions of particulate matter species can be assessed using these technologies. Sampling approaches that require long sampling time periods, however, limit the use of the technology because the changing meteorologic conditions during that time period will affect the emission rate estimation capability of the model. Sampling media and analysis techniques are not discussed here.

Indirect emission measurement technologies generally do not provide significant data on the emission rate variability for different locations across a site. This is because the emission concentration is measured downwind of the site after some atmospheric mixing. They generally do not allow for the evaluation of individual contaminated areas at the site unless the areas are separated from one another and are not located upwind of one another.

The costs of the indirect emission measurement technologies vary considerably. The indirect technologies are complex and require considerable equipment, labor, and analysis costs. All of these technologies are subject to data loss or sampling delay due to inappropriate meteorologic conditions.

C. Air Monitoring/Modeling Approach

Air monitoring technologies can be combined with air dispersion modeling to calculate the area source emission rate. The primary difference between indirect measurement technologies and air monitoring technologies is the distance at which measurements are made downwind from the source. Indirect measurements are made near the source (usually onsite). Air monitoring is generally performed at a distance downwind from the source and can measure emissions from the entire site. Air monitoring technologies typically measures lower concentrations because the contaminant plume is subject to additional air dispersion as compared to indirect technologies.

The first step in using the ambient air sampling data to develop emission rate estimates is to select an air dispersion model which accurately reflects the site-specific conditions, including regional and local terrain, typical wind stability, etc. Guidance for selecting an appropriate model is given in the EPA guideline on air quality models.[11] Air monitoring and air dispersion models are used to determine the emission rate through an iterative process. An emission rate is first estimated for the area source. This estimated emission rate, along with meteorologic data collected during air monitoring, is used to calculate a predicted downwind concentration. The predicted concentration is then compared to the measured downwind concentration. Based on this comparison, the estimated emission rate is adjusted appropriately, and the process is repeated until acceptable agreement is reached between the measured and predicted downwind air concentrations. The emission rate estimates made in this manner generally have a very large uncertainty associated with them.

D. Predictive Modeling Assessment Approach

Emissions models have been developed to predict emission rates for a variety of area source types.[2] Each model is generally applicable to a specific type of area emission source.

The predictive models can be used as screening tools or for estimating emission rates. Emission models used to provide a crude estimate of emission rate employ data that can be obtained or calculated from information available in the literature, or can be assumed with some level of confidence. Emissions models used to provide an emissions rate estimate with higher confidence require site-specific and waste characterization data. The selection of model input sources (site-specific, literature value, or assumed) should be based on the requirements of the decision-making process and the level of resources available. Site-specific data should be used whenever possible to increase the accuracy of emission rate estimates.

The emission models for soil (nonliquid) sources are diffusion controlled. The key input variables are area and size of the emission source, types of contaminants present, concentration of contaminants, depth of any soil cover, air-filled porosity of soil, and vapor pressure of contaminants.

III. SURFACE ISOLATION EMISSION FLUX CHAMBER TECHNOLOGY

The emission isolation flux chamber is a device used to make direct emission flux measurements from land or liquid surfaces such as landfills, spill sites, and surface impoundments. The enclosure approach has been used by researchers to measure emission fluxes of a variety of gaseous species including sulfur-containing species and volatile organic species.[12-20] The approach uses an enclosure device (flux chamber) to sample gaseous emissions from a defined surface area. Clean dry sweep air is added to the chamber at a fixed controlled rate. The volumetric flow rate of sweep air through the chamber is recorded and the concentration of the species of interest is measured at the exit of the chamber.

A. Theory of Operation

The flux chamber technology for hazardous waste applications was developed and tested under contract to EPA-EMSL (Las Vegas) by Radian Corporation. The design, theory of application, and user information is documented in the EPA User's Guide.[6] The intent of this research effort was to develop an area source emissions assessment technology that was accurate, precise, relatively easy to use, field practical, and inert or nonreactive when in contact with chemicals/waste materials. Several designs were constructed and tested leading to the current design. The flux chamber design has satisfied the original design objectives and has been tested for mixing properties (or performance as a continuously stirred tank reactor), precision (repeatability), accuracy, inertness (recovery of standards), and reproducibility.[21] The User's Guide reports a precision of about 5% and a variability (reproducibility) of 9.5% on land surfaces. Precision was determined by laboratory studies using gas standards; and reproducibility was determined from bench-scale studies on controlled, simulated land surfaces emitting volatile species. Accuracy was established by measuring recovery of gas standards. As reported in Table 1, the average recovery for 40 test compounds was about 103%. Good recoveries were observed for most species tested with the exception of low levels (i.e., less than 10 ppbv) of some compounds including reduced sulfur-containing species. Precision and accuracy under field conditions will be higher. Field data for precision vary from 10 to 20%.[5]

Table 1. Compounds Tested in the Emission Isolation Flux Chamber and the Measured % Recovery

Compound	% Recovery	Compound	% Recovery
Total C_2	100	3-Methylhexane	106
Total C_3	108	2,2,4-Trimethylpentane	106
Isobutane	109	n-Heptane	103
1-Butene	108	Methylcyclohexane	103
n-Butane	106	Toluene	103
t-2-Butene	107	Ethylbenzene	94.7
c-2-Butene	109	m,p-Xylene	88.5
Isopentane	112	o-Xylene	97.3
1-Pentene	105	n-Nonane	99.4
2-Methyl-1-Butene	124	n-Propylbenzene	95.5
n-Pentene	103	p-Ethyltoluene	92.5
c-2-Pentene	105	1,3,5-Trimethylbenzene	93.5
Cyclopentene	105	1,2,4-Trimethylbenzene	88.7
Isohexane	107	2-Methyl-2-butene	103
3-Methylpentane	106	Methyl mercaptan	107
Methylcyclopentane	105	Ethyl mercaptan	107
Benzene	106	Butyl mercaptan	101
1,2-Dimethylpentane	105	Tetrahydrothiopene	115

The sensitivity of the technology is dependent on the analytical method used for application. A typical sample collection includes real-time monitoring; grab sample collection by syringe or evacuated canister; and integrated sample collection by solid sorbent, absorbing solution/impinger, or evacuated canister. With canister sample collection and offsite cryofocus/gas chromatography (GC)/mass spectrometry (MS) analysis, sensitivities in the tens of picograms per square meter per minute can be attained. For instance, a chamber gas concentration of 1 ppbv with a sweep air flow rate of 5.0 liters per minute (Lpm) is an emission rate of about 0.1 $\mu g/m^2$ min^{-1}. For halogenated species and detection by GC/electron capture detector (ECD), emission rates of 1 pg/m^2 min^{-1} for TCE can be achieved. Range of detection, however, is as important as low-end sensitivity. Again, like sensitivity, range is determined by sample collection and analytical method. Using evacuated canisters as a collection media, range is not limited since any desired dilution or sample volume can be obtained from the canister and injected in the GC.

Since the chamber is constructed from relatively inert materials and operated in the dynamic mode, chamber interaction with sample constituents is usually not a concern. Blank system testing should be conducted per application by placing the chamber on a clean, inert surface and sampling per project protocol for species of interest. Since the chamber dome is acrylic and sample lines/air supply can off-gas organic species, low levels (parts per billion volume levels of toluene and some halogenated species) are not uncommon. Crossover contamination during field testing is possible and can be avoided by scheduling (measure low emission sources first, high emission sources last), allowing for adequate chamber purging (minimum of 5 residence times); and/or by cleaning the chamber with soap and water, flushing the sample lines, drying the chamber/lines, and retesting blank levels as needed. Species loss by wall adsorption is more difficult to detect, but can be tested by determining the recovery of target compounds. Wall effects are typically minimal because most wall surfaces are stainless steel or Teflon®* and the system is typically well purged prior to sample collection which passivates the wall surface and minimizes sample loss.

Additional information on the use of the flux chamber technology including sensitivity analysis and emission rate data from a variety of sources is found in the EPA database of emission rate measurement projects.[5]

B. Standard Operation

The standard operation of the flux chamber includes mobilizing the test equipment to the site, performing setup quality control (system checks, instrument calibration, blank testing), screening the site and/or selecting test locations, and measuring emission rates. The user must have knowledge of the site and have developed a strategy for assessing emissions from the area source. The objective is to measure the emissions from each similar zone and then sum emissions from each zone on a surface area basis and estimate emissions from the site. Distinct zones of emission rate can often be distinguished by chemical content, surface cover, oil layers, or other visible/obvious indications. The EPA User's Guide provides a prescription for representative testing which includes gridding the site, identifying zones of similar emissions, and then testing until project objectives for variability in these data (coefficient of variation [CV]) are achieved. The level of testing, thus, is determined by the consistency of emissions from grid point to grid point per zone and the predetermined data quality objectives. Oftentimes, useful data can be obtained from a few tests within a zone for applications such as a detailed screening study where high confidence in these data is not required. Full characterization of an area source, however, can include intense testing.

A standard sampling protocol that can be used for most applications follows:

1.   Locate the flux chamber, sweep air, real-time gas analyzer(s), sample collection equipment, and field documents at the test location.

2.   Document site information, location information, equipment information, name of sampler, date, and time on the field data sheet (Figure 1).

3.   Select the exact test location and place the chamber on the surface. Do not attempt to seal or force the chamber to the emitting surface. Sealing is generally not necessary and in doing so, the emission rate may be affected by disturbing the surface. Place thermocouples to monitor soil/air temperature inside and outside of the chamber as needed. Temperature data are used to show that the emission event was not disturbed during the measurement or to correlate emission rate to temperature.

---

* Registered trademark by E.I. duPont de Nemours and Company, Inc., Wilmington, Delaware.

FIGURE 2

SURFACE EMISSIONS MEASUREMENT DATA FORM

DATE _____   SAMPLES _____

LOCATION _____

SURFACE DESCRIPTION _____

CURRENT ACTIVITY _____

INSTRUMENT TYPE _____ ID NO. _____ TYPE _____ ID NO. _____

AMBIENT _____

Time	Sweep Air (L/min)	Residence Number	Air Temp (C°)		Real-Time (ppmv)		Sample Number	Comments
			Chamber	Ambient	FID	PID		
		0						
		1						
		2						
		3						
		4						
		5						
		6						
		7						
		8						

TEST COMMENTS _____
_____
_____

SUMMARY: PEAK EMISSIONS _____ TIME _____ TAU _____

STEADY STATE EMISSION _____ TIME _____ TAU _____

FIGURE 1.   Surface emissions measurement data form.

4.   Initiate the sweep air flow rate and set the rotameter at 5.0 Lpm. Constant sweep air flow rate is critical. Record time.

5.   Collect instrument background data (gas analyzers, thermocouples) and record time.

6.   Connect real-time gas analyzer(s) to the exhaust manifold or purge pump as needed. Do not exceed an exhaust gas sample/purge rate of 2.5 Lpm.

7.   Operate the chamber at 5.0 Lpm and record data every residence time (6 min) for at least four residence times or 24 min. Record data. The chamber is at steady state.

8.   Interface the evacuated canister to the purged sample line and collect the gas sample. Do not exceed a collection rate of 2.5 Lpm at any time. This will prevent unwanted dilution of chamber exhaust gas by ambient air. Try to keep sample collection to less than a 30-min time interval, preferably 2–10 min.

9.   Label samples; record sample collection or real-time monitoring data on the data sheet.

10. If collected, properly store/condition gas samples as appropriate.

11. Document sample collection in field master log book (if collected).

12. Discontinue the flux measurement, shut off the sweep air, remove chamber, and secure equipment.

13. Decontaminate the chamber where contact was made with the surface using appropriate cleaning supplies. Purge the sample lines with sweep air for 1–2 min, depending on species and concentrations.

14. Relocate equipment to the next test location and follow Steps 1 through 13.

The emission rate is expressed as:

$$E_i = C_iR/A \qquad (1)$$

where E_i = emission rate of component i, $\mu g/m^2$ min^{-1}
         C_i = concentration of component i in the air flowing from the chamber, $\mu g/m^3$
         R = low rate of air through the chamber, m^3/min
         A = surface area enclosed by the chamber, m^2

All parameters in Equation 1 are measured directly. For standard application (room temperature) and concentration data reported in $\mu g/m_3$, Equation 1 reduces to concentration ($\mu g/m_3$) times 0.0385.

A diagram of the flux chamber apparatus used for measuring emission rates from area sources is shown in Figure 2. (Note that the sweep air addition line should be located near the bottom of the chamber to avoid clean air short-circuiting problems and poor mixing.) The design and operating protocol has been validated through testing at a controlled emission source. The sampling equipment consists of a stainless steel/acrylic chamber, ultrahigh purity sweep air (normally 5 Lpm), a rotameter for measuring flow into the chamber, a thermocouple for measuring the air temperature within the chamber, and a sampling manifold for monitoring and/or collection of the species of interest. For wastes containing volatile organic compounds (VOC), concentrations of total hydrocarbons can be monitored continuously in the chamber outlet gas stream using portable flame ionization detector (FID)- and/or photoionization detector (PID)-based analyzers.

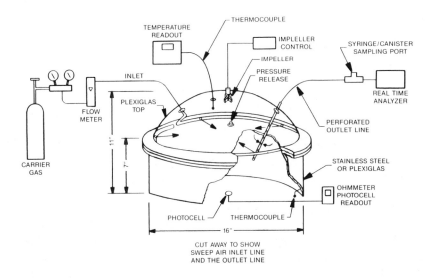

FIGURE 2.   Schematic diagram of isolation emission flux chamber.

A modified design of the emission isolation flux chamber has also been used to measure emission rates during coring of wastes.[22] The downhole flux chamber is lowered down a hollow stem auger in the same manner as a core sampler. This technique makes it possible to measure emission rates as a function of depth from waste bodies which may be exposed at a later date during remedial action activities.

IV. APPLICATION OF THE FLUX CHAMBER TO VARIOUS AREA SOURCES

The EPA recommended flux chamber technology as described in the literature can be applied to a variety of area sources with only minor adaptions and equipment modification. This section describes the application of the technology to land surfaces; nonaerated, nonmixed liquid surfaces, aerated and mixed liquid surfaces, and fugitive process emission sources.

A. Land Surfaces

The flux chamber was first designed and tested for land surfaces where the volatile/semivolatile source (waste material) is on the surface or is subsurface. The difference regarding surface vs subsurface applications is that for direct contact with the waste, emission rates are generally greater; those rates are often subject to influences such as solar heating, ambient temperature, surface moisture, and physical disturbance of the waste material. Where the waste material is subsurface and the volatile species must migrate through a layer of soil, the emission rates can be lower and often are not affected by surface conditions or activity.

Assessing volatile (including inorganic species)/semivolatile species emissions from surface waste (i.e., uncovered RCRA landfill) and land surfaces (farm, sludge lagoon, surface spill) is a straightforward and common application of the technology. Surface conditions are ge nerally recorded during the test including air/soil temperatures inside and outside the chamber. Solar cover data may also be significant for surfaces that absorb solar radiation and increase emission rate as a function of surface temperatures. Since the chamber will be in contact with the waste, contamination of equipment and cross contamination is a concern. This concern can be addressed by cleaning the chamber walls where contact with waste has been made and back-flushing the chamber exhaust or sample line. One approach is to avoid contact by wrapping the chamber lip with wide, disposable Teflon® tape per test location and preventing contact with solid waste material. Care must be taken to not introduce materials into the chamber, such as adhesive tape, that may off-gas contaminate species. All wrapping must be secured outside the chamber. In addition to cleaning and preventing limiting contact with waste, more frequent (10–20%) blank testing is recommended for direct waste contact testing.

Physical disturbance of surface waste must be avoided for undisturbed or baseline emissions assessment. This can be accomplished by suspending the chamber from a tripod or overhead support and lowering the chamber to the surface. There is no need to "seal" the chamber to a surface for the undisturbed test; the chamber should "grace" the surface. All emission measurements should be conducted with atmospheric pressure in the chamber via the open port atop the chamber. Small leaks at the chamber lip will not significantly affect the measurement. Disturbing an exposed, weathered waste surface may cause a significant increase in emission rate since the exterior may be acting somewhat like a vapor cap. Disturbing the surface and increasing emissions would invalidate the undisturbed or baseline test.

The subsurface waste application (i.e., buried waste, contaminated groundwater) is straightforward, and lower emission rates can be expected with greater depth-to-source distances. Surface coverings greatly affect emission rates, and these must be considered when designing the testing approach. Oftentimes emissions are channeled between surface barriers, and higher emissions can be observed next to building foundations and parking lots. Barometric pressure can affect soil gas diffusion and emissions of volatile species at the surface, as well as soil type, soil compaction, surface water (rain events), and soil moisture. Any condition or process that may affect soil porosity (i.e., water) and the migration of soil gas (i.e., nearby vapor extraction well) must be considered in the testing strategy.

B. Nonaerated, Nonmixed Liquid Surfaces

Application of the technology to liquid surfaces requires floating or suspending the chamber on/over the liquid surface. For nonaerated and nonmixed liquid surfaces such as abandoned lagoons and ponds, quality assurance testing has demonstrated that lower emission rates will be found if the chamber lip penetrates the liquid surface and prevents communication of the isolated surface with the waste body.[23] In nonmixed liquid systems, natural mixing occurs by heating and cooling of the surface layer and the resulting connective mixing. If the chamber contains a layer of trapped liquid, the volatile species may diffuse and the resulting liquid layer may retard emissions. This could result in a bias in test data (in this case, a negative bias). This bias can be prevented by suspending the chamber and keeping the chamber lip floating just under the liquid surface. Flotation/suspension systems have been designed and used successfully.[24] Aside from this concern, liquid testing can be accomplished following the testing protocol as described for land surfaces including spatial and temporal test strategies.

C. Aerated, Mixed Liquid Surfaces

Aerated and/or mixed surfaces, usually associated with a treatment process such as municipal sewage or industrial wastewater treatment, are tested as nonaerated, nonmixed liquid surfaces.[25,26] Mixed liquid surfaces usually prevent the low bias situation described in the previous section by continually renewing the liquid surface. For vigorous mixing, it may be necessary to attach flotation devices to the chamber to improve the stability of the equipment preventing chamber upset.

Aerated surfaces are unique in that bulk air flow from the system strips volatile species and carries contaminants into the chamber. This air flow must be measured and used in the calculation for determining emission rate. Aeration flow can be measured by attaching a volume-calibrated, deflated plastic bag to the chamber pressure port, sealing all other ports, and timing the bag filling.[25] Dividing bag volume by filling rate affords the aeration flow rate assessment. Aeration rate can also be measured using a pump and a manometer to match flow at zero pressure difference and a mass flow measuring device such as a rotameter.

In assessing emissions from an aerated treatment process, both aerated and nonaerated zones must be assessed independently. Average unit emission rate $(g/m^2 min^{-1})$ from each zone and estimated surface areas (m^2) are needed to assess emissions (g/min) from each zone. Soil biofilters with forced aeration, which are aerated systems similar to aeration basins, have also been tested using this technology.[27]

D. Process Fugitive Seam/Leak Assessment

The flux chamber technology can also be used to assess emissions from passive vents, seams, leaking valves, ports, and cracks in control devices ranging from fixed and floating roofs to clay caps on landfills.[28] These applications all have the same two modifications to the standard protocol:

- The chamber must be adapted to the fugitive source.
- The process or source must be well understood in order to properly design a testing strategy and assess the area source.

Adapting the chamber to these fugitive emission sources can be as simple as placing the chamber on a flat seam or as involved as constructing an adapter to interface between the port, valve, or process opening. When possible, these adapters should be made from inert materials and the entire system should be blank-tested. Operating conditions such as flow rate and residence time parameters may need to be changed due to increased enclosure volume.

The requirement of representative testing, however, is a bigger challenge for this application. Typically, process seam/fugitive emissions are first surveyed with real-time analyzers and all fugitive emissions are identified, organized into zones or ranges of similar emission potential, and tested as zones of emissions potential as described for land surfaces. An estimate of emissions can be obtained by averaging emission rate per zone and calculating emissions per zone by knowing the number of sources, area of source, or linear feet of source. Unit emission rate data for this type of source can have units such as mass per time per vent or foot of seam leak.

V. EMISSION RATE DATA USAGE

Emission rate data from area sources can be used to satisfy a variety of project needs including waste site management investigations, support for feasibility studies, air pathway analysis, and input to health risk assessment. The mass unit emission rate data per volatile/semivolatile organic and volatile inorganic species data are useful for predicting, via modeling, downwind (fenceline) concentrations under a variety of dispersion and transport conditions. These data can be used to satisfy air pathway analysis requirements for screening and in-depth assessment.

Emission rate data can also be used as input to dispersion modeling supporting health risk assessment. Having emission rate data, as opposed to a limited air monitoring data set, offers the advantage of providing a better estimate of yearly exposure since a range of dispersion and transport conditions affecting exposure can be modeled. In some cases, when only air monitoring data are available, it is advantageous to estimate emission rate via modeling to obtain modeled emission rate data and then to estimate exposure by conducting modeling.[2] Direct measurement of emission rate data is, however, preferred over indirect monitoring and modeling approaches for most applications.

VI. SUMMARY

The EPA recommended surface emission isolation flux chamber technology has many uses and applications for assessing emission rate from area sources. Current limitations of the technology include a lack of understanding of those parameters that affect the emissions event per application and a limited amount of quality assurance data for diverse applications of the technology. Users of the technology must be aware of the limitations of the technology and attempt to collect emission rate data that represent the area source and that are not perturbed or biased in the process of applying the measurement technology. Typical errors in using the technology include improper assessments due to a lack of understanding of the emission source and/or the measurement technology, and attempts to define area sources with only a limited set of data (i.e., data extrapolation). All emission sources should be screened with real-time analyzers and the area source should be divided into zones of similar emissions potential. Each zone must then be studied at a level of testing that will satisfy program specifications. In addition to spatial differences, diurnal and seasonal effects should also be included in the testing strategy.

Finally, in order to ensure high quality, representative emission rate data, the technology including testing strategy, equipment design and fabrication, and data use/interpretation, all should adhere to the EPA recommended protocols as described in the EPA User's Guide. Using the technology as described in this guide can provide meaningful and high quality area/fugitive source emission rate data.

ACKNOWLEDGMENTS

I would like to thank John Clark and Bart Eklund of Radian Corporation for reviewing/editing this chapter. Furthermore, I would like to acknowledge their active participation in developing, testing, and applying these area source assessment technologies, in particular, the surface emission isolation flux chamber technology.

REFERENCES

1. Air/Superfund National Technical Guidance Study Series, Vol. 1 — Application of Air Pathway Analyses for Superfund Activities, Interim Final, EPA-450/1-89-001 (July 1989).
2. Air/Superfund National Technical Guidance Study Series, Vol. 2, — Estimation of Baseline Air Emissions at Superfund Sites, EPA-450/1-89-002, Radian Corporation (August 1990).
3. Air/Superfund National Technical Guidance Study Series, Vol. 2 — Application of Air Pathway Analyses for Superfund Activities, Interim Final, EPA-450/1-89-001. (July 1989).
4. Air/Superfund National Technical Guidance Study Series, Vol. 4 — Application of Air Pathway Analyses for Superfund Activities, Interim Final, EPA-450/1-89-001 (July 1989).
5. Air/Superfund National Technical Guidance Study Series, Database of Emission Rate Measurement Projects — Technical Note, EPA-450-1-91-003, Radian Corporation (June 1991).
6. Measurement of Gaseous Emission Rates From Land Surfaces Using an Emission Isolation Flux Chamber User's Guide, EPA Contract No. 68-02-3889, Radian Corporation (February 1986).

7. Cowherd, C., K. Axetell, C. M. Guenther, and Jutze. Development of Emission Factors for Fugitive Dust Sources EPA 450/3-74-037, U.S. EPA, Cincinnati, OH (1974).

8. Scotto, R. L., T. R. Minnich, and M. R. Leo. "A Method for Estimating VOC Emission Rates from Area Sources Using Remote Optical Sensing," presented at the 1991 AWMA/EPA International Symposium on the Measurement of Toxic and Related Air Pollutants, Durham, NC (May 1991).

9. Thibodeaux, L. J., D. G. Parker, and M. M. Heck. "Measurements of Volatile Chemical Emissions from Wastewater Basins," U.S. EPA, Hazardous Waste Engineering Research Laboratory, EPA 1600/5-2-82/095, Cincinnati, OH (1982).

10. Esplin, G. J. "Boundary Layer Emissions Monitoring." *J. Air Pollut. Control Assoc.* 38(9):1158–1161 (September 1988).

11. U.S. EPA. "Superfund Exposure Assessment Manual," EPA/540/1-88/001, Washington, DC (April 1988).

12. Hill, F. B., V. P. Aneja, and R. M. Felder. "A Technique for Measurements of Biogenic Sulfur Emission Fluxes," *J. Environ. Sci. Health* AIB(3): 199–225 (1978).

13. Adams, D. F., M. R. Pack, W. L. Bamesberger, and A. E. Sherrard. "Measurement of Biogenic Sulfur-Containing Gas Emissions from Soils and Vegetation." Proceedings of 71st Annual APCA Meeting, Houston, TX (1978), 76–78.

14. Adams, D. F. "Sulfur Gas Emissions from Flue Gas Desulfurization Sludge Ponds." *J. Air Pollut. Control Assoc.,* 29(9): 963–968 (1979).

15. Schmidt, C. E., W. D. Balfour, and R. D. Cox. "Sampling Techniques for Emissions Measurement at Hazardous Waste Sites," Proceedings of 3rd National Conference and Exhibition on Management of Uncontrolled Waste Sites, Washington, DC (1982).

16. Eklund, B. M., W. D. Balfour, and C. E. Schmidt. "Measurement of Fugitive Volatile Organic Compound Emission Rates with an Emission Isolation Flux Chamber," presented at the AIChE 1984 Summer National Meeting, Philadelphia, PA (August 19–22, 1984).

17. Pearson, J. E., D. H. Rimbey, and G. E. Jones. "A Soil-Gas Emanation Measurement System Used for Radon-222," *J. Appl. Meteorol.* 4: 349–356 (1965).

18. Ryden, J. C., L. J. Lund, and D. D. Focht. "Direct In-Field Measurement of Nitrous Oxide Flux from Soils," *Soil Sci. Soc. Am. J.* 42: 731–737 (1978).

19. Denmead, O. T. "Chamber Systems for Measuring Nitrous Oxide Emission from Soils in the Field," *Soil Sci. Soc. Am. J.* 43: 89–95 (1979).

20. Matthias, A. D., A. M. Blackmer, and J. M. Bremner. "A Simple Chamber Technique for Field Measurement of Emissions of Nitrous Oxide from Soils," *J. Environ. Qual.* 9(2): 251-256 (1980).

21. Radian Corporation. "Validation of Flux Chamber Emission Measurements on a Soil Surface," EPA Contract Number: 68-02-3889, Work Assignment 18, work in progress.

22. Radian Corporation. "User's Guide for the Measurement of Gaseous Emissions from Subsurface Wastes Using a Downhole Flux Chamber," U.S. EPA, Work Assignment 0-13, Task 3, Contract No. 68-CO-0003, U.S. EPA Office of Research and Development, Cincinnati, OH (1991).

23. Gholson, A. R., J. R. Albritton, K. M. Jayanty, and J. E. Knoll. "Evaluation of the Flux Chamber Method for Measuring Volatile Organic Emissions from Surface Impoundments," Final Report for U.S. E.P.A. Contract No. 68-02-4550 (1988).

24. Eklund, B. M., M. R. Keinbusch, D. Ranum, and T. Harrison. "Development of a Sampling Method for Measuring VOC Emissions from Surface Impoundments," Radian Corporation, Austin, TX (No date), p. 7.

25. Schmidt, C. E., W. R. Faught, and J. Nottoli. "Using the EPA Recommended Surface Emission Isolation Flux Chamber to Assess Emissions from Aerated and Non-Aerated Liquid Surfaces," presented at the 1991 EPA/AWMA Measurement of Air Toxics, Raleigh, NC (May 1991).

26. Schmidt, C. E., and W. R. Faught. "Review of Direct and Indirect Emission Measurement Technologies for Measuring Fugitive Emissions From Waste Water Treatment Facilities," American Institute of Chemical Engineers Symposium, San Diego, CA (August, 1990).

27. Berry, R. S., C. E. Schmidt, and R. C. Wells. "An Evaluation of Sampling Strategies for Determining Soil Filter Control Efficiencies," presented at the 1991 AWMA Symposium, 91-17.4, Vancouver, Canada (June 1991).

28. Schmidt, C. E., and J. Clark. "Use of the Surface Isolation Flux Chamber to Assess Fugitive Emissions from a Fixed-Roof on an Oil-Water Separator Facility," presented at the 1990 EPA/AWMA Measurement of Air Toxics, Raleigh, NC (May, 1990).

CHAPTER 4

A Combustion Wind Tunnel for Direct Determination of Emission Factors from Open Burning of Agricultural Wastes

B. M. Jenkins, S. Q. Turn, R. B. Williams, D. P. Y. Chang, O. G. Raabe, J. Paskind, and M. Ahuja

TABLE OF CONTENTS

0-87371-606-0/93/$0.00+$.50

© 1993 by Lewis Publishers

I. INTRODUCTION

California legislation enacted in 1983 under Assembly Bill (AB) 1223 directed the California Air Resources Board (ARB) to develop procedures to determine the magnitudes of emission offsets available to facilities that burn biomass for the generation of steam or electricity. AB 1223 added Section 41605.5 to the California Health and Safety Code requiring districts to include incremental emission benefits in considering any offset requirements for projects in which biomass is used as fuel.

This legislation was introduced in recognition of the potential reduction in air emissions from open burning of crop and forest wastes in the state. Increased activity in the development of biomass fueled facilities occurred after the enactment of the federal Public Utilities Regulatory and Policy Act of 1978 that increased incentives provided by utility companies for the purchase of energy and capacity from third-party developers.

In satisfaction of AB 1223, in 1984 the ARB created a guideline procedure for calculating the magnitude of available offsets. This procedure was later modified in 1988 in response to AB 2158.[1] The ARB procedure for calculating the offsets is of the form:

$$P_i = \frac{4x}{365} \sum_{j=1}^{n} B_j\, EF_j\, HBF_j\, QDF_j \qquad (1)$$

where

x = fraction corresponding to allowable offset as determined by proximity of biomass to facility

B_j = quantity (metric ton) of biomass of type j used during year i

EF_j = emission factor of pollutant in kg/t of biomass j which is open burned

HBF_j = fraction of total biomass j that is open burned,

P_i = pollutant offset credit in kilograms per day in quarter i

QDF_j = quarterly distribution factor for biomass of type j

n = number of different types of biomass used by facility

A value P is computed for each type of pollutant which may be offset. The parameter x takes on the value 1/2 for biomass originating more than 24 km away from the plant, and 1/1.2 for biomass from within 24 km of the facility.

This study was concerned with methods to determine the emission factors, EF_j, of Equation 1. Other work was concerned with the development of factors HBF_j pertaining to the fraction of biomass actually field burned, but this matter is not further discussed here.[2] The work undertaken in this project was designed to elucidate the effects of fuel and environmental conditions on emission factors required for Equation 1. To this end, a procedure for conducting the burning and the emission sampling was required to reduce uncertainty associated with uncontrolled internal and external parameters of the fire.

Open burning of agricultural and forestry wastes is principally done in one of two forms: spreading fires and pile burns. For field crop residues which are typically spread behind the crop harvesting equipment in a layer distributed over the soil and stubble surface, the burning is done as a fire front which is allowed to propagate naturally into the fuel bed. Current regulations in California call for this burning to be done in either backing or strip-lighted mode. For backing fires, the direction of fire propagation is directly opposite the direction of the wind, and this is largely true for strip-lighted fires as well. The benefits of these techniques in reducing particulate matter emissions have been shown to be substantial when compared to the emissions from so-called heading fires, in which the direction of fire propagation is concurrent with the direction of the wind. In backing fires, the fire propagation velocity is thought to be less sensitive to the wind velocity; whereas in heading fires, the fire velocity is directly proportional to the wind velocity due to heat transfer from the flame overlying the unreacted fuel bed. Volatiles emitted from the fuel during heating and pyrolysis prior to ignition are also more likely to escape the flame in a heading fire. For these reasons, burning against the wind is the adopted practice for field crop residues.

Pile burns are typically conducted for orchard prunings and forest slash. These fires are set after the biomass has been pushed into a pile, generally at the edge of the field. Ignition may be initiated at several locations around the pile such that the fire propagates into the pile from the periphery, and involves much of the fuel simultaneously throughout the duration of the burn. The pile represents a point source of emissions, as compared to the line or area sources occurring with spreading fires of field crop residues.

In both situations, the fire will induce its own draft. It will also respond to an impressed wind which is almost always present to some extent. Also, emissions continue after the passage of the initial flame front, generally as a smoldering afterburn. In spreading fires of field crop residues, these secondary emissions may continue for some time as a result of nonuniform deposition of residue. Where deep piles of residue have accumulated or been deposited, the short residence time of the fire and the reduced diffusion of atmospheric oxygen into the pile contribute to the prolonged emission. The structure of the fuel bed therefore has an influence on both the primary and secondary emissions from the fire. In addition to the direction and magnitude of the wind and the fuel bed structure, other major parameters that influence emissions from the fire include fuel type, fuel moisture, fuel morphology, and condition of the soil surface beneath the fire. This surface can serve as a source of extinction and as a source of moisture. The influence is probably greater on spreading fires than on large pile burns, but may have some effect nonetheless.

To simulate the major influences on the fire, a wind tunnel was constructed that was capable of supporting both spreading- and pile-type fires. This tunnel was in some respects similar to other wind tunnels used elsewhere to study the fire propagation velocity in spreading type fires. Most of these tunnels had stationary floors and rather limited duration testing capabilities, which was seen to be undesirable from the standpoint of emissions monitoring because of the very small sample sizes involved. The UCD tunnel includes a moving floor that can be used when conducting spreading fires. By translating the floor, the fire can be made to propagate indefinitely into the fuel bed introduced into the upstream end of the tunnel. Emissions sampling can then occur for as long as necessary to capture representative samples or to integrate the emission load over the anticipated duration of an actual field fire. Primary emission sampling is done in a stack some 8 m above the fire. When testing spreading fires with an impressed wind, the gas temperature in the stack is about 15 K above the temperature of ambient air entering the tunnel. The data collected therefore are representative of emission levels in the plume shortly after quenching, and do not include any longer term effects due to atmospheric reactions occurring far behind the fire. These may be understood to some degree from the primary emissions. The tunnel provides a means to control the factors influencing the fire and thus permits the study of how changes in the condition of the fire affect the emission levels observed, a study which is virtually impossible to conduct in the field. Because the wind and fuel conditions can be controlled in the wind tunnel, it also serves as a means to calibrate emission models for application to the field.

Described here are the design of the wind tunnel and initial experiments conducted to determine if repeatable emission rates could be obtained under separate tests. Subsequent phases of the program are intended to carry out the actual determination of the emission factors for various fuels of importance, and to understand how changes in configuration of the tunnel and condition of the fuel, the wind, and the soil influence the emission rates of pollutants from the fire.

II. Wind Tunnel Design

Design of the combustion wind tunnel was based on the need to determine and control variables of the fuel and the fire environment in order to understand the effects of these variables on the observed emission rates. The use of a wind tunnel was intended to enable well-prescribed conditions that could not be obtained under similar field experiments where spatial and temporal variation of the major parameters influencing the fire are large, and difficulties of sampling a moving emission stream in the case of a spreading fire are severe. The major variables to be controlled are freestream wind speed, wind velocity profile, turbulence intensity, fuel loading, and fuel moisture content.

The main concern in designing the combustion wind tunnel was that the fires produced in the tunnel and the resulting emissions be physically representative of those that occur in the field. In addition, a balance had to be found between blower capacity needed to generate model wind speed, tunnel cross-sectional area, and heat exchange between the fire and the wind tunnel structure. Consequently, important design considerations were identified, namely, quantity of fuel to be burned during each test, ability to reproduce the velocity profile and turbulence intensity of the wind, simulation of the fuel bed structure of the field, configuration of the tunnel to allow for the free expansion of the plume, and minimization of the effect of heat exchange between the physical tunnel structure and the fire. Unfortunately, compromises among these parameters reduced the size of fire that can be permitted in the tunnel, so that large pile fires cannot be directly tested. Because these types of fires can be accommodated with simpler structural considerations, however, the wind tunnel design was dominated by the potential to measure long-term emission rates in spreading-type fires.

For the purposes of sampling the emissions from spreading fires, the use of stationary fuel beds requires extremely long combustion test sections to permit burns of sufficient duration. The concept employed here instead was to translate the fuel bed relative to the tunnel floor with a conveyor system. By moving the fuel bed at the fire propagation velocity, the flame could be held stationary in space. In principle, a fire could be sustained indefinitely with this approach. By fixing the position of the flame, sampling instruments could also be fixed without the need to follow the fire.

The major difficulty arising from translation of the bed is the need to configure the bed so that the structure remains essentially intact during transport, and thus an accurate simulation of the fuel bed structure occurring in the field becomes more difficult. The structure used to convey the fuel also needs to provide minimum quench surface so as not to unduly influence emission of volatile organics. The wind profile and turbulence characteristics are also problematic. In fact, complete simulation of the atmospheric boundary layer in a wind tunnel cannot be achieved. As discussed by Plate,[3] however, sufficient similarity in the lower part of the boundary layer can be obtained to warrant use of the wind tunnel in this kind of investigation. Because scaling of the fire itself appeared undesirable from the standpoint of emissions testing, the tunnel was designed with the intent of generating a sufficiently thick boundary layer above the fuel surface with turbulence intensities of the same order as the field. As mentioned earlier, sampling was to be done from the immediate vicinity of the fire, without attempting to model plume processes.

Another design consideration was to allow for the free expansion of the plume behind the fire in the wind tunnel. The use of a closed channel wind tunnel leads to a reduced flame angle and recirculation behind the flame, neither of which appears to occur to any large extent in the field. An open channel tunnel would lead to deceleration of the jetlike flow at the exit of the flow development section.[4] To deal with this problem, a semiopen channel was devised consisting of a large volume combustion test section incorporating a movable ceiling which was positioned at the leading edge of the plume in a manner similar to that used by Fleeter et al.[4] The plume was then free to expand beyond the ceiling, and the problems of recirculation and turbulent mixing effects generated by closed and open channel tunnels were avoided. The adjustable ceiling need not be moved once the position of the flame has been established in the case of a spreading fire.

The width of the tunnel was selected to reduce the thermal dissipation at the walls to a reasonable fraction of the estimated energy release rate of the fire. Consideration was also given to the blower capacity required to reach the intended maximum wind velocity of 5 m/sec, the ability to maintain a uniform fire profile

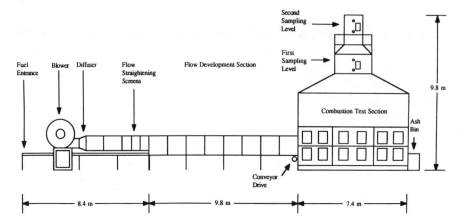

FIGURE 1.   Exterior schematic of the combustion wind tunnel.

across the width of the tunnel in the case of a spreading fire, and the physical strength requirements of the conveyor elements supporting the fuel across the width of the tunnel. Using the results of prototype studies, the transverse thermal energy loss from the flame adjacent to the walls in a wind of 2.5 m/sec and a loading rate similar to that of a rice straw fuel in the field was estimated to be approximately 17.5 kW. From the heating value of fuel and the heating value of residual char following the fire, the fireline intensity (energy released per unit width of fire) was estimated to be 130 kW/m. As the width of the tunnel increases, the fractional wall dissipation rate (the ratio of wall loss to energy release of the fire) declines. There is a rapid decline in the wall dissipation rate up to 1-m width. The incremental improvement is reduced past this point. An increase in width from 1.5 to 2.0 m, for example, results in a decline of only 2% in the wall dissipation rate, but requires an increase of 33% in blower capacity to match the wind velocity. The final width was selected as 1.2 m and provided a wall thermal dissipation rate of approximately 10% under the assumptions made.

III. Wind Tunnel Description

The combustion wind tunnel is comprised of three major functional parts: the flow development section, the combustion test section, and the emissions sampling stack. An exterior design view of the wind tunnel is shown in Figure 1. The entire tunnel extends nearly 26 m in length and stands approximately 10 m high. The tunnel is a forced draft open circuit type, using a 45-kW centrifugal blower capable of generating an average 5 m/sec wind speed approaching the fire in the combustion test section. A series of conveyors extend the length of the tunnel and are used to transport fuel to the fire at a speed matching that of the fire propagation rate so as to produce a flame standing stationary in space.

The flow development section consists of the blower, the flow straightening elements, and the primary fuel feed conveyor used when testing spreading-type

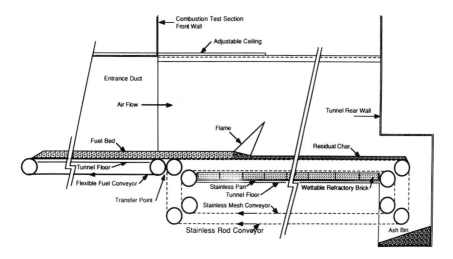

FIGURE 2. Arrangement of fuel conveyors for maintaining a stationary flame zone.

fires. Fuel is loaded onto the primary conveyor behind the blower and passes beneath the blower, diffuser, and flow straightening elements before entering the airstream 10 m upstream from the combustion test section. The primary fuel conveyor consists of a continuous flexible solid belt 36.5 m long. The belt travels along the inner floor of the flow development section on its delivery pass, and returns on the outside below the tunnel. Fuel can be loaded continously onto the belt as the experiment proceeds. The centrifugal blower is of fixed velocity, but has variable dampers at the inlet and outlet to adjust the air flow rate. The blower is connected to the diffuser by a flexible coupling to reduce transmission of vibration from the blower to the tunnel structure. Flow straightening is done with five screens located downstream of the diffuser. Two 40 mesh stainless screens are followed by three 60 mesh stainless screens. Just following the last screen, the fuel enters the flow development section, providing a roughened lower surface to aid velocity profile development. At a distance of 2.4 m upstream of the combustion test section, a wire mesh fence was installed to trip the flow shortly ahead of the fire.

The combustion test section extends 7.4 m in length. The configuration of the conveyors, floor, and ash accumulation bin are shown in Figure 2. At the entrance to the combustion test section, the fuel is transferred from the primary conveyor to a stainless rod conveyor consisting of 12.7-mm diameter stainless tubes spaced every 100 mm along the tunnel and connected at the ends to continuous roller chains. The floor of the combustion test section consists of porous refractory brick that can be wetted to provide a moist surface below the fuel. The brick was placed in stainless pans connected to a manifold supplying water to saturate the brick. The brick is intended to remain wet in the presence of a flame, as in the case of a flame moving across a saturated soil surface.

Just above the brick and 150 mm below the rod conveyor is a third conveyor of 3-mm stainless mesh screen. This conveyor is used to carry fine fuel material which drops through the rod conveyor downstream so that it will not serve as a source of ignition to incoming fuel. All three conveyors are driven at the same linear velocity by a variable speed DC motor regulated to match the fire propagation velocity. The rod and mesh conveyors also return on the outside of the tunnel, and are cooled to near ambient before they enter the tunnel. An ash collection bin is located at the discharge end of the conveyors to receive and accumulate residual char as a test proceeds. This bin is sealed so as to reduce air intrusion or loss of char. Located on the walls of the tunnel are 25 glass windows, 12 on each side and 1 in the end wall of the tunnel. These windows permit observation of the flame. Each window is 508 mm wide, 660 mm high, and 5 mm thick. During operation, the flame is held on the window surfaces to reduce catalytic reactions, rather than on the sheet steel surfaces making up the remainder of the tunnel wall. Each window is held against a silicon gasket by a spring-loaded frame to allow the windows to expand upon heating, although the glass used is reported to have zero thermal expansion at 800°C. The windows have not failed under thermal-mechanical stress generated by operating with the flame held on the glass. The lower six windows on each side of the tunnel are installed on three doors, hinged at the top, which may be opened for inspection purposes. The doors are also designed to be opened when conducting pile fires under no wind conditions when air should be drawn from around the fuel bed. An adjustable ceiling can be extended into the combustion test section at a height of 1.2 m above the rod conveyor. A water manifold is installed at the entrance to the combustion test section to extinguish any uncontrolled fires approaching the flow development section.

At a height of 3.7 m above the fuel bed the combustion test section begins to narrow into the stack sampling section. Sampling ports are located at two levels in the stack. The first level is in a section of duct 2.4 m long extending 1.2 m upward from the initial transition. Above this is another transition leading into the second level sampling position situated in the vent stack of 1.2 m square cross section. This second level is the principal sampling area and is situated nearly 8 m above the fuel bed. Windows are also provided at both levels to view the sampling instruments and the conditions of the exhaust. A ladder and catwalks provide operator access to the sampling ports.

IV. Flow and Fire Characterization

Investigations of the velocity profiles and turbulence intensities generated at the end of the flow development section and in the stack at the second-level sampling position under cold conditions were conducted by traversing the flow in these regions with a single element hot film anemometer. The anemometer output was connected to a high speed data acquisition system. The anemometer was sampled at 7 kHz, and ensemble averages were collected over a period of at least

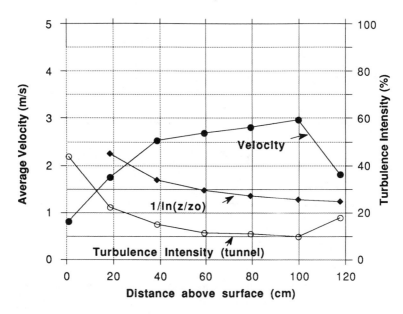

FIGURE 3.   Vertical velocity profile and turbulence intensity at the entrance to the combustion test section. Shown for comparison is the computed turbulence intensity of the field estimated for a grass surface.

1 min in each position of the traverse. Turbulence intensities at each point were computed from 15,000 data points according to:

$$I_1 = \frac{(\overline{u'u'})^{0.5}}{u} = \frac{\sigma_u}{u} \tag{2}$$

where

I_1 = turbulence intensity normalized to mean velocity, u, in streamwise direction
u' = streamwise velocity fluctuation
$(\overline{u'u'})^{0.5}$ = root mean square of fluctuation, i.e., standard deviation (SD), σ_u.

Vertical velocity profiles and turbulence intensities in the approach flow for a mean velocity of 2.35 m/sec are shown in Figure 3. The duct Reynolds number at this velocity was 1.9×10^5. The profiles were determined in the combustion test section with the ceiling extended at a position approximately 250 mm upstream from the edge of the ceiling. Data were collected without a fire in the tunnel, but with a layer of straw (600 g/m²) extending the full length of the flow development section and into the combustion test section. The vertical profile at the entrance

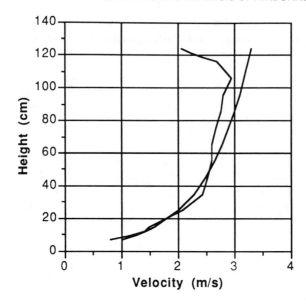

FIGURE 4.   Comparison of the actual inlet velocity profile to a logarithmic law of
the wall velocity profile based on a 5-cm high grass surface (u* =
0.32, z_0 = 0.02, d_0= 0).

shows the boundary layer extending some 35–40 cm above the fuel surface as well
as a boundary layer of some 10– to 15-cm thickness developed on the ceiling of
the tunnel. A comparison of the boundary layer over the fuel to that computed for
a natural grass surface estimated to be of similar roughness to the field shows
similarity up to about 35-cm depth (Figure 4). The latter profile was computed
from a logarithmic law of the wall:[5]

$$u = \frac{u*}{k} \ln\left(\frac{z - d_0}{z_0}\right)$$
(3)

where

u = wind velocity (m/sec) at elevation z (m) normal to the surface
k = von Karman's constant (here taken numerically equal to 0.4)
$u*$ = friction velocity (m/sec)
d_0 = zero plane displacement (m)
z_0 = surface roughness parameter

Values used in developing Figure 4 were $u*$ = 0.32, z_0 = 0.02, and d_0 = 0. The
35-cm thickness corresponds roughly to the flame height of a field fire in rice
straw propagating in opposition to the wind as shown later. Plate[3] has shown that
Equation 3 is valid with roughness height z_0 which depends only on the geometry

of the roughness elements of the surface, and is independent of any Reynolds number for roughness $u^*z_0/v > 5$ (where v is the kinematic viscosity, m^2/sec), which is the situation of Figure 4.

The velocity profile across the horizontal direction was fairly uniform except for boundary layers of some 20-cm thickness at the wall. Such boundary layers do not appear in the stack, and the stack velocity profiles are uniform out to the walls. Conditions of uniformity for sampling the stack appear to be satisfactory.

Of primary importance is to match the gross velocity profile and the turbulence characteristics of the flow within the immersion depth of the flame as closely as possible. Wichman[6] and Wichman et al.[7] have demonstrated dependence of fire propagation characteristics on the velocity profile of air entering the fire. In fact, Wichman[6] has used this result to relate experiments conducted under different external flow conditions which yield similar flame spread rates. Comparing fires conducted under forced convection in wind tunnels with open flames in a free convection boundary, he found the results to be similar based on velocity profile similarity over nearly an order of magnitude change in air velocity.

The vertical inlet velocity profile appears reasonably well matched to the field up to 35–40 cm above the fuel surface. The turbulence intensities are also of the proper order, although a comparsion with atmospheric turbulence intensities suggests the wind tunnel intensities may be somewhat low. The logarithmic profile developed from the Monin-Obukhov similarity theory and K theory yields a streamwise turbulence intensity:[3,8-12]

$$\frac{\sigma_u}{u} = \frac{1}{\ln(z/z_0)} \tag{4}$$

For $z_0 = 0.02$ m (Figure 4), this yields estimated turbulence intensities of the magnitude shown in Figure 3. Estimates of intensities are about twice as large as those measured in the tunnel, but decrease upward from the surface in the same manner. Turbulence intensities in the range of both the estimates of Equation 4 and those measured in the wind tunnel (40% just above the fuel surface decreasing to about 10% at 60-cm height) have been reported in the literature. Plate[3] summarizes field data for $z_0 = 0.05–0.5$ m which are nearly identical to those measured in the wind tunnel. Cionco[13] measured horizontal turbulence intensities of 30–100% in the region immediately above a crop canopy. Maitani[14] measured turbulence intensities over various surfaces and found values of 40–120% in the region just above plant canopies. Ohtou et al.[15] consistently measured turbulence intensities of about 50% at an elevation of 65 cm above a rice crop. Ohtaki[16] measured turbulence intensities of about 40% at 80 cm above a rice crop.

Tests were also conducted to determine whether the time-temperature relationships of fires in the field are adequately matched in the tunnel. Species emitted from a combustion process are directly dependent on the residence time of reactants at elevated temperature. Field data were obtained for a fire spreading in opposition to the wind (backing fire) and compared against wind tunnel data.

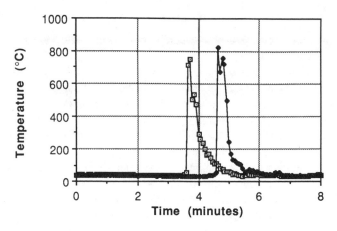

FIGURE 5a. Temperature at the fuel surface for a field backing fire in rice straw.

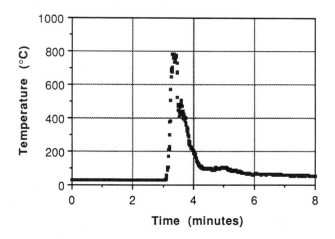

FIGURE 5b. Temperature at the fuel surface for a backing fire in rice straw conducted in the wind tunnel.

Results of field and tunnel experiments in spread rice straw are shown in Figure 5a and b. Temperature data were obtained by positioning 0.5-mm type K thermocouples a known distance apart in the fuel bed. The wind velocity in the field ranged from 1 to 2 m/sec at 0.5- to 0.6-m elevation. The wind tunnel experiments were conducted at an average duct wind speed of 2.3 m/sec. The spreading velocity for the backing fires in the field was 0.95 m/min. The period of elevated temperature at the fuel surface lasted about 60 sec. Measured peak temperatures were 800–900°C. The thermocouple time constant was too large to measure true peak flame temperatures, but the results are consistent from wind tunnel to field.

Comparable tests in the wind tunnel show similar trends in temperature and duration (Figure 5b). The peak temperature is about 800°C, and the duration is

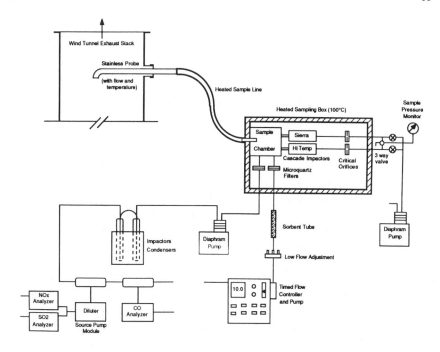

FIGURE 6.   Sampling train used in preliminary emission tests.

about 60 sec. Spreading velocity was 0.94 m/min. This is in reasonable correspondence with the field data.

V. Emissions Testing

A number of experiments were conducted to determine whether repeatable data could be obtained on emissions from fires conducted under similar conditions of wind speed, fuel type, fuel loading, fuel moisture, air temperature and relative humidity, and moisture status of the floor. Experiments on September 1 and September 7, 1988, were conducted with a mean wind speed of 2.35 m/sec and a fuel-loading rate to match that of the field at 7% moisture, or approximately 600 g/m^2. Rice straw was used for both tests. Each test was designed to accumulate information on fuel properties and fuel bed structure, residual solid-phase composition, wind tunnel characteristics, and solid- and gas-phase emissions (particulate matter concentration and size distribution, CO_2, CO, NO_2, NO, SO_2, CH_4, C_2H_2, C_2H_4, C_2H_6, C_6H_6. Qualitative assessments of other organic compounds were made as well.

The sampling train used in conducting these experiments is illustrated in Figure 6. Samples were drawn from the second level sampling position through one of the ports in the sidewall of the stack. The sample was removed isokinetically with a stainless steel probe, 9.5 mm in diameter, located in the center of the stack. The end of the probe was bent and cut 90° from the probe axis, such that it faced into the exhaust stream. The probe flow rate was matched to the stack velocity measured by a hot film anemometer positioned near the probe. The probe was connected to a 25-mm i.d. heated line of flexible stainless steel tubing lined with Teflon® and wrapped with heating tape. The line was controlled to $100 \pm 2°C$ by a temperature controller using an embedded type K thermocouple. The sample line was connected to a heated aluminum manifold having a volume of 2.5 L from which various samples were drawn via stainless steel pipe connectors tapped into the wall. Filters and impactors were housed within an insulated heated chamber also maintained at $100 \pm 2°C$.

A high temperature, high pressure (HTHP) impactor (Model 02-300, In-Tox Products, Albuquerque, NM) operated at 13.3 liters per minute (Lpm) average flow rate was used to determine particulate matter size distribution. For the tests reported here, Gelman Micro-Quartz filters were used (Part No. 66089, 47 mm, Gelman Sciences Inc., Ann Arbor, MI). The filters were hand cut to 37-mm diameter for all stages except the final filter stage which was left at 47-mm diameter.

Two other samples were drawn from the manifold. One stream was used for collection of volatile organics on sorbent tubes. The other was used for gas analysis. Each line was filtered through a Micro-Quartz filter. Flow through each sorbent tube was pulled by a timed flow controller and pump at a rate of 100 mL/min for a period of 10 min yielding a total flow of 1 L. The sorbent tubes used were 6-mm o.d. packed with 50 mm of Tenax-TA 20×30 mesh and 38 mm of Ambersorb XE-340 (tubes supplied by T.R. Associates, Lewisburg, PA). The tube was placed immediately after the heated manifold and connected with a soft graphite ferrule on Teflon® tubing. The gas sample line was connected via 6.4-mm Teflon® tubing to a diaphragm pump and an ice bath impinger train before descending to ground level. Samples were pulled from this line for each of the NO_x, SO_2, and CO analyzers located in an instrument trailer next to the tunnel. Sorbent tubes were analyzed by gas chromatography/mass spectrometry (GC/MS).

The NO_x and SO_2 analyzers (Model 8840 and 8850, Monitor Labs, San Diego, CA) were supplied with sample from a source pump and dilution module (Monitor Labs Model 8730) which diluted the sample 20 to 1 with zero air by means of a critical orifice. Sample dilution was specified on the basis of results from the prototype tunnel that indicated levels within the source range of the instruments, which are designed as ambient monitors. Carbon monoxide was first monitored using an electrochemical polargraphic analyzer, but was later monitored with an infrared gas analyzer (Anarad AR-500). Grab samples were collected in glass containers at regular intervals throughout each test. These were analyzed by GC/thermal conductivity detector (TCD) for CO_2 and low molecular weight hydrocarbons. Ambient air background samples were also collected and analyzed.

Other instrumentation included various transducers for monitoring the conditions of the test, and the data acquisition system to read, process, and store the data. Inlet air temperature and relative humidity were monitored downstream of

Table 1. Fuel and Ash Analysis for Combustion Trials of September 1 and 7, 1988, in Rice Straw

	September 1		September 7	
	Fuel	Ash	Fuel	Ash
Proximate analysis				
Moisture (% wet basis)	6.5	2.8	7.3	3.3
Ash (% dry basis)	13.59	78.41	13.16	77.68
Volatiles (% dry basis)	69.44	11.11	69.94	10.34
Fixed carbon (% dry basis)	16.97	10.48	16.90	11.98
Higher heating value (MJ/kg, dry basis)	15.91	4.99	15.99	5.34
Ultimate analysis (% dry basis)				
Carbon	40.41	11.17	40.29	15.23
Hydrogen	5.42	0.49	5.39	0.56
Oxygen (by difference)	38.69	4.05	38.98	3.18
Nitrogen	0.53	0.24	0.69	0.31
Sulfur	0.05	0.09	0.08	0.05
Chlorine	0.20	0.06	0.27	—
Residual	14.7	83.9	14.3	80.7

the blower by a thermistor and polymeric grid-type humidity sensor (Campbell Scientific Model 207, Logan, UT). Pressure differential across the five screens in the flow development section was monitored with a strain gauge-type pressure transducer. Temperatures in the flame, in the stack, and at the outlet of the impingers were monitored with type K and type T thermocouples. Velocity in the stack was monitored using a hot film air velocity sensor (Kurz Instruments, Carmel Valley, CA) capable of withstanding the particulate load in the gas stream. All sensors except the velocity transducer were connected to an electronic data logger (Campbell Scientific Model CR21X, Logan, UT) which was serially interfaced to a microcomputer (IBM PC-AT) running a custom data acquisition program. Outputs from the NO_x, SO_2, and CO analyzers were also connected to the data logger and computer. Results were written to disk every 3 sec throughout the test. Five 0.5-mm type K thermocouples were suspended from the leading edge of the adjustable ceiling into the flame. These were positioned at 44, 70, 127, 146, and 216 mm above the surface of the rod conveyor.

VI. RESULTS

Results of the fuel and ash analyses for both tests are included in Table 1. Fuel moisture was similar in both cases (6.5 and 7.3% wet basis).

Material balances and characteristics of the fire appear in Table 2. The fire propagation velocities were 0.96 and 0.93 m/min. These values are consistent with values reported by Mobley[17] for backing fires in grass fuels at similar moisture, and to the velocities obtained in the field tests described previously. Total ash

Table 2. Material Balances and Fire Characteristics for Tests of September 1 and 7, 1988, in Rice Straw

	September 1	September 7
Fire characteristics		
Effective fuel loading (g/m^2 wet basis)	584	577
Fuel consumption (g wet basis)	23,190	22,896
Fuel moisture (% wet basis)	6.5	7.3
Dry fuel consumption (g)	21,683	21,225
Burn duration (min)	34	35
Dry fuel consumption rate (g/sec)	10.63	10.11
Fire Propagation Velocity (m/min)	0.96	0.93
Mean wind velocity (m/sec)	2.35	2.35
Mean air temperature (°C)	37	32
Mean relative humidity (%)	15	15
Air density (kg/m^3)	1.140	1.159
Air flow rate (g/sec)	3,982	4,048
Overall air/fuel ratio (dry basis)	375	400
Stack gas mass flow rate (g/sec)	3,991	4,057
Stack gas temperature (°C)	53	48
Stack gas flow rate at temperature (m^3/sec)	3.68	3.68
Ash recovery		
Total ash recovered (g)	3,993	3,538
Moisture content (% wet basis)	2.8	3.3
Dry ash recovery (g)	3,881	3,421
Fraction of dry fuel consumption (%)	17.9	16.1
Inorganic fraction of fuel (%)	14.7	14.3
Inorganic fraction of ash (%)	83.9	80.7
Expected ash recovery (% of fuel)	17.5	17.7
Nitrogen		
Fuel nitrogen (% dry basis)	0.53	0.69
Ash nitrogen (% dry basis)	0.24	0.31
NO concentration in stack (ppm)	3.0	2.8
NO$_2$ concentration in stack (ppm)	0.9	0.6
Fuel N (g)	115	146
Ash N (g)	9	11
Gas phase N from NO, NO$_2$ (g)	15	14
Excess fuel N (g)	91	121
Sulfur		
Fuel sulfur (% dry basis)	0.05	0.08
Ash sulfur (% dry basis)	.09	0.05
SO$_2$ concentration in stack (ppm)	0.12	0.25
Fuel S (g)	11	17
Ash S (g)	3	2
Gas phase S from SO$_2$ (g)	1	2
Excess fuel S (g)	7	13
Carbon		
Fuel carbon (% dry basis)	40.41	40.29
Ash carbon (% dry basis)	11.17	15.23

Table 2. Material Balances and Fire Characteristics for Tests of September 1 and 7, 1988, in Rice Straw (continued)

	September 1	September 7
Stack gas concentrations (ppm, excludes background)		
CO_2	2,000	2,000
CO	168	155
CH_4	8	4
C_2H_2	0.3	0.2
C_2H_4	1.7	1.7
C_2H_6	0.2	0.13
C_6H_6	0.137	0.082
Fuel C (g)	8,762	8,552
Ash C (g)	436	521
Gas phase C (g)	7,346	7,617
Particulate C (g)	ND	ND
Excess fuel C (g)	980	414
Closure on Carbon Balance (%)	89	95

recovery was similar to what would be expected on the basis of the inorganic component of the fuel and char. The amount recovered from the first test was somewhat higher than expected and the amount from the second test was somewhat lower, but both were similar to anticipated values and were within the expected experimental error. There was some loss of char to the floor of the tunnel which was lost during cleaning in preparation for the next experiment.

Inlet air conditions for both tests were 32–37°C and 15% relative humidity. Stack temperatures at the sampling point were 15–20 K above inlet air temperature. The flue gas heat loss was approximately 80 kW, or half of the energy released by the fire (160 kW) determined on the basis of difference in heating values of the fuel and char. Roughly 20 kW was transferred from the flame through the windows by radiation, conduction, and convection. The remainder was lost in the structure of the tunnel and by cooling of the char after deposition in the collection bin.

Temperatures measured in the flame at 44 mm above the conveyor surface ranged from 400 to 800°C. At 216 mm above the surface, average temperature measured 100°C, with excursions to 400°C as flamelets burned past the thermocouple.

Average concentrations of the gas-phase emissions observed during each test are listed in Table 2. As shown by the material balance for nitrogen, the concentrations of N in the char and in the gas-phase emissions of NO and NO_2 account for only 20% of the fuel nitrogen.

SO_2 emissions were extremely difficult to measure, the signal from the analyzer being only slightly above the noise level of the instrument. The values reported are uncertain, but there did not appear to be a significant quantity of SO_2 in the gas. The discrepancy in the sulfur concentrations in the char by ultimate analysis makes it difficult to compare the closure values on sulfur, but the

Table 3. Particulate Matter Concentrations and Size Distributions for Combustion Trials of September 1 and 7, 1988, in Rice Straw

| | | September 1 | | September 7 | |
| | | | Cumulative | | Cumulative |
Stage	Effective diameter (μm)	Mass (mg)	fraction (%)	Mass (mg)	fraction (%)
1	12.80	0.11	100.0	0.12	100.0
2	7.64	0.18	97.1	0.17	96.7
3	4.31	0.16	92.3	0.17	92.1
4	1.97	0.15	88.1	0.17	87.5
5	1.30	0.15	84.1	0.27	82.9
6	0.75	0.35	80.1	0.43	75.5
7	0.46	1.16	70.9	0.91	63.8
8	Filter	1.52	40.2	1.42	38.9
Concentration (mg/m³)		9.47		9.17	
Aerodynamic diameter (μm)		0.47		0.51	
σ_g		1.63		2.49	

gas-phase emission of SO_2 was below that computed on the difference of sulfur in the fuel and the ash as reported by Darley.[18] The low readings obtained on the SO_2 analyzer indicate that an ambient monitor should have been used because the actual concentrations are so much lower than those estimated from the fuel-bound sulfur.

Carbon balances show reasonable closure levels, being 89% for the test of September 1 and 95% for the test of September 7. The carbon balance is extremely sensitive to the concentration of CO_2 determined for the stack gas, and represents the critical problem in determination of emission factors from field studies where carbon balance is used to determine emitted fraction per unit mass of fuel.

The predominant hydrocarbon was methane at an average of 6 ppmv, although the concentration measured on September 7 was half that of the value on September 1. Concentrations of other light hydrocarbons were comparable. Ethene was present at the next highest concentration after methane.

Benzene concentrations determined from the sorbent tubes were in the range of 100 ppbv. The analysis of benzene by trapping on the sorbent tubes was compared to grab sample analyses using Tedlar bags by analyzing a calibration standard with a concentration of 112 ppbv. Chromatograms from GC/MS exhibit a large number of peaks, some of which were tentatively identified using the library of the instrument and comparing boiling points and mass spectrograms. Tentative qualitative identification has been made for toluene, ethylbenzene, ethenylbenzene, benzaldehyde (possible artifact from tube packing), phenol (possible artifact), and napthalene. Work is continuing to both qualitatively and quantitatively determine volatile organics emitted from the fire. There is some concern that quench effects could be contributing to the concentrations of volatile organics observed. Further investigation is required in this respect as well.

Particle concentrations and size distributions measured by the HTHP cascade impactor are given in Table 3. The size distributions were matched to an equivalent

Table 4. Emission Factors (kg/ton)[a] for Preliminary Combustion Trials of
September 1 and 7, 1988, in Rice Straw (7% Moisture), with
Comparison to Darley[18] at 14.9% Average Moisture Wet Basis

	September 1	September 7	Average	Darley[18]
Particulate matter[b]	3.28	3.34	3.31	1.06
NO	1.16	1.16	1.16	1.88
NO_2	0.54	0.38	0.46	0.65
NO_x (as NO_2)	2.32	2.16	2.24	3.53
SO_2	0.10	0.22	0.16	1.00[c]
CH_4	1.66	0.88	1.27	0.82
C_2H_2	0.10	0.07	0.09	0.11
C_2H_4	0.62	0.66	0.64	0.41
C_2H_6	0.08	0.05	0.07	0.11
C_6H_6	0.14	0.09	0.12	0.05
CO	60.89	59.99	60.44	89.62[d]
CO_2	1139	1216	1178	NR[e]

[a] All values dry basis (zero moisture). Multiply by 2.0 to obtain lb/ton.
[b] Based on results of HTHP impactor. Note that Darley[20] reports 3.47 kg/ton
particulate matter (dry basis) for a backing fire in 9.3% moisture rice straw on a
25° slope, and 4.64 kg/ton at 10.5% moisture and 15° burning slope.
[c] Darley's value for SO_2 computed from the difference of sulfur in the fuel and ash.
[d] Darley.[20] Corrected for average moisture of 9.9% wet basis. CO for rice straw
is not reported by Darley.[18]
[e] Not reported.

mass median aerodynamic diameter using a lognormal distribution with the stage
efficiency least squares method.[19] About 90% of the particles were below 4 μm
in size, and 80% were less than 1 μm.

Preliminary emission factors computed on the basis of concentrations obtained
during the two experiments of September 1 and 7 are listed in Table 4 for rice
straw. The results of Darley[18] are listed for comparison, because these values are
in current regulatory use in California. There exist significant discrepancies in the
particulate matter and SO_2 concentrations. Earlier work by Darley[20] does not
support the particulate matter emission rate reported by his later studies. This
uncertainty suggests the need for more direct measurements of the type conducted
here.

VII. SUMMARY AND CONCLUSIONS

The combustion wind tunnel provides a means of conducting controlled experiments to directly quantify emissions from open burning of biomass materials.
Emission data are therefore not subject to the uncertainty which accompanies field
data determined by elemental balance methods. By permitting direct control over
all fuel and oxidizer conditions, the wind tunnel experiments are also useful for

understanding the combustion dynamics. The ability to translate the fuel bed relative to the floor allows long-term testing of spreading fires to be conducted under the same controlled conditions, a capability not available earlier. The scale of the tunnel is not sufficient to conduct fires of the same size that occur in forest burning. Data collected from the tunnel, however, can provide calibrations for model predictions of how fire emission is influenced by fuel and environment conditions, and serve to validate larger scale field measurements.

REFERENCES

1. "A Procedure Relating to the Determination of Agricultural/Forestry Waste Emission Offset Credits," California Air Resources Board, Sacramento, CA (1988).
2. Jenkins, B.M., S.Q. Turn, R. Williams, S. Hickman, E. Stanghellini, and L. Bounds. "An Assessment of Burn Fractions and Seasonal Burn Profiles for Agricultural Crop Residues in the San Joaquin Valley of California, " California Air Resources Board Final Report, ARB Interagency Agreement No. A847-110, Sacramento, CA (1990).
3. Plate, E. *Engineering Meteorology* (Amsterdam: Elsevier Scientific, 1982).
4. Fleeter, R.D., F.E. Fendell, L.M. Cohen, N. Gat, and A.B. Witte. "Laboratory Facility for Wind Aided Firespread Along a Fuel Matrix," *Combust. Flame* 57:289–311 (1984).
5. Sutton, O.G. *Atmospheric Turbulence* (New York: John Wiley & Sons, Inc., 1960).
6. Wichman, I.S. "Flame Spread in an Opposed Flow with a Linear Velocity Gradient," *Combust. Flame* 50:287–304 (1983).
7. Wichman, I.S., F.A. Williams, and I. Glassman. "Theoretical Aspects of Flame Spread in an Opposed Flow Over Flat Surfaces of Solid Fuels," *Nineteenth International Symposium on Combustion* (1982), 835–845.
8. Lumley, J.L., and H.A. Panofsky. *The Structure of Atmospheric Turbulence* (New York: John Wiley & Sons, Inc., 1964).
9. Panofsky, H.A. and J.A. Dutton. *Atmospheric Turbulence* (New York: John Wiley & Sons, Inc., 1984).
10. Pasquill, F. *Atmospheric Diffusion* (New York: John Willey & Sons, Inc., 1974).
11. Nieuwstadt, F.T.M., and H. van Dop. *Atmospheric Turbulence and Air Pollution Modelling* (Dordrecht, Holland: D. Reidel Publishing Co., 1982).
12. McBean, G.A. "The Variations of the Statistics of Wind, Temperature and Humidity Fluctuations with Stability," *Boundary-Layer Meteorol.* 1:438–457 (1971).
13. Cionco, R.M. "Intensity of Turbulence Within Canopies with Simple and Complex Roughness Elements," *Boundary-Layer Meteorol.* 2:453–465 (1972).
14. Maitani, T. "A Comparison of Turbulence Statistics in the Surface Layer Over Plant Canopies with Those Over Several Other Surfaces," *Boundary-Layer Meteorol.* 17:213–222 (1979).
15. Ohtou, A., T. Maitani, and T. Seo. "Direct Measurement of Vorticity and Its Transport in the Surface Layer Over a Paddy Field," *Boundary-Layer Meteorol.* 27:197–207 (1983).
16. Ohtaki, E. "Turbulent Transport of Carbon Dioxide Over a Paddy Field," *Boundary-Layer Meteorol.* 19:315–336 (1980).
17. Mobley, H.E., Ed. "Southern Forestry Smoke Management Guidebook," USDA Forest Service, Southeastern Forest Experiment Station, Asheville, NC and Southern Forest Fire Laboratory, Macon, GA (1976).

18. Darley, E.F. "Hydrocarbon Characterization of Agricultural Waste Burning," Final Report, CAL/ARB Project A7-068-30, Statewide Air Pollution Research Center, University of California, Riverside, CA (1979).

19. Raabe, O.G. "A General Method for Fitting Size Distributuions to Multicomponent Aerosol Data Using Weighted Least-Squares," *Environm. Sci. Technol.* 12(10):1162–1167 (1968).

20. Darley, E.F. "Emission Factors from Burning Agricultural Wastes Collected in California," Final Report, CAL/ARB Project 4-011, Statewide Air Pollution Research Center, University of California, Riverside, CA (1977).

CHAPTER 5

Comparison of a Charcoal Tube and a Passive Sampling Device for Determination of Low Concentrations of Styrene in Air

Y. Z. Tang, P. Fellin, and R. Otson

TABLE OF CONTENTS

0-87371-606-0/93/$0.00+$.50

I. INTRODUCTION

Public concern has recently increased substantially regarding indoor air quality (IAQ) in residential and workplace environments. Largely due to the energy conservation practices that lead to the construction of energy-efficient buildings with low air exchange rates, considerable accumulation of airborne pollutants can occur in indoor environments. These pollutants can be produced by human activities such as smoking, fireplace use, and cooking using gas stoves or can originate from household products and materials used in construction and furnishings. Volatile organic compounds such as methylene chloride, toluene, and benzene are common solvents and are found in the indoor environment.[1] Styrene vapor is also found in the indoor environment because it can be released from polystyrene used in packaging, containers and molded household wares, toys, and thermal insulation.[1,2]

The complete characterization of organic pollutants in the indoor environment is a formidable task because of the complexity of organic mixtures and the fact that many compounds are present at trace levels ($\mu g/m^3$).[1] Methods available for measurement normally rely on either passive diffusion-based sampling or on active sampling by means of a mechanical pump. Application of passive samplers to the characterization of IAQ offers particular advantages such as low cost, convenience, and simplicity of deployment. Gas chromatographic analysis of extracts from the samplers is usually employed to quantify pollutant concentrations.

We recently evaluated the performance of a commercially available passive sampler (3M OVM 3500) for determination of a number of organic compounds under controlled conditions. Styrene was among the listed target compounds, because it is a central nervous system depressant and may cause dermatitis after repeated exposures.[3] The OSHA[3] and NIOSH[4] methods prescribe collection of styrene using active sampling with activated charcoal sorbent. Carbon disulfide extraction of the sorbent and gas chromatography-flame ionization detector (GC-FID) analysis are employed for quantitation. However, the NIOSH method is validated only for styrene as a single component with concentrations in the range of 426–1710 mg/m^3 with a 5-L air sample, and the reliable quantitation limit of the OSHA method is 13 mg/m^3 based on a 10-L air sample. These concentrations are encountered in some occupational/industrial environments, but are higher than those expected in typical residential environments. In addition, styrene is rarely the major constituent in IAQ situations. The OSHA method also indicates that the extraction efficiency of styrene from the charcoal is dependent on the loading, but this effect at loadings less than 4.7 μg is not documented. It is anticipated that passive samplers (such as the 3M OVM 3500), which contain activated charcoal as the collection medium, may exhibit similar nonlinear behavior with respect to the recovery of styrene. Thus, performance of the 3M passive sampler for determination of styrene deserves evelution under conditions similar to IAQ situations.

II. EXPERIMENTAL

A. Materials and Standard Solutions

Passive samplers (3M OVM 3500, lot No. 7043-007) were obtained from 3M Centre (St. Paul, MN). Activated charcoal tubes (400-mg front and 200-mg backup sections, lot 120, catalog no. 226-09) were supplied by SKC Inc. (334 Valley View Road, Eighty-Four, PA), and the air was drawn through the tubes at a flow rate of 30 mL/min by Gilian Model LFS 113D pumps (Gilian Instrument Corp., Wayne, NJ). Styrene (99.8%, monomer, inhibited) was from Fisher Scientific Company (Fair Lawn, NJ). Carbon disulfide used for preparation of standard solutions and for extraction was supplied by Caledon Laboratories (Georgetown, Ontario). The five styrene standard solutions (0.208, 1.04, 5.20, 26.0, and 130 µg/mL) also contained 23 other compounds (Figure 1) at concentrations similar to those for styrene.

Diffusion tubes custom made by Bohemia Glass Blowing (Agincourt, Ontario) were used for generating vapors of styrene and the 23 other compounds. The diffusion path dimensions required for target compound emission rates and concentrations in air were calculated based on the mathematical relationship:[5]

$$R(ng / min) = 2.216 \times 10^6 \times \left(\frac{DMP_A}{T} \right) \times \frac{A}{L} \times \log \left(\frac{P_A}{P_A - P} \right) \qquad (1)$$

where

R = diffusion tube emission rate (ng/min)
D = diffusion constant (cm^2/sec)
M = molecular weight (g/mol)
P_A = ambient pressure (mmHg)
P = vapor pressure of compound at temperature T
T = temperature of liquid (K)
A = cross-sectional area of diffusional path (cm^2)
L = length of diffusional path (cm)

The calculation of diffusion constants was based on the method given in 3M "Diffusional Air Sampling Reference Manual"[6] and the calculation of vapor pressures was based on the formula and constants provided in the CRC *Handbook of Chemistry and Physics*.[7] Each diffusion tube was loaded with a compound, and all tubes were placed in the diffusion oven of the test atmosphere generation system described next.

B. Test Atmosphere Generation System

The samplers were exposed in a custom-designed test atmosphere chamber with associated vapor generation systems.[8] Two thermostated ovens were used to allow two different temperature zones for diffusion tubes so that vapors in desired

GC Conditions		Peak Identification			
Initial oven temperature:	= 60°C	1.	n-Hexane	13.	d-Limonene
Initial hold time:	= 1 min	2.	Dichloromethane	14.	1,3,5-Trimethylbenzene
First ramp:	= 10°C/min to 110°C	3.	Benzene	15.	Styrene
Second ramp:	= 25°C/min to 135°C	4.	n-Decane	16.	p-Cymene
Second hold time:	= 2 min	5.	Chloroform	17.	1,2,4-Trimethylbenzene
Third ramp:	= 25°C/min to 210°C	6.	a-Pinene	18.	1,3-Dichlorobenzene
Split on:	= 0 min	7.	Toluene	19.	Pentachloroethane
Injector temperature:	= 200°C	8.	1,2-Dichloroethane	20.	Hexachloroethane
Detector (FID) temperature:	= 250°C	9.	Ethylbenzene	21.	1,4-Dichlorobenzene
Carrier (He) Flow rate:	= 1 mL/min	10.	p-Xylene	22.	1,1,2,2-Tetrachloroethane
Air flow rate:	= 300 mL/min	11.	m-Xylene	23.	1,2,4-Trichlorobenzene
Hydrogen flow rate:	= 30 mL/min	12.	o-Xylene	24.	Naphthalene
Aux (He) flow rate:	= 30 mL/min				
Splitting ratio:	= 1/10				

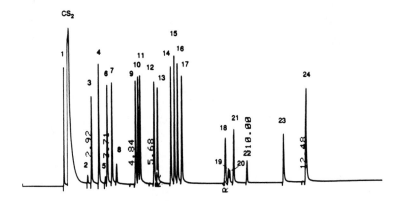

FIGURE 1.   GC separation of styrene from 23 copresent compounds.

concentration ranges could be generated by selecting diffusion tube dimensions and/or the temperature. The low temperature (30°C) apparatus was an Analytical Instrument Development Inc. Model 350 diffusion permeation system modified by replacing the Teflon® tubing with 316 stainless steel tubing with similar dimensions. The second temperature (100°C) zone consisted of a Precision Model No.16EG gravity convection oven fitted with a custom-made glass diffusion chamber. A J-type thermocouple was used to measure the oven temperature settings. Potential deposition losses of generated vapors on the connecting tubing were reduced by heating the tubing with heater tapes manually controlled by a variable (0–120 V) transformer (Staco Energy Products Co., Dayton, OH).

The chamber featured a 304 stainless steel body, which had dimensions of 1 m long, 0.64 m wide, and 0.35 m high in the exposure zone and had an inlet cone tapered to 0.025 m and an outlet cone tapered to 0.075 m. A 316 stainless steel rotating tray allowed the evaluation of passive samplers under different face velocities. The air delivered to the chamber was cleaned by a series of filters and conditioned by a Miller/Nelson 301 temperature-relative humidity controller.[8] This cleaned and conditioned air flow was mixed with the nitrogen carrier from the vapor generation system before entering into the chamber. The total gas flow was 100 ± 1 liters per minute (Lpm), which allowed this dynamic dilution system to equilibrate immediately after sampler deployment. The vapors of the other 23 compounds (Figure 1) were generated simultaneously with styrene vapor.

C. Gas Chromatographic Analysis

A Shimadzu Mini-II gas chromatograph (GC) equipped with a flame ionization detector (FID) and a Carle six-port gas sampling valve (Carle Mini MK-II. P/W 72303-00, 1-mL sampling loop) was used for on-line monitoring of chamber concentrations. The valve was activated by a Chrontrol timer (Cole-Parmer Instrument Company, Chicago, IL) at 30-min intervals and automatically sampled the air from the chamber. A 15-m J&W DB-Wax Megabore column (1.0 µm film) at 50°C isothermal with a helium carrier flow rate of 10 mL/min was used for separating the target compounds. This GC was calibrated with standard solutions and the sensitivity stability was checked by injection of gas standards containing hexane, benzene, and toluene.

The extracts from exposed samplers were analyzed using a HP5840 GC equipped with a FID. The column used in this GC was a 30 m x 0.32mm x 0.5 µm J&W DB-Wax column, and the column was temperature programmed. The GC operation conditions are listed in Figure 1. Under these conditions, styrene is chromatographically resolved from the 23 co-occurring compounds. Sample injection (1 µL) was executed by a Varian Model 8000 autosampler. The system was calibrated by injection of standard solutions.

D. Recovery Test

Extraction efficiencies were established by using the phase equilibrium technique.[9-11] Aliquots (2 mL) of standard solutions were added to the 3M badges and to vials containing 400-mg SKC charcoal sorbent. The sorbents were then exposed to the solutions for 30 min with frequent shaking. Overnight exposure of the sorbents to the solutions did not reveal any difference from the 30-min exposure. Aliquots (1 µL) of the solvent phase were then analyzed, and the results were compared with those of standard solutions analyses to allow calculation of extraction efficiencies based on the formula:

$$R = 100 \times E / S_s \qquad (2)$$

where

R = extraction efficiency (%)
E = amount of styrene found in the extract
S_s = amount of styrene in the standard solution

Triplicate samples were used for each concentration level tested.

E. Procedures

Four charcoal tubes were connected to the sampling ports of the chamber, two upstream and two downstream of the rotating tray, with air drawn through by pumps at flow rates of 30mL/min. Three 3M badges were placed near the charcoal tubes upstream of the tray, and two sets of triplicate 3M badges were suspended from the rotating tray at two different distances from the rotation center. Therefore, the three sets of 3M badges were deployed at three face velocities. The samplers were exposed to the test atmosphere for 24 h and were then extracted with 2-mL carbon disulfide for 30 min. Since the diffusion tube dimensions were not sufficiently precise and losses of styrene due to adsorption and other mechanisms occurred in the system, the styrene concentration calculated based on the theoretical emission rate was not accurate and served only as a guide for generating styrene vapor in the desired concentration range. Therefore, the concentration of styrene in the chamber was determined by: (1) analysis of the extracts from charcoal tubes and application of extraction efficiencies and (2) analysis of air samples on the on-line GC. The chamber concentrations then were used to calculate the 3M badge sampling rates:

$$S = W / (C \times T) \tag{3}$$

where

S = sampling rate (mL/min)
W = weight of styrene collected by the badge (μg)
C = styrene concentration in the chamber (μg/mL)
T = sampling time (min)

III. RESULTS AND DISCUSSION

It has been noted that the recovery of styrene from charcoal adsorbents is dependent on the compound load,[3,12] i.e., the extraction efficiency for a compound depends on the amount of compound on the sorbent. However, this information was obtained in the presence of styrene only and at styrene loadings greater than 4.7 μg. In addition, the loading effect on styrene recovery from the 3M badge has not been documented. Also, in typical residential environments, a variety of compounds are expected to be present in the air. It was, therefore, necessary to examine the extraction of styrene by carbon disulfide from the SKC charcoal and the 3M sorbent at conditions similar to those anticipated in IAQ situations, i.e., at low concentrations with consequent low sampler loadings and in the presence of other organics.

The dependence of the extraction efficiency for styrene on loading was studied. Control blank experiments indicated that no detectable amount of styrene was found in the samplers and the solvent. The extraction efficiencies for styrene at five loading levels were then obtained and are listed in Table 1. The lowest loading level (0.416 μg) was less than one-tenth of that previously reported.[3]

Table 1. Styrene Extraction Efficiency

Samplers	Loading (µg)														
	0.416			2.08			10.4			52.0			260		
	REC (%)	SD	RSD (%)	REC (%)	SD	RSD (%)	REC (%)	SD	RSD (%)	REC (%)	SD	RSD (%)	REC (%)	SD	RSD (%)
3M sorbent	67.8			74.0	2.7	3.7	81.3	0.8	1.0	89.1	1.7	1.9	96.8	1.2	1.2
SKC charcoal	47.7	3.8	6.9	58.0	2.1	3.7	70.2	2.3	3.3	74.7	2.0	2.7	88.1	4.1	4.7

Note: Triplicate measurements except 3M at 0.416 µg; 400-mg SKC charcoal.

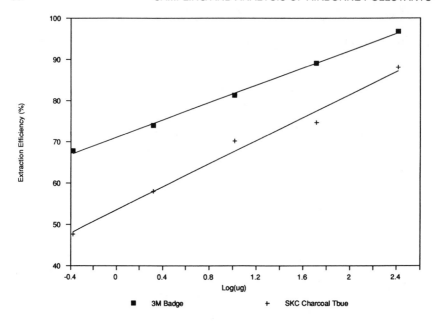

FIGURE 2.   Extraction efficiency of styrene vs logarithm of loading.

However, the amount of styrene in a 1-µL aliquot of extract was, due to the low
recovery, below the limit of detection (0.2 ng) when the sampler was loaded with
0.416 µg of styrene. Therefore, an aliquot of 2 µL was used to achieve the
necessary GC response at this loading level. Even so, the amount of styrene
injected was still near the limit of detection, and consequently the quantitation was
difficult and the results were variable. Therefore, the data at low loading levels
should be treated with caution. Nevertheless, at three tested loading levels the
relative standard deviations for the SKC charcoal results are greater than those for
the 3M sorbent results.

The dependence of extraction on loading is graphically demonstrated in Figure
2, where the extraction efficiencies for the charcoal tube and the 3M badge are
plotted against the logarithm of styrene loadings (µg). For SKC charcoal tubes the
relationship is formulated as:

$$R = 53.53 + 13.95 \times \log(L) = 100 \times W / L \qquad (4)$$

where

R = extraction efficiency (%)
W = amount (µg) of styrene found in extract
L = amount (µg) of styrene loaded on sampler

or

$$R = 71.15 + 10.46 \times \log(L) = 100 \times W / L \qquad (5)$$

for 3M badges. By solving Equations 4 and 5, it is possible to determine the actual
amounts (L) of styrene collected on the SKC sorbent and the 3M sorbent, respec-

Table 2. Comparison of Styrene Data Obtained by On-Line GC and Charcoal Tube

Chamber concentration (μg/m^3)		3M badge sampling rates (mL/min)	
On-line GC	Charcoal tube	On-line GC	Charcoal tube
22	20	37.9	40.8
27	19	39.3	55.0
117	45	33.7	87.9
136	88	36.2	56.0
138	8	31.4	567.6
172	66	27.1	62.1
182	119	33.8	51.9
306	397	46.0	35.4
348	436	39.0	35.6
471	399	33.3	39.4
447	389	32.0	39.2
500	512	34.5	41.9
809	779	34.9	36.2
814	779	36.8	38.5
2374	2042	33.6	39.0
	Average	35.3	47.0[a]
	SD	4.3	14.6
	RSD%	12.3	31.0

Note: 3M badge results: averages of triplicate measurements. SKC tube results: averages of quadruple measurements. On-line GC results: averages of 48 measurements (interval: 30 min.)

[a] 567.6 Value removed from data set. All data average 81.8 mL/min; SD 131 and RSD 159%.

tively. However, the constants in Equations 4 and 5 are not universally valid and thus should be determined for individual sets of experimental parameters.

Although the extraction efficiencies for the 3M sorbent increased with loading, they varied with loading to a lesser extent than those for the SKC charcoal. In addition, the extraction efficiencies for styrene from the 3M sorbent were substantially higher than those measured for the SKC charcoal under the test conditions. A minimum recovery of 75% is required by the NIOSH method validation criteria.[13] Therefore, 3M badges and SKC charcoal tubes should (according to Equations 4 and 5) collect at least 2.4 and 35 μg of styrene, respectively, to meet the recovery criteria.

Styrene vapor concentrations from 22 to 2400 μg/m^3 were generated together with 23 other compounds at concentrations similar to those of styrene for testing 3M badges (Table 2). These concentrations were determined based on the on-line GC data and corresponded to sorbent loadings from 0.97 to 106 μg of styrene, assuming 24-h samplings at ca. 30 mL/min. The 3M sampling rates calculated

Table 3. Effect of Face Velocity on 3M Sampling Rate

Face Velocity (m/sec)	0.01	0.5	1.7
Sampling Rate[a] (mL/min)	34.1 ± 4.1	34.7 ± 4.6	37.1 ± 4.3

[a] Triplicate measurements.

Table 4. Temperature and Humidity Effects on 3M Sampling Rate (mL/min)

Temperature (°C)	10	23	36
RH 20%	33.3 ± 1.9	36.0 ± 1.4	39.0 ± 1.5
RH 90%	32.0 ± 0.6	34.5 ± 2.9	35.4 ± 1.0

Note: Averages of triplicate measurements.

Table 5. Storage Effect on 3M Sampler

First Analysis	7-day Storage		21-day Storage	
μg/Sample	μg/Sample[a]	% Loss	μg/Sample[a]	% Loss
22.3	19.9	10.8	18.8	15.7

[a] Averages of triplicate measurements.

based on this set of concentrations vary from 27.1 to 46.0 mL/min with an average of 35.3 ± 4.3 mL/min. The overall relative standard deviation (RSD) of 12.3% indicates the consistency in measurement. The chamber concentrations calculated from the charcoal tube results are also listed in Table 2. They deviate at low concentrations from values based on the on-line GC data. Also, 3M badge sampling rates determined according to concentrations based on the charcoal tube results vary over a wide range: from 35.4 to 87.9 mL/min with an average of 47.0 ± 14.6 mL/min and an RSD of 31.0% (excluding the data point of 567.6 mL/min). At higher concentrations, (300 μg/m^3 and above), i.e., when samplers are at higher loadings, the charcoal tube data are similar to those from the on-line GC.

It is proposed that deviations of the two sets of data at low concentrations arise partly from low loadings and thus the difficulty in accurate determination of extraction efficiency. However, this seems insufficient to fully account for all deviations observed. Polymerization of styrene collected in the 3M badge and the SKC tube may have contributed to deviations of the two sets of data. However, it has been reported that polymerization of styrene in charcoal tubes was not evident.[3] In view of the fact that 3M sampling rates based on the on-line GC data fall in a consistent and reasonable range, it seems that the on-line GC data are more reliable than the charcoal tube results.

It is possible to increase the air sample volume to cope with the low loading problem. With the 43-L sample volume (a sampling rate of ca. 30 mL/min for 24

h) used, no breakthrough of styrene from the front section (400 mg) to the backup section (200 mg) of the charcoal tube was observed under test conditions. However, higher volumes may lead to breakthrough. The OSHA method[3] suggests a sample volume of 10 L, about one-quarter of that we used, but the amount of adsorbent (100 mg/50 mg) used in the OSHA method is also one-quarter of that used in our tests.

The effects of sampling conditions on the 3M sampling rates were also studied. The data in Table 2 do not demonstrate an apparent relationship between the sampling rate and the airborne styrene concentration. The effect of face velocity is illustrated by Table 3. The trend in the data of Table 3 suggests a small effect of face velocity: increase in the face velocity leads to a slight increase in the sampling rate. Worthy of particular comment is the fact that the 3M sampler performed well at a face velocity of 0.01 m/sec, which is below the minimum face velocity (0.13 m/sec) suggested by 3M for proper performance of the badges.[6] It is generally believed that starvation will occur at face velocities below 0.1 m/sec[6,14,15] and will cause apparent reduction in sampling rate, because the external resistance to mass transfer becomes a significant fraction of the total diffusional resistance.

The data in Table 4 show a slight increase in the sampling rate with temperature. This is consistent with diffusion theory which predicts an increase in molecular diffusivity with temperature. The decrease of sampling rate with elevated relative humidity, also indicated by Table 4, is less than 10%. The effect is likely due to the modification of sorbent surface by moisture. Table 5 summarizes the results of storage effect study. Three sets of 3M samplers exposed in the chamber under the same condition were collected; one set was analyzed immediately after sampling; and the other two sets were analyzed after storage in the 3M badge container[6] at room temperature for 7 and 21 days, respectively. The losses observed were less than 16%. This allows sufficient time for transportation of the samplers from the field to an analytical laboratory without significant sample losses.

IV. CONCLUSION

It was found that the extraction efficiency of carbon disulfide for styrene from both the SKC charcoal and 3M sorbent depended on the amount of styrene loaded on the samplers. The relationships between extraction efficiencies and styrene loadings were established for both samplers. It is more precise to determine the extraction efficiency for styrene from the 3M badge than to determine that from the SKC charcoal. The 3M badge is reliable, in addition to other advantages of passive samplers, for sampling airborne styrene at low concentrations. The 3M badge sampling rate for styrene was slightly affected by sampling conditions such as humidity, temperature, and face velocity. The variations were less than 10% over the ranges of relative humidity 20–90%, temperature 10–36°C, and face velocity 0.01–1.7 m/sec. The average sampling rate was 35.3 ± 4.3 mL/min based on the on-line GC data.

REFERENCES

1. "Indoor Air Quality Environmental Information Handbook: Building System Characteristics," U.S. DOE Report DOE/EV/10450 - H1 (1987).
2. *The Condensed Chemical Dictionary* (New York: Van Nostrand Reinhold Co., 1987), p. 943
3. OSHA Manual of Analytical Methods, Method #09, U.S. Department of Labor, Occupational Safety & Health Administration (1985).
4. NIOSH Manual of Analytical Methods, Method #1501, U.S. Department of Health, Education, and Welfare, NIOSH (1984).
5. Analytical Instrument Development, Inc. "Operations Manual for AID Model 350," (Undated).
6. 3M Company. "Diffusional Air Sampling Reference Manual," T-C-1872-E (Undated).
7. *Handbook of Chemistry and Physics,* 51st ed. (Cleveland, OH: CRC Press, Inc., 1970–71).
8. Fellin, F., R. Otson, and K. A. Brice. "A Versatile System for Evaluation of Organic Vapour Monitoring Methods," in *Proceedings of the 8th World Clean Air Congress,* (Society for Clean Air in the Netherlands, 1989), 675–680.
9. Dommer, R. A., and R. G. Melcher. "Phase Equilibrium Method for Determination of Desorption Efficiencies," *Am. Ind. Hyg. Assoc. J.* 39(3):240–246 (1978).
10. Krajewski, J., J. Gromiec, and M. Dobecki. "Comparison of Methods for Determination of Desorption Efficiencies," *Am. Ind. Hyg. Assoc. J.* 41(7):531–534 (1980).
11. Rodriguez, S. T., D. W. Gosselink, and H. E. Mullins. "Determination of Desorption Efficiencies in the 3M 3500 Organic Vapour Monitor," *Am. Ind. Hyg. Assoc. J.* 43(8):569–574 (1982).
12. Evans, P. R., and S. W. Horstman. "Desorption Efficiency Determination Methods for Styrene using Charcoal Tubes and Passive Monitors," *Am. Ind. Hyg. Assoc. J.* 42(6):471–474 (1981).
13. "Development and Validation of Methods for Sampling and Analysis of Workplace Toxic Substances," NIOSH Contract #210-76-0123, U.S. Department of Health, Education, and Welfare, NIOSH (1979).
14. Lautenberger, W. J., E. V. Kring, and J. Morello. "A New Personal Badge Monitor for Organic Vapours," *Am. Ind. Hyg. Assoc. J.* 41(10):737–747 (1980).
15. Fowler, W. K. "Fundamental of Passive Vapour Sampling," *Am. Lab.* (12):80–87 (1982).

CHAPTER 6

Dynamic Gas Chromatographic Techniques for the Characterization of Carbon-Based Adsorbents Utilized in Air Sampling

W. R. Betz, K. S. Ho, S. A. Hazard, and S. J. Lambiase

TABLE OF CONTENTS

0-87371-606-0/93/$0.00+$.50
© 1993 by Lewis Publishers

I. INTRODUCTION

The use of gas-solid chromatography (GSC) as a tool for the characterization of adsorbents, as well as an analytical separations tool evolved in the late 1940s and early 1950s.[1] In this era, the physicochemical measurements that could be obtained by GSC were established, and the technique remains a viable analytical tool. In the 1960s, considerable attention was given to the development of adsorbents with physical (i.e., structural) and chemical (i.e., homogeneous or inert surface) properties that enabled them to be used in both GSC and sample enrichment modes of operation. This improvement in adsorbent characteristics was responsible for the application of solid adsorbents in the field of environmental monitoring, specifically in airborne contaminant sampling for industrial workplace and ambient air atmospheres, and for enrichment of contaminants present in aqueous media.

The adsorbents typically used in these sample enrichment devices were and remain: activated charcoals, activated silica gels, porous polymers, carbon molecular sieves, and graphitized carbon blacks. Characterization of these adsorbents in the 1960s led to a classification scheme for both adsorbents and adsorbates, and laid the groundwork for understanding adsorption phenomena at the gas-solid interface.[2] The use of this classification scheme and the principles characterizing these adsorbate/adsorbent interactions have assisted in constructing adsorbent devices currently used in sample enrichment procedures.

Kiselev[2] first categorized the interactions between adsorbate and adsorbent, and generated a scheme that categorizes the adsorbents into three classes, or types: Type I, Type II, and Type III. Type I adsorbents are defined in this scheme as adsorbents that possess no ionic charges (i.e., positive or negative) on the surface. This lack of ionic interaction, or nonspecific interaction, allows for predictable behavior between the adsorbate and adsorbent surfaces, and also allows the sampling professional to choose sampling parameters based on the molecular size and shape of the molecule(s) of interest.[3] The nonspecific surface characteristic of a Type I adsorbent is also hydrophobic. Examples of Type I adsorbents include the graphitized carbon blacks.

Type II adsorbents possess localized, positive charges that interact specifically with the adsorbates. This specificity allows for a strong or weak (depending on the adsorbent) electrostatic interaction between the adsorbate and the adsorbent. This type of interaction can be effective when sampling of adsorbates with specific, similar functionalities is desired. A characteristic shortcoming of these specific adsorbents is their affinity for water (i.e., hydrophilicity) which can negate the specificity of the surface. An example of a Type II adsorbent is activated silica gel.

Type III adsorbents possess localized, negative charges which, as with the Type II adsorbents, interact specifically with the adsorbates. This specific interaction has similar positive and negative surface characteristics (i.e., specificity and hydrophilicity). Examples of Type III adsorbents include activated charcoal and porous polymers such as Amberlite® XAD®-2 and Tenax.®

Table 1. Classification of Adsorbents and Adsorbates and Their Subsequent Interactions

Adsorbate types	Adsorbent types		
Molecular groups	Type I without ions or active groups	Type II localized positive active groups	Type III localized negative active group
Group A (n-alkanes) Spherically symmetrical shells σ-Bonds	Nonspecific interactions		
Group B (aromatic hydrocarbons, chlorinated hydrocarbons, ketones) Electron density concentrated on bonds/links π-Bonds			
Group C (organometallics) (+) Charge on peripheral links	Nonspecific interactions	Nonspecific and specific interactions	
Group D (organic acids,organic bases, aliphatic alcohols) Concentrated electron densities (+) Charges on adjacent links			

Classification of Adsorbents

Adsorbent	Surface	Classification (Kiselev)
Graphitized carbon blacks	Graphitic carbon	Class I
Carbon molecular sieves	Amorphous carbon	Weak Class III (can approach Class I)
Activated silica gel	Oxides of silica gel	Class II
Activated charcoal	Oxides of amorphous carbon	Class III
Porous polymers	Organic "plastics"	Weak → Strong Class III

Kiselev[2] also categorized adsorbates, according to electronic activity, into four groups: Group A — n-alkanes; Group B — aromatic hydrocarbons, chlorinated hydrocarbons, ketones; Group C — organometallics; and Group D — organic acids, organic bases, and aliphatic alcohols. Table 1 summarizes the interactions between the three types of adsorbents and the four groups of adsorbates.

Characterization evaluations for the three types of adsorbents interacting with adsorbates typically encountered in environmental air sampling techniques have been and remain the focus of activity in this laboratory.

II. METHODS

Initially, experiments to evaluate the adsorbate/adsorbent interactions occurring in an adsorbent tube of dimensions typically utilized in indoor air sampling studies

U-Tube Configuration for Adsorbent Characterization Evaluations

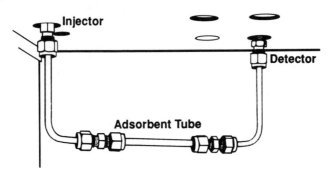

©1989 SUPELCO
84-62

FIGURE 1.

were performed using an apparatus which was constructed to parallel previous studies.[4] This apparatus, illustrated in Figure 1, consisted of a 4.0-mm i.d. packed adsorbent tube/column connected to a gas chromatograph through two L-shaped silanized glass tubes. The tube was utilized to determine specific retention (breakthrough) volumes, adsorption capacities, and equilibrium sorption capacities.[4]

The specific retention volume is defined as the calculated volume of carrier gas (i.e., nitrogen) per unit mass of adsorbent necessary to cause a mass of adsorbate molecules, introduced into the front (i.e., injector end) of the adsorbent tube, to migrate to the back (i.e., detector end) of the tube. The adsorption coefficient is defined as the specific retention volume divided by the adsorbent surface area, and can provide information concerning the degree of specific interactions occurring between the adsorbate and adsorbent surfaces. The equilibrium sorption capacity is a measure of the adsorbent capacity for the adsorbate in low surface coverage sampling modes. The specific equations and subsequent reduction of terms for these three parameters are extensively covered in previous documents.[4,5]

Additional, fundamental experiments were performed using the column/adsorbent tube approach mentioned here. However, a reduction in the internal diameter of the apparatus from 4.0- to 2.0-mm i.d. was necessary. This reduction in internal volume improved the determination of adsorbent capacities (i.e., monolayer capacities) and adsorption isotherms.[6] The adsorption isotherm is the relationship between the amount of adsorbate adsorbed and the equilibrium pressure of the carrier gas at a constant temperature.[7]

Current efforts have focused on the use of narrower (i.e., 0.75 mm i.d.) column/tube configurations to provide further fundamental insights. The choice of 0.75-mm i.d. tubing has allowed for the generation of adsorption isotherms that can be correlated with those generated using static methods (i.e., nitrogen isotherms via Micromeritics®* instrumentation). A Brunnaeur, Emmet, Teller (BET) plot was

BET Isotherm Plots Carbopack™ B / Pentane

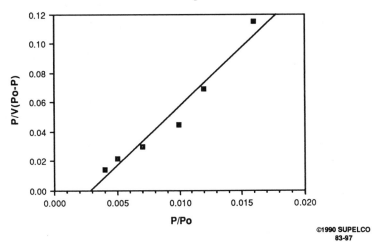

FIGURE 2.

Table 2. Breakthrough Volume Data for Carbotrap, Amberlite XAD-2, and Tenax

	Breakthrough volume (mL/g)		
Adsorbate	Carbotrap	Tenax	Amberlite XAD-2
n-Decane	4.79×10^9	1.56×10^7	3.36×10^7
Benzylamine	2.23×10^7	3.57×10^6	1.63×10^7
Chlorobenzene	1.58×10^6	1.51×10^5	4.84×10^5
p-Xylene	4.24×10^7	3.88×10^5	7.95×10^6
p-Cresol	2.06×10^7	1.50×10^7	4.96×10^6
n-Pentanoic acid	4.31×10^5	9.78×10^5	1.01×10^5
Cyclohexanone	2.04×10^6	1.06×10^6	6.27×10^5
2-Methyl-2-propanol	6.52×10^3	6.86×10^2	5.42×10^3

also constructed to determine the adsorbent surface area used by the adsorbate during air sampling, as shown in Figure 2.

III. RESULTS AND DISCUSSION

A. Studies Involving 4.0-mm i.d. Tubes

Results obtained from the 4.0-mm i.d. tube studies have allowed for comparisons between the adsorbents. Table 2 provides a comparison between Carbotrap® (B), Amberlite XAD-2, and Tenax.® These data indicate that Carbotrap, a

Table 3. Breakthrough Volume Data for Carbotrap

Adsorbate	Breakthrough volume (mL/g, 20°C)
Ethane	1.73×10^1
n-Propane	5.49×10^1
n-Butane	4.06×10^2
Ethanol	4.93×10^2
Acetic acid	7.16×10^2
Propionic acid	1.66×10^3
1,2-Dichloroethane	1.94×10^3
2-Butanone	3.76×10^3
n-Pentane	5.89×10^3
2-Methyl-2-propanol	6.52×10^3
Benzene	1.17×10^4
1,1,2-Trichloroethylene	1.27×10^4
n-Butanol	1.92×10^4
1,1,2-Trichloroethane	2.47×10^4
n-Hexane	7.99×10^4
n-Pentanoic acid	4.31×10^5
Phenol	6.16×10^5
Toluene	6.50×10^5
Chlorobenzene	1.58×10^6
Cyclohexanone	2.04×10^6
n-Butylamine	2.08×10^6
4-Heptanone	2.44×10^6
1,4-Dichlorobenzene	1.34×10^7
n-Octane	1.61×10^7
Ethylbenzene	2.03×10^7
p-Cresol	2.06×10^7
Benzylamine	2.23×10^7
p-Xylene	4.27×10^7
Acetophenone	6.40×10^7
Isopropylbenzene	1.70×10^8
n-Propylbenzene	1.72×10^9
n-Decane	4.79×10^9
n-Butylbenzene	5.83×10^9
Diphenyl	3.74×10^{12}
n-Hexylbenzene	7.00×10^{12}
n-Dodecane	1.63×10^{14}
n-Octylbenzene	1.31×10^{15}
n-Tetradecane	8.32×10^{16}

graphitized carbon black, generally possesses a greater affinity for the eight chosen adsorbates. The greater affinity of Tenax for 2-methyl-2-propanol is due to the strong electrostatic interactions occurring between the phenylene oxide

Table 4. Breakthrough Volume Data for Carbon Molecular Sieves and Activated Charcoal

Adsorbent	Breakthrough volume dichloromethane (L)	Water
Carbosieve S-III	66.2	0.32
Carboxen-569	43.2	0.06
Activated coconut charcoal	39.2	2.44
Carboxen-564	31.5	0.10
Carbosieve S-II	31.5	1.02
Purasieve	5.05	0.24
Carboxen-563	1.56	0.80
Spherocarb	1.05	0.22

functional groups present on the adsorbent surface and the alcohol group present on the adsorbate surface. Table 3 illustrates the ordered (i.e., molecular size and shape, or molecular volume) breakthrough volumes obtained with a Type I adsorbent such as the graphitized carbon black.

Table 4 provides data obtained from a comparison evaluation of several carbon molecular sieves and activated charcoal. These data indicate an ordering based on the adsorption strength of the pores of these porous solids and on the ability of the adsorbate molecules to access the pores, specifically the smaller pores. These and other data obtained over a 4-year period have allowed for a classification of the adsorbents according to the functional range of all the adsorbates, based on molecular size or carbon number. Table 4 illustrates this ordering for several carbon-based adsorbents currently used in multibed adsorbent tubes.

B. Studies Involving 2.0-mm i.d. Tubes

Results obtained from 2.0-mm i.d. tube studies provided data that suggest the entire external surface of nonporous, nonspecific adsorbents such as the graphitic carbons is utilized during air sampling modes of operation.[8] These suggestions were exemplified by the behavior of the isotherm plots, which indicate Type II behavior, or behavior indicative of a nonporous or macroporous solid (Figure 3). Further work using the 0.75-mm columns confirmed these hypotheses.

C. Studies Involving 0.75-mm i.d. Tubes

Results obtained from the 0.75-mm i.d. tube studies are providing valuable insight into characteristics of both the nonporous and porous carbon-based adsorbents. The data presented in Table 5 confirm the statements mentioned above concerning the surface usage of a graphitic carbon. Table 6 provides insight into the lack of correlation between the surface area data provided by using two differing probe molecules, such as nitrogen and dichloromethane, with several

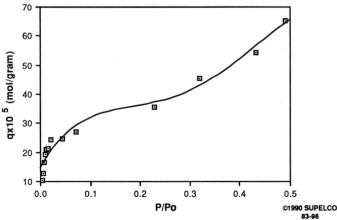

FIGURE 3.

Table 5. Surface Area Data for Graphitized Carbon Blacks

Adsorbent	Adsorbate	Surface area (m²/g)
Carbopack B	Nitrogen	86
	n-Pentane	84
	Toluene	83
	Benzene	86
	n-Octane	85
Carbopack C	Nitrogen	10
	n-Pentane	10
	Toluene	10
	n-Decane	10
Carbopack F	Nitrogen	6.0
	Toluene	5.2
	n-Octane	5.7
	n-Decane	5.7
	n-Dodecane	5.4

carbon molecular sieves and activated charcoal. This discrepancy can be explained as follows. Data obtained from a static nitrogen isotherm study provide micropore (i.e., pores less than 20 Å in diameter) volume data rather than micropore surface area data. In contrast, the generation of surface area data using the dynamic gas-solid chromatographic approach allows for an interaction of the probe molecules in the low coverage, or Henry's law, region where the probe molecule interacts only with the internal surface that is accessible to the molecule of interest. This adsorbate/adsorbent surface interaction provides practical surface

Table 6. Surface Area/Breakthrough Volume Data for Carbon Molecular Sieves and Activated Coconut Charcoal

Adsorbent	Breakthrough volume (L)	Surface area (m²/g) Dichloromethane	Nitrogen
Carbosieve S-III	66.2	697	820
Carboxen-569	43.2	466	485
Activated Charcoal	39.2	526	1070
Carbosieve S-II	31.5	506	1060
Carboxen-564	31.5	380	400
Purasieve	5.05	364	950
Carboxen-563	1.56	291	510
Spherocarb	1.05	291	880

Multi-Bed Adsorbent Tube

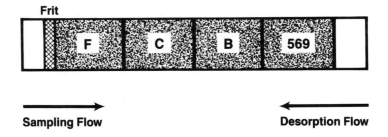

FIGURE 4.

area data and correlates with the breakthrough volume data generated using the 4.0-mm i.d. column/tube approach.

IV. CONCLUSION

Several multibed adsorbent tubes have been constructed based on data obtained from these dynamic characterization studies (i.e., Figure 4). These tubes are currently being utilized to effectively adsorb and desorb adsorbates having a wide range of molecular sizes in sample enrichment modes of operation.[9] Further work is currently ongoing to investigate other carbon-based adsorbents and multibed adsorbent tubes, and applications of the carbon-based (i.e., Type I) adsorbents for a variety of sample enrichment modes of operation.

REFERENCES

1. Conder J.R., and C.L. Young. *Physicochemical Measurement by Gas Chromatography* (New York: John Wiley & Sons, Inc., 1979).
2. Kiselev, A.V., and Y.I. Yashin. *Gas Adsorption Chromatography* (New York: Plenum Press, 1969).
3. Betz, W.R., and W.R. Supina. "Use of Thermally Modified Carbon Black and Carbon Molecular Sieve Adsorbents in Sampling Air Contaminants," *Pure Appl. Chem,* 61(11): 105–112 (1989).
4. U.S. Environmental Protection Agency. "Characterization of Sorbent Resins for Use in Airborne Environmental Sampling," EPA Document No. 600/7-78-054/NTIS Document No. PB284347, Springfield, VA, National Technical Information Service, (1978).
5. Betz, W.R., S.G. Maroldo, G.D. Wachob, and M.C. Firth. "Characterization of Carbon Molecular Sieves and Activated Charcoal for Use in Airborne Contaminant Sampling," *Am. Ind. Hygiene Assoc. J.* 50(4): 181–187 (1989).
6. Betz, W.R., and W.R. Supina. "Determination of the Gas Chromatographic Performance Characteristics of Several Graphitized Carbon Blacks," *J. Chromatog.* 471: 105–112 (1989).
7. Gregg, S.J., and K.S.W. Sing. *Adsorption, Surface Area and Porosity.* (New York: Academic Press, 1982).
8. Bruner, F., G. Bertoni, and G. Crescentini. "Critical Evaluation of Sampling and Gas Chromatographic Analysis of Halocarbon and Other Organic Air Pollutants," *J. Chromatog.* 167: 399–407 (1978).
9. Betz, W.R., S.A. Hazard, and E.M. Yearick. "Characterization and Utilization of Carbon-Based Adsorbents for Adsorption and Thermal Desorption of Volatile, Semi-Volatile and Non-Volatile Organic Contaminants in Air, Water, and Soil Sample Matrices," *Int. Labmate* 15(1): 41–44 (1990).

TRADEMARKS

Amberlite — Rohm and Haas
Carbopack — Supelco
Carbosieve — Supelco
Carbotrap — Supelco
Carboxen — Supelco
Purasieve — Union Carbide
Spherocarb — The Foxboro Co.
Tenax — Enka Research Institute Arnhem
XAD — Rohm and Haas

Volatile Organic Compound Analysis

CHAPTER 7

Detector Selection for the Analysis of Volatile Organic Air Toxic Compounds Using U.S. EPA Compendium Method TO14

Larry D. Ogle, David A. Brymer, and Ruth L. Carlson

TABLE OF CONTENTS

0-87371-606-0/93/$0.00+$.50

© 1993 by Lewis Publishers

I. INTRODUCTION

U.S. EPA Compendium Method TO-14[1] describes the analysis of volatile
organic compounds (VOCs) using SUMMA™ polished stainless steel canisters
for sample collection and cryogenic concentration of organics before analysis.
However, the analytical techniques to be used are not strictly defined. This
method allows for the use of gas chromatography with multiple detectors (GC/
MD) or gas chromatography coupled with mass spectrometry (GC/MS). Selection
of the appropriate detector(s) is predicated by the target compounds, equipment
availability, budgetary restrictions, and required sensitivity and selectivity. The
goal of this chapter is to present a logical approach to detector selection and usage
by presenting advantages and disadvantages of each detector or combination of
detectors.

As a general rule, high resolution gas chromatography utilizing bonded-phase
capillary columns is desirable for VOC separation. Packed columns may be used,
but often will not provide the resolution required to analyze ambient organic
compounds. Sample preconcentration can be achieved by concentrating the organics into a trap cooled to $-150°C$ or colder. Solid sorbents may also be used for
sample concentration. Sorbents generally are suitable for a particular compound
volatility range or polarity range. They cannot trap the wide range of compounds
that can be concentrated with cryogens.

Cryogens colder than liquid argon, such as liquid nitrogen, should be avoided
because oxygen in the sample can be enriched in the trap. This oxygen could result
in column and detector degradation. Water may be removed through the use of a
Nafion®* membrane, but these membranes also result in the loss of polar organic
compounds, as pointed out in Method TO-14. If water is concentrated in the
cryotrap, bonded-phase columns must be used to avoid degradation of the liquid
phase.

Detector selection for the analysis of VOCs generally depends on the class(es)
of compound(s) to be detected. The flame ionization detector (FID) responds to
almost all carbon-containing compounds. The photoionization detector (PID)
responds to compounds with unsaturation. Electron capture detectors (ECD) and
Hall electrolytic conductivity detectors (HECD or ElCD) both respond to halogenated compounds. Table 1 presents a listing of possible detectors and the types of
compounds for which they are useful. Some detectors respond to only one or two
compound classes, but generally provide excellent and specific response to those
compounds.

Multiple detector systems can be configured to utilize a combination of detectors to provide selectivity and sensitivity for different compound classes. Use of

Table 1. Detector Types and Compound Classes Detected

Detector	Compound Classes
Flame ionization	All organic
Electron capture	Halogenated, nitrogen containing, some ketones and alcohols
Photoionization	Ionizable compounds at lamp voltage (aromatic and unsaturated hydrocarbons, mercaptans, ketones, esters, ethers)
Atomic emission	All compounds
Mass spectrometry	All compounds

multiple detector responses and a ratio of those responses can provide increased certainty in compound identification on nonspecific detectors. This approach was described by Cox and Earp[2] using a combination of FID and PID detectors. A multidetector system utilizing FID, PID, and ElCD detectors has also been described.[3]

II. DETECTORS

A. Flame Ionization Detectors

The FID is the most common type of nonspecific detector. It responds to virtually all carbon-containing compounds. However, the FID responds very poorly to some compounds, such as CO, CO_2, and CS_2. FID response is relatively linear with the number of carbon atoms in the compound. Therefore, FID response can be used as a quantitative tool (on a carbon basis) for unknown compounds and for determining a total nonmethane response in samples because methane is not trapped on most sorbents or at liquid nitrogen temperatures or warmer. However, response is diminished with the addition of a heteroatom to the compound. For example, the response of methanol is 0.87 times that of methane, the response of carbon tetrachloride 0.67 times methane, and freon-12 is 0.42 times methane.[4]

A summary of the advantages and disadvantages of the FID is shown in Table 2. The FID offers a linear response to organic compounds, wide linear range, good sensitivity, and stability with little maintenance required. In addition, most laboratories have one or more of these detectors available. Disadvantages of the FID include the fact that it responds to most organic compounds and can make identifications difficult due to the large number of compounds present. Figure 1 is an FID chromatogram of an ambient air sample with a number of peaks. Compound identification is difficult due to the number of coeluting compounds. The FID is also a destructive detector which means it must be used as the last detector in series when multiple detectors are used. One other important consideration is the effect of moisture on the FID. In undried samples, moisture can cause spikes in the chromatogram which can obscure components of interest or, if present in large enough quantities, can extinguish the flame.

Table 2. Advantages and Disadvantages of a Flame Ionization Detector

Advantages	Disadvantages
Universal response to carbon	Nonspecific
Excellent linear range ($>10^6$)[a]	Destructive detector
Good sensitivity (<1 ng)[a]	Poor sensitivity for some compounds, i.e., halocarbons
Low maintenance	Generally requires 3 to 4 different support gases (N_2, H_2, He, and air)
Small analytical variation	Can be affected by moisture in the sample

[a] Information taken from manufacturer's instrument manuals.

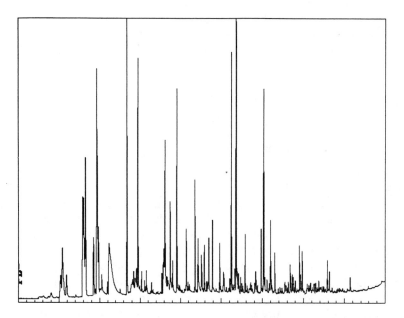

FIGURE 1.   FID chromatogram of ambient air from a petrochemical complex conditions: 60 m x 0.32-mm i.d., 1.0-μm film DB-1 column; –50°C for 2 min, 6°C/min to 175°C then 25°C/min to 225°C hold 5 min. Load volume 0.234 L.

B. Photoionization Detectors

Another common detector used for ambient organics is the PID, which provides excellent response for those compounds with an ionization potential at or below the lamp energy. The most common lamp has an energy of 10.2 eV and ionizes compounds with unsaturated bonds. It has many of the advantages of the FID, as can be seen in Table 3, and is not significantly affected by moisture from the sample. However, it responds only to a limited number of compounds and by itself is not useful in estimating the concentration of an unknown compound. The response is not linear to any property of a class of compounds in the same manner the response of the FID is linear to carbon.

Table 3. Advantages and Disadvantages of a Photoionization Detector

Advantages	Disadvantages
Very sensitive to compounds with unsaturated bonds, i.e., aromatics (2 pg of benzene with 10.2 eV lamp)[a]	Responds to a limited number of compounds
Nondestructive detector	Difficult to predict response from molecular structure
Low maintenance	High maintenance costs compared to FID
Requires only one support gas	
Excellent linear range ($>10^7$)[a]	
Unaffected by moisture	

[a] Information taken from manufacturer's instrument manuals.

Table 4. Advantages and Disadvantages of an Electron Capture Detector

Advantages	Disadvantages
Excellent sensitivity to electron capturing compounds, i.e., polychlorinated	Small linear range ($<10^3$)[a]
Low maintenance	Difficult to predict response from molecular structure
Requires only one support gas and a carrier gas	Does not respond well to some compounds of environmental interest, e.g., vinyl chloride, methylene chloride
Nondestructive detector	
Unaffected by moisture	
Very stable detector	

[a] Information taken from manufacturer's instrument manuals.

C. Electron Capture Detectors

Method TO-14 lists the ECD as a detector of choice in a multidetector system. Table 4 lists the advantages and disadvantages of the ECD. Its major advantage is the reliability and stability of the detector. It requires very little maintenance as long as dry, contaminant-free support gases are utilized. It is generally nondestructive, but cannot be used in series with a FID if argon/methane is used as the carrier or makeup gas. Linearity is generally better with Ar/CH$_4$ than with nitrogen. The ECD is also unaffected by moisture in the sample. It is extremely sensitive to some compounds containing multiple halogen atoms, such as chloroform or tetrachloroethylene. Therefore, it can be used to achieve parts per trillion level detection limits for these compounds.

The ECD suffers from the fact that it has a very limited linear range. This fact is a disadvantage due to the need to dilute samples containing high concentrations of electron-capturing compounds so that they are within the linear range of the detector. For most ambient samples, this is not a large problem because concentrations are generally low (parts per billion levels).

Table 5. Advantages and Disadvantages of an Electrolytic Conductivity Detector

Advantages	Disadvantages
Excellent sensitivity for halogenated compounds (2.5 pg heptachlor)[a]	Complex detector
Readily available	More analytical variation than other detectors such as ECD and PID
Requires one or two support gases	Affected by moisture from the sample
Excellent linear range (10^6 for Cl)[a]	Destructive detector
Response linear with the number of halogens in the molecule	Considerable maintenance required
Responds to all halogenated compounds of environmental interest, i.e., vinyl chloride	Can be affected by samples containing large amounts of acidic or basic components, e.g., CO_2, NH_3

[a] Information taken from manufacturer's instrument manuals.

D. Hall Electrolytic Conductivity Detectors

The HECD or ElCD is often used as an alternative to the ECD for the detection of halogenated compounds. It can also be used for nitrogen and sulfur compounds by switching the mode of operation. The ElCD exhibits linear response for the number of halogens in a molecule, within the same halogen group, since the halogen is converted to the corresponding acid for detection. For example, dichloroethane will give twice the response of chloromethane and dibromoethane will give twice the response of bromomethane. This characteristic can be used to estimate the amounts of unknown compounds when a boiling point column, such as methyl silicone, is used for separation. In addition, environmentally sensitive compounds, such as vinyl chloride, respond much better on the ElCD than on the ECD. The ElCD is also more selective than the ECD and does not respond to nonhalogenated compounds.

Shortcomings of the ElCD include the complexity of the detector. It has a nickel reaction tube, a solvent, a high temperature furnace, an ion-exchange resin, and a reaction gas any of which can affect the detector operation. Troubleshooting can be very difficult with this many potential sources of problems. The ElCD is adversely affected by water from environmental samples and should be used with a Nafion® drier. Fortunately, most halogenated components are not removed by the drier. It is a destructive detector and must be used last in series with another detector. It is often used in series with a PID as described in EPA Method 502.2 for water samples.[5] Other advantages and disadvantages of the ElCD are presented in Table 5.

Figure 2 presents a comparison of the same 17-component standard analyzed on the ECD and the ElCD. Some components, such as trichloroethylene and tetrachloroethylene, have a much greater response on the ECD than on the ElCD. A greater response provides a lower detection limit for these compounds, but can

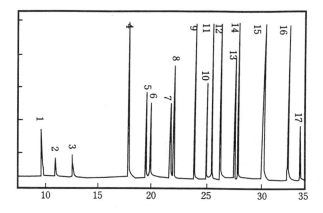

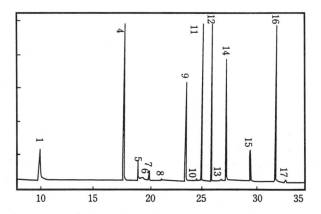

FIGURE 2.   Comparison of chromatograms from a 17-component standard on the ECD and EICD. (Peak numbers correspond to the compounds listed in Table 6.)

be complicated by a rather narrow linear range for the ECD. Table 6 provides a comparison between responses of the ECD and the EICD normalized to the FID response. This comparison was for one set of conditions and does not represent the comparison of a number of different detectors or manufacturers.

The EICD provided a factor of 10 times greater sensitivity than the FID and ECD for chloromethane, methylene chloride, vinyl chloride, 1,1-dichloroethane, 1,2-dichloroethane, and 1,2-dichloropropane. The ECD provides greater than a factor of 10 better sensitivity than the EICD for fluorotrichloromethane and carbon tetrachloride. Basically, the decision to use the ECD or EICD for the detection of halogenated compounds depends on the components to be detected and the experience of the analyst with the EICD.

Table 6. Relative Responses of the ECD and EICD to the FID for a Select Group of Compounds

Compound	EICD	ECD
1. Dichlorodifluoromethane	25	130
2. Chloromethane	15	NR
3. Vinyl chloride	10	0.2
4. Fluorotrichloromethane	240	4900
5. 1,1-Dichloroethylene	15	9
6. Methylene chloride	25	2
7. t-1,2-Dichloroethene	15	13
8. 1,1-Dichloroethane	20	1
9. Chloroform	50	100
10. 1,2-Dichloroethane	15	1
11. 1,1,1-Trichloroethane	22	135
12. Carbon tetrachloride	90	1100
13. 1,2-Dichloropropane	10	1
14. Trichloroethylene	20	80
15. 1,1,2-Trichloroethane	20	13
16. Tetrachloroethylene	20	60
17. Chlorobenzene	2	1

Note: Response relative to FID for a 2.4 nL per species injection. NR = no response.

E. Flame Photometric and Nitrogen/Phosphorus Detectors

Table 7 provides the advantages and disadvantages for the nitrogen/phosphorus (NPD) and the flame photometric detectors (FPD). Both of these detectors are of use for specific applications in the analysis of air toxics. The NPD would be applicable to low molecular weight amines and phosphorus-containing compounds. The FPD would be applicable to sulfur compounds such a mercaptans, disulfides, and thiophenes. However, the problem with using either of these detectors in such applications is that the stability of the compounds of interest have not been demonstrated in canisters. Brymer et al.[6] reported that methyl and ethyl mercaptans displayed poor stability in SUMMA™ polished canisters.

F. Atomic Emission Detector

A relatively new detector, the Hewlett-Packard (HP) 5921A atomic emission detector (AED) shows great promise for the detection of air toxic compounds since it detects specific elements and can monitor as many as 15 at one time. According to HP literature,[7] it has sensitivity in the picogram range. A recent article[8] reported using this detector with supercritical fluid extraction for the analysis of sulfur compounds in spices. Sensitivity was five times better than in the FPD, and linear response was reported.

Great advantages could be obtained by the ability to monitor carbon for a FID-type response at the same time as chlorine, nitrogen, sulfur, and oxygen. An oxygen specific detector would be very useful in detecting several classes of

Table 7. Advantages and Disadvantages of Nitrogen/Phosphorus and Flame Photometric Detectors

Advantages	Disadvantages
Nitrogen/phosphorus	
Specific for nitrogen and phosphorus containing compounds	Large analytical variation
Good sensitivity	Destruction detector
	Limited applicability for volatile air toxics
	Affected by water from the sample
Flame Photometric	
Good specificity for sulfur or phosphorus containing compounds	Poor linearity
Good sensitivity	Destructive detector
Small analytical variation	May be affected by water from the sample

compounds such as ethers, esters, epoxides, aldehydes, ketones, and alcohols. Such a system could easily replace multidetector systems utilizing all of the detectors mentioned here. The major drawback to such a system is its price (around $88,000), which places it in the price range of many GC/MS instruments.

G. Other Detectors

Other detectors are available for use in the detection of air toxic compounds. For example, Rice et al.[9] reported the use of the helium discharge detector in the detection of volatile halogenated compounds in water. Di Sanzo[10] reported the use of an oxygen-specific detector for the determination of oxygenates in gasoline range hydrocarbons. However, these types of detectors are not commonly found in most laboratories and therefore have limited usefulness in the measurement of air toxics.

III. Gas Chromatography/Mass Spectrometry (GC/MS) Instrumentation

As described in Method TO-14, use of mass spectrometry in conjunction with high resolution gas chromatography allows for positive compound identification. The price paid for the selectivity obtained using GC/MS in the full-scan mode is less sensitivity than can be obtained with most of the detectors used in multidetector systems. Operation of the GC/MS in the specific ion monitoring (SIM or MID) mode generally provides greater sensitivity than the full-scan mode, but can limit the number of compounds that can be monitored and provides information only on the compounds of interest. Sensitivity increases of 2 to 10 times greater than the full-scan mode are common. An example of increased sensitivity in the SIM mode would be that observed for 1,4-dioxane in ambient samples. The sensitivity of this compound increases by a factor of approximately 10 in the SIM mode over

Table 8. GC/MD and GC/MS System Selection: Advantages and Disadvantages

Advantages	Disadvantages
GC/MD System	
Lower capital investment	Possible coelution
Good sensitivity	Labor intensive data reduction
Detects low molecular weight compounds	Different detectors required for different compound classes
Very reliable and stable detectors	
GC/MS System	
Positive identification for most compounds	Difficulty detecting low molecular weight compounds
Good sensitivity in SIM mode	Lower sensitivity in full-scan mode
	Inability to analyze for a large number of compounds in the SIM mode
	Larger analytical variability

that observed in the full-scan mode on a Finnigan 4500 GC/MS system. Other compounds, such as toluene, chloroform, and trichloroethylene, analyzed on the same instrument exhibit sensitivity increases of approximately a factor of 2 in the SIM mode over that observed in the full-scan mode.

Table 8 presents a guideline for the selection of GC/MS or GC/MD as the analytical system. It should be noted that the mass spectrometer can be used in conjunction with other detectors in a multidetector system if specificity and sensitivity are desired in the same analysis.

The advantages and disadvantages of GC/MS systems in the full-scan and SIM modes are listed in Table 9. GC/MS has a very distinct advantage in that it can generally distinguish between coeluting compounds providing identification and quantitation of both components. Data systems are generally more automated than those available with other detectors allowing automated quantitation and search routines. Operation in the full-scan mode provides screening and identification information for all compounds detected in an ambient sample.

GC/MS does have some limitations for the analysis of ambient level VOC compounds. In most cases, positional isomers cannot be distinguished from the mass spectrum. Some compounds, such as methanol and ethylene, cannot be easily determined by GC/MS due to their molecular structure and interferences from the major components in air. This shortcoming may be of concern to those interested in the use of ethylene concentrations in the generation of ozone precursor models.

Water can also have a very detrimental effect on the operation of the GC/MS systems as reported by Ogle et al.[11] Nafion® driers are often used to remove the water, but these also remove lower molecular weight polar compounds. Concentration of the water in the cryotrap results in the injection of 2–3 mg of water, based on 0.25 l of sample at 70% relative humidity. This amount of water can cause GC/MS systems having turbomolecular pumps to shut down due to pressure

Table 9. Advantages and Disadvantages of GC/MS Systems

Advantages	Disadvantages
MS in full-scan mode	
Positively identifies a large number of compounds at one time	Cannot distinguish structural isomers
Few interferences	Low sensitivity
Universal detector	Larger analytical variability
Good linear range	Affected by moisture in the sample
	Interferences with low molecular weight compounds
	High maintenance costs
MS in SIM mode	
Good sensitivity	Sensitivity decreases with number of ions (compounds) monitored
Limits interferences	Will only detect compounds of interest or others containing the mass being monitored

increases in the system, and can cause differentially pumped systems to lose sensitivity and reproducibility. Smaller sample sizes may be used for the analysis of wet samples, but compound detection limits will be affected accordingly. Use of a jet separator can also be used to remove most of the water, but a sensitivity loss is observed due to the transfer efficiencies across the separator. Larger samples must then be concentrated to achieve equivalent sensitivities.

Figure 3 shows a total ion chromatogram of an ambient air sample analyzed by GC/MS. This sample was analyzed on a GC/MD system first and identifications were verified by GC/MS. A large number of air toxic compounds can be detected and quantitated with a high degree of certainty using a combination of multidetector systems coupled with GC/MS confirmation of a percentage of the samples. This approach is used by Radian Corporation for the Urban Air Toxics Monitoring Program conducted for the U.S. EPA.[12] The primary analytical system consists of a FID, PID, and ECD multidetector system and is used to determine 38 compounds. Approximately 10% of the samples are then confirmed by GC/MS analysis in the full-scan mode. For this program, confirmation of the compounds observed on the GC/MD system by GC/MS is approximately 90%.

Radian also utilizes dual column multidetector systems containing FID, PID, and ElCD detectors. The eluent from one column is directed to the PID and FID in series and the eluent from the second column goes to the ElCD. Identifications are determined by retention times relative to toluene and by use of a PID to FID ratio normalized to the PID/FID ratio determined for toluene (toluene normalized response or TNR). Coupled with 60-m, 0.32-mm i.d., 1.0-μm film methyl silicone columns, this approach can provide considerable information about unknown compounds. Unknown compounds are quantitated on a carbon basis: boiling range of the compound can be determined from the elution time, information on the level of unsaturation can be determined from the TNR, and ElCD response can

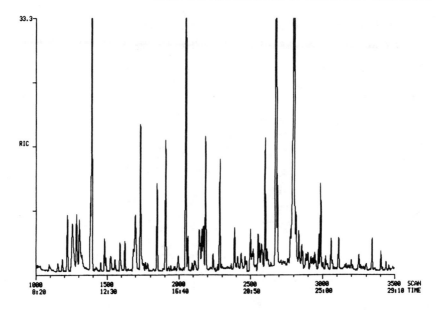

FIGURE 3.   GC/MS chromatogram of an ambient air sample. Conditions: 30 m
x 0.32-mm i.d. 1.0-μm film DB-1 column to direct coupling source;
−50°C for 2 min 6°C/min to 220°C; Finnigan 4500 GC/MS in full-scan
mode 35–350 amu; 70 eV.

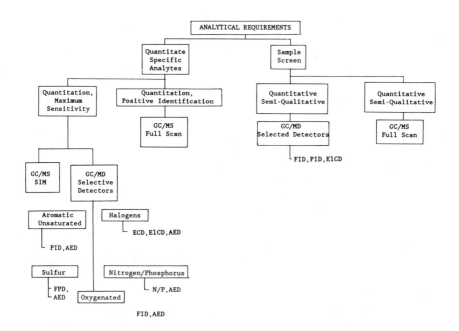

FIGURE 4.   Detector selection flow chart.

be used to determine whether the compound contains halogens. The carbon response factor can also be used to determine a total nonmethane hydrocarbon level in the sample based on the FID response. Coupled with GC/MS confirmation, such an analytical system can provide a very powerful, sensitive, and cost-effective tool for the analysis of VOCs.

IV. CONCLUSIONS

The selection of detector systems for the analysis of VOC air toxics depends on the sensitivity required, the availability of detectors, the list of compounds to be monitored, and the degree of specificity required to meet the program objectives. Figure 4 presents a flow chart to help determine the best detector or combination of detectors for a specific application. GC/MS detection and quantitation is highly recommended when absolute identification is necessary or when target analytes are not specified and a general screen to identify all components is desired. Multidetector systems are recommended when extreme sensitivities, lower analytical costs, or quantitative and semiqualitative analyses are desired.

REFERENCES

1. "Compendium Method TO-14, The Determination of Volatile Organic Compounds (VOCs) in Ambient Air Using SUMMA Passivated Canister Sampling and Gas Chromatographic Analysis," U.S. EPA, Quality Assurance Division, Environmental Monitoring Systems Laboratory, Research Triangle Park, NC (1988).
2. Cox, R. D., and R. F. Earp. "Determination of Trace Level Organics in Ambient Air by High Resolution Gas Chromatography with Simultaneous Photoionization and Flame Ionization Detection," *Anal. Chem.* 54:2265–2270 (1982).
3. Earp, R. F., and R. D. Cox. "Identification and Quantitation of Organic in Ambient Air Using Multiple Gas Chromatographic Detection," in *Identification and Analysis of Organic Pollutants in Air*, L. Keith, Ed. (Woburn, MA: Butterworth Publishers, Inc., 1984), pp. 159–169.
4. Berezkin, V. G., and V. S. Tatarinskii. *Gas-Chromatographic Analysis of Trace Impurities* (New York: Consultants Bureau, 1973), p. 49.
5. Ho, J. S. "Method 502.2 Volatile Organic Compounds in Water by Purge-and-Trap Capillary Column Gas Chromatography with Photoionization and Electrolytic Conductivity Detectors in Series, Revision 2.0, *Methods for the Determination of Organic Compounds in Drinking Water,* U.S. EPA Publication EPA-600/4-88/039 (1988), pp. 31–62.
6. Brymer, D. A., L. D. Ogle, W. L. Crow, M. J. Carlo, and L. A. Bendele. "Storage Stability of Ambient Level Volatile Organics in SUMMA™ Polished Canisters," 81st Annual APCA Conference, Dallas, TX (1988).
7. Hewlett-Packard Company. "HP 5921A Atomic Emission Detector Specification Guide," (1990).
8. Miles, W. S., and B. D. Quimbly. "Characterization of Sulfur Compounds in Spices Using SFE-GC-AED," *Am. Lab.* 28F–28L (July 1990).

9. Rice, G. W., D. A. Ryan, and S. M. Argentine. "Helium Discharge Detector for Quantitation of Volatile Organohalogen Compounds," *Anal. Chem.* 62(8): 853–857 (1990).
10. Di Sanzo, F. P. "Determination of Oxygenates in Gasoline Range Hydrocarbons by Capillary Column Gas Chromatography: A User's Experience with an Oxygen-Specific Detector (O FID)," *J. Chrom. Sci.* 28:73–75 (1990).
11. Ogle, L. D., R. B. White, D. A. Brymer, and M. C. Shepherd. "Applicability of GC/MS Instrumentation for the Analysis of Undried Air Toxic Samples," 1990 EPA/AQMA Symposium on Measurement of Toxic and Related Pollutants, Raleigh, NC (1990).
12. Dayton, D. P., and J. Rice. "Development and Evaluation of a Prototype Analytical System for Measuring Air Toxics," Final Report, Radian Corporation, prepared for U.S. EPA EMSL Research Triangle Park, NC, EPA Contract No. 68-02-3889, WA No. 120 (1987).

CHAPTER 8

The Chemiluminescent Detection of Volatile Sulfur Compounds

Raul Dominguez, Jr. and Margil W. Wadley

TABLE OF CONTENTS

0-87371-606-0/93/$0.00 + $.50

© 1993 by Lewis Publishers

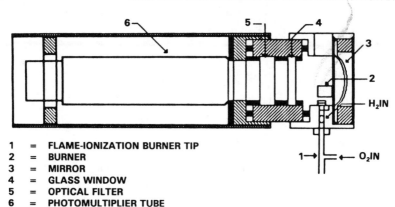

1 = FLAME-IONIZATION BURNER TIP
2 = BURNER
3 = MIRROR
4 = GLASS WINDOW
5 = OPTICAL FILTER
6 = PHOTOMULTIPLIER TUBE

FIGURE 1.   Schematic diagram of a flame photometric detector (FPD). (From
             Brody, S. S., and J. E. Chaney. *J. Gas Chromatogr.* 4: 42–46 (1966).
             With permission.)

I. INTRODUCTION

The analysis of air samples for volatile sulfur compounds such as H_2S, COS, and SO_2 has become an important concern of environmental and process control laboratories. Gas chromatography (GC) is generally used to separate sulfur compounds prior to delivery to a detector. Although a number of detectors are available for sulfur quantitation, the single-flame photometric detector (FPD) is the most commonly used.[1-3] This is due to its high selectivity, ease of operation, low maintenance, and relative low cost.

The FPD, despite its popularity, is hampered by a number of disadvantages.[4-8] Response for the single-flame FPD is only approximately proportional to the square of the mass flow rate of sulfur. The FPD is sensitive to matrix effects, changes in flame geometry, changes in burner block temperature, and environmental factors such as barometric pressure. The dual-flame FPD is an improvement on the single-flame instrument (Figures 1 and 2).[9-11] The response for this detector follows a strict dependence on the square of the sulfur content. The dual-flame FPD is also less susceptible to hydrocarbon-induced quenching than the single-flame detector. It is, however, still sensitive to changes in flame characteristics and environmental factors. Extinction of the first flame can occur during analysis of hydrocarbon rich samples. Although this flame is quickly relit, results from the dual-flame FPD are rendered suspect. The dual-flame FPD is also less sensitive than the single-flame FPD.[10]

In addition to the chemiluminescence detectors (CLDs) to be discussed here, a number of alternative detection systems for sulfur have been investigated. These include a nonflame source-induced sulfur fluorescence detector,[8] a vacuum ultraviolet plasma atomic emission spectrometer,[12] a microwave-induced plasma detector,[13] and the Hall electrolytic conductivity detector.[14,15] Each of these systems are viable alternatives to the FPD; however, each suffers from its own characteristic limitations.

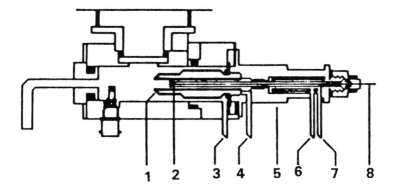

1. Secondary flame jet	5. Modified detector base
2. Primary flame jet	6. Makeup gas inlet
3. Secondary air inlet	7. Primary air inlet
4. H₂ gas inlet	8. Column (restrictor

FIGURE 2. Schematic diagram of a dual-flame photometric detector. (From Markides, K. E., E. D. Lee, R. Bolick, and M. L. Lee. *Anal. Chem.* 58: 740–743(1986). With permission.)

Chemiluminescence detection of sulfur compounds has recently come into vogue with a number of instruments commercially available. We chose to investigate the utility of using three of the available instruments in the analysis of air samples. Results for detection limits, linearity, and sample analyses are compared to those obtained using an FPD.

II. EXPERIMENTAL SECTION

Operating conditions, sample loops, and columns were selected for optimal sensitivity and peak quality for each detection system. Parameters used in the analyses of sulfur compounds for these systems are collected in Table 1.

A. Instrumentation

A Hewlett-Packard FPD, factory installed to a HP5890 GC was used without modification. The FPD was interfaced to a HP3890 integrator. A Sievers Model 300 sulfur chemiluminescence detector (SCD 300) was attached to a HP5890 GC and HP3392 integrator according to the manufacturer's instructions. A Sievers Model 350 sulfur chemiluminescence detector (SCD 350) was attached to a HP5890 GC via a heat-conditioned ceramic probe inserted into the exit port of a HP flame ionization detector (FID). Hydrogen and air flow rates of 205 and 275 mL/min were established using Veriflo Corporation flow control valves (MIR

Table 1. Parameters Used in the Analysis of Sulfur Compounds

Parameter	FPD	SCD 300	SCD 350	RCD
Column	J & W Scientific GSQ 30M	J & W Scientific GSQ 30M	Chromosil 310/Teflon® 6 ft x 1/8 in. o.d.	Chromosil 310/Teflon® 6 ft x 1/8 in. o.d.
Sample loop volume (mL)	1	5	5	5
Injector temperature (°C)	125	30	50	30
Detector temperature (°C)	200	150	—	—
Oven program	Initial temperature 40°C for 2.5 min Ramp at 6.0°C/min hold at 110°C for 10 min			50°C isothermal
Carrier flow	40 mL/min nitrogen	28.5 mL/min helium	33 mL/min helium	15 mL/min helium
Detector supply gas	150 mL/min air	—	205 mL/min oxygen	20 mL/min helium
	150 mL/min hydrogen	—	257 mL/min hydrogen	—
Integration Time(s)	—	0.12	0.012	0.25

701). An ozonator supply gas pressure of 8 psi was established. The ceramic probe height was adjusted until the peak height of a 10.1 ppmv carbonyl sulfide standard was optimized. A Sievers Model 207 redox chemiluminescence detector (RCD) with a gold-on-glass bead catalyst was attached to a HP5890 GC and HP3392 integrator according to the manufacturer's instructions. Replacement nitrogen dioxide permeation tubes (No. 110-100-0081) were obtained from VICI Metronics. Permeation tube supply gas was maintained at 20 mL/min.

B. Standards and Supply Gases

Scott Specialty Gas mixtures containing low parts per million by volume levels of sulfur gases in nitrogen were used as stock standards. Dilutions were prepared in either a 100-mL ground glass syringe or in 5-L Tedlar®* bags with house nitrogen as the diluent.

Helium (UHP grade, Scott Specialty Gases) was used as carrier for the CLDs and as the RCD permeation tube supply gas. FPD nitrogen carrier was obtained from liquid nitrogen (Cryogenic Distributors Inc.) evaporation. Hydrogen, air, and oxygen (MG Industries) were used as received.

Natural gas was obtained from the El Monte Municipal Pipe System via the plumbing in the South Coast Air Quality Management District (SCAQMD) Laboratory. Environmental air samples were obtained from a variety of sources and were analyzed without modification.

III. RESULTS AND DISCUSSION

A. Principles

A number of chemistries have been used in the chemiluminescence detection of sulfur compounds.[16] Five of these are presented in Table 2. The three CLDs studied are based upon the chemistries in Schemes 1–3.

---

* Registered trademark of E. I. du Pont de Nemours & Co.

Table 2. Chemiluminescence Reactions Used in GC Detection of Sulfur Compounds

1.	$RCH_2-S-H + F_2$	$\longrightarrow$	$RCH=SHF + HF^{*a}$
2.	NO_2 + analyte	$\xrightarrow{\text{Au/glass}}$	NO + oxidized analyte
	$NO + O_3$	$\longrightarrow$	$O_2 + NO_2^{*b}$
3.	Analyte + $H_2 + O_2$	$\xrightarrow{\text{FID}}$	SO + other products
	$SO + O_3$	$\longrightarrow$	$SO_2 + O_2 +{}^{c}$
4.	$2H_2S + ClO_2$	$\longrightarrow$	$S_2^{*} + 1/2\ Cl_2 + 2H_2O^{d}$
5.	$O_3 + R_2S$	$\longrightarrow$	$O_2 + SO_2^{*e}$

* Excited state.
a Nelson, J.K., Getty, R.H., and Birks, J.W., *Anal. Chem.*, 55: 1767-1770 (1983).
b Nyarady, S.A., Barkley, R.M., Sievers, R.E., *Anal. Chem.*, 57: 2074-2079 (1985).
c Shearer, R.L., O'Neal, D.L., Rios, R., Baker, M.D., *J. Chromatogr. Sci.*, 28: 24-28, (1990).
d Spulin, S.R., Yeung, E.S., *Anal. Chem.*, 54: 318-320 (1982).
e Hutte, R.S., Sievers, R.E., and Birks, J.W., *J. Chromatogr. Sci.*, 24: 499-505 (1986).

The SCD 300 responds to emission of photons from excited-state hydrogen fluoride produced in the reaction of fluorine with a reducible sulfur compound (Scheme 1).[17,18] The RCD responds to photons produced in a two-step reaction series (Scheme 2).[19-20] The first step involves reaction of nitrogen dioxide with analyte on a heated gold catalyst. The generated nitric oxide is then reacted with ozone at low pressure to produce molecular oxygen and excited-state nitrogen dioxide. Relaxation of this molecule produces a photon that is detected by the RCD.

Operation of the SCD 350 is based upon the chemistry depicted in Scheme 3.[21] A sulfur compound is converted to sulfur monoxide in a hydrogen-rich FID flame. The sulfur monoxide is transferred to a reaction chamber where it is reacted with ozone producing sulfur dioxide, oxygen, and light.

The chemistries depicted in Schemes 4 and 5 have been reported to yield promising results; however, they have not yet attained widespread use.

B. Detection Limits and Linearity

A viable alternative detector to the FPD should possess sensitivity in the low parts per billion by volume range, linearity over several orders of magnitude, and tolerance to matrix effects. With this in mind, all detector systems examined in this study were optimized with respect to sensitivity and ability to separate hydrogen sulfide, carbonyl sulfide, and methyl mercaptan. This has resulted in systems with slightly different operating parameters.

Detection limits determined for a variety of sulfur compounds, using the CLDs and FPD, are collected in Table 3. Detection limits are based on three times the average chromatogram baseline noise observed in low parts per billion by volume standards analyzed, at least, in triplicate.

Table 3. Sulfur Compound Detection Limit Comparison. Concentration in Parts per Billion by Volume

Compound	FPD	SCD 300	RCD	SCD 350
Carbonyl sulfide	90	—	1000	13
Hydrogen sulfide	130	—	130	61
Carbon disulfide	40	—	1800	16
Sulfur dioxide	430	—	1710	90
Methyl mercaptan	240	7	—	40
Ethyl mercaptan	290	4	—	55
1-Propanethiol	100	8	—	84
2-Propanethiol	100	4	—	47
Dimethyl sulfide	120	3	—	25
Dimethyl disulfide	180	1	—	45

The SCD 350 has detection limits of 10–90 ppbv and is generally more sensitive than the FPD, which typically responds to from 40 to 430 ppbv of analyte. The SCD 300 is greater than an order of magnitude more sensitive to mercaptans, dimethyl sulfide (DMS), and dimethyl disulfide (DMD); however, it does not respond to carbonyl sulfide, hydrogen sulfide, carbon disulfide, or sulfur dioxide. The RCD, with the notable exception of hydrogen sulfide, is generally more than two orders of magnitude less sensitive than the SCD 350.

Response linearities for the CLDs were determined from analysis of three concentrations of analyte spanning at least one order of magnitude from the detection limit. Linearity for the SCD 350 is maintained over at least two orders of magnitude with r^2 values greater than 0.99. A typical plot of linearity data is presented in Figure 3. For the SCD 300, linearity is maintained over 2.5 orders of magnitude from the detection limit. Linearity for the RCD was verified for one order of magnitude.

C. Analysis of Environmental Samples

A number of environmental samples were analyzed using one of the CLDs and an FPD. The results are reported in Table 4. Results for analysis of landfill and refinery inlet gas using the SCD 300 and RCD varied considerably from those obtained using the FPD. This was probably due to the high hydrocarbon content of the samples. Landfill gas can contain up to several percent hydrocarbons and carbon dioxide while fuel gas is composed almost exclusively of hydrocarbons. These levels of hydrocarbons are expected to adversely affect FPD response to coeluting sulfur compounds. Results obtained using the SCD 350 generally compare well with those obtained using the FPD, with only two entries conflicting by more than a few percent.

D. Analysis of Natural Gas

Municipal natural gas was analyzed using both the FPD and the SCD 350. The FPD did not detect sulfur compounds in these samples. In contrast, the SCD 350

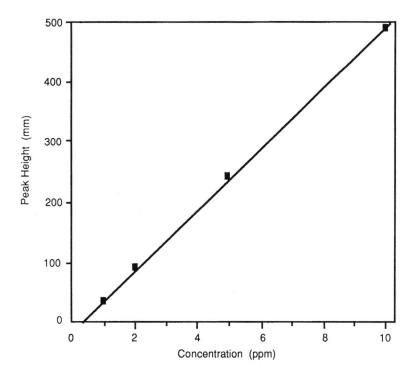

FIGURE 3.   SCD 350 linearity: hydrogen sulfide

detected sulfur compounds in the 0.1–1.0 ppmv range (Table 5). Daily hydrogen sulfide levels were found to fluctuate from 260 up to 650 ppbv. Concentrations of methyl mercaptan, dimethyl sulfide, and dimethyl disulfide exhibited similar variability. A typical chromatogram obtained in the analysis of natural gas is illustrated in Figure 4.

The concentration of sulfur compounds found using the SCD 350 are well within the detection limits of the FPD. The failure of this instrument to detect sulfur compounds in natural gas was thought to be due to hydrocarbon quenching of the sulfur response.

In order to determine whether this discrepancy was due to a matrix effect, the method of standard additions was applied to the analysis of natural gas. Standard additions of a standard containing hydrogen sulfide (10.1 ppmv) and methyl mercaptan (10.8 ppmv) was applied in ratios of 1:1, 1:5, and 1:10 to a natural gas sample. This sample was analyzed in parallel using the FPD and SCD 350. The SCD 350 results obtained are reported in Table 6, and FPD results are presented in Table 7. The SCD 350 consistently generated results differing by less than 8% from the predicted amounts of hydrogen sulfide and methyl mercaptan. In contrast, the FPD gave results differing by as much as 26% from the theoretical values. In the case of municipal natural gas, the FPD results were affected by the sample matrix.

Table 4. Environmental Sample Analyses Comparison

Sample	Detector	Compound	(ppmv)
Bag 404	FPD	H_2S	14.00
Interior landfill well	SCD 350	H_2S	18.78
Bag 422	FPD	H_2S	9.30
Perimeter landfill well	SCD 350	H_2S	9.36
Bulb 112	FPD	H_2S	6.98
Fuel gas	CD 350	H_2S	11.04
	FPD	CH_3SH	1.79
	SCD 350	CH_3SH	1.94
	FPD	CH_3SCH_3	<0.12
	SCD 350	CH_3SCH_3	0.14
	FPD	CH_3SSCH_3	0.39
	SCD 350	CH_3SSCH_3	0.36
Bulb 171	FPD	COS	0.20
Fuel outlet gas	SCD 350	COS	0.21
Bulb 193	FPD	COS	<0.09
Flare outlet gas	SCD 350	COS	0.15
Bag 400	FPD	SO_2	<0.43
Flare outlet gas	SCD 350	SO_2	0.16
Bag 431	FPD	H_2S	3.32
Landfill raw gas	SCD 350	H_2S	3.30
	FPD	CH_3SH	0.46
	SCD 350	CH_3SH	0.58
	FPD	COS	<0.09
	SCD 350	COS	0.04
Bag 1025	FPD	CH_3SH	5.49
Land reclamation	SCD 300	CH_3SH	27.57
Dehumidified gas	FPD	CH_3SCH_3	34.93
	SCD 300	CH_3SCH_3	140.57
Bag 754	FPD	CH_3SH	2.88
Land reclamation	SCD 300	CH_3SH	9.94
Raw gas	FPD	CH_3SCH_3	68.90
	SCD 300	CH_3SCH_3	78.20
Bag 01	FPD	CH_3SH	<0.24
Landfill inlet gas	SCD 300	CH_3SH	0.35
	FPD	CH_3SCH_3	2.69
	SCD 300	CH_3SCH_3	1.97
Bag 04	FPD	CH_3SH	1.74
Landfill flare outlet	SCD 300	CH_3SH	0.75
	FPD	CH_3SCH_3	3.28
	SCD 300	CH_3SCH_3	3.08
Bulb MS	FPD	H_2S	4.78
Refinery outlet gas	RCD	H_2S	20.45
Bulb 1	FPD	H_2S	4.55
Refinery outlet gas	RCD	H_2S	37.60
Bulb 197	FPD	H_2S	8.11
Landfill gas	RCD	H_2S	4.00

Table 4. Environmental Sample Analyses Comparison (continued)

Sample	Detector	Compound	(ppmv)
Bulb 206	FPD	H_2S	9.06
Landfill gas	RCD	H_2S	2.13
Bulb 208	FPD	H_2S	9.01
Landfill gas	RCD	H_2S	1.20
Refinery flare outlet	RCD	SO_2	29.01
Bulb 122	FPD	SO_2	42.88
Refinery flare outlet	RCD	SO_2	29.99

Table 5. SCD 350 Analysis of Natural Gas

Compound	Date	Concentration (ppbv)
Hydrogen sulfide	8/28/89	650
	9/18/89	260
	9/19/89	430
	9/20/89	490
Methyl mercaptan	8/28/89	740
	9/18/89	730
	9/19/89	980
	9/20/89	800
Dimethyl sulfide	8/28/89	190
	9/18/89	130
	9/19/89	120
	9/20/89	110
Dimethyl disulfide	8/28/89	360
	9/18/89	310
	9/19/89	270
	9/20/89	330

E. Interferences

The results obtained for analysis of environmental samples and natural gas prompted a further investigation into matrix effects on the analysis for sulfur compounds. The SCD 350 was selected for comparison with the FPD due to its general applicability to sulfur analysis and relatively low detection limits. A mixed sulfur compound standard containing about 10-ppmv levels of hydrogen sulfide, methyl mercaptan, dimethyl sulfide, and dimethyl disulfide was diluted 1:1 with nitrogen containing parts per million volume levels of selected compounds. These mixtures generally gave results differing from those obtained using house nitrogen as the diluent by 5% or less (Table 8). Only two significant exceptions were observed; unsaturated hydrocarbons gave a result for dimethyl sulfide 19% higher than observed in a nitrogen matrix alone while analysis of dimethyl disulfide in a matrix containing aromatic hydrocarbons gave a result 10% higher than observed for dilution with house nitrogen.

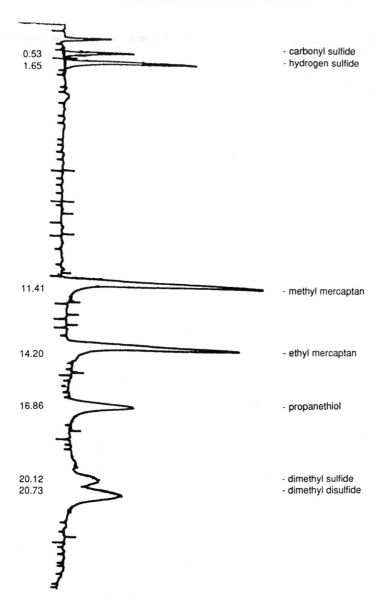

FIGURE 4.   SCD 350 analysis of natural gas

The FPD was not adversely affected by aromatic hydrocarbons, halocarbons, alkanes, unsaturated hydrocarbons, carbon monoxide, or carbon dioxide at parts per million by volume levels. Results differed by less than 5% from those obtained from analysis of the sulfur standard diluted in nitrogen alone (Table 9). These results, which are in sharp contrast to those found in the analysis of environmental

Table 6. SCD 350 Analysis of Natural Gas: Standard Additions

Sample	Compound	Concentration (ppmv)	Theory (ppmv)	% Difference
Natural gas 9/20/89	H_2S	0.49	—	—
	CH_3SH	0.80	—	—
Std/natural gas = 1:1	H_2S	4.90	5.30	−7
	CH_3SH	5.69	5.80	−2
Std/natural gas = 1:5	H_2S	2.30	2.42	−5
	CH_3SH	2.61	2.80	−7
Std/natural gas = 1:10	H_2S	1.40	1.45	−4
	CH_3SH	1.69	1.80	−6
Natural gas	H_2S	0.51	0.49	+3
	CH_3SH	0.78	0.80	−2

Table 7. FPD Analysis of Natural Gas: Standard Additions

Sample	Compound	Concentration (ppmv)	Theory (ppmv)	% Difference
Natural gas 9/20/89	H_2S	<0.13	—	—
	CH_3SH	<0.24	—	—
Std/natural gas = 1:1	H_2S	6.05	5.05	+20
	CH_3SH	4.59	5.40	−15
Std/natural gas = 1:5	H_2S	2.14	2.02	+6
	CH_3SH	1.61	2.16	−26
Std/natural gas = 1:10	H_2S	0.96	1.01	−5
	CH_3SH	0.80	1.08	−26

and natural gas samples, are thought to be due to the levels of hydrocarbons in these samples. Hydrocarbon content at low parts per million volume levels does not appear to adversely affect FPD response. However, at percent or greater concentrations, these species have a significant effect on FPD response. Analyses of sulfur compounds in a matrix containing oxygenates generally resulted in good agreement with results obtained from analysis of the nitrogen diluted standard. Only dimethyl disulfide was adversely affected by this matrix, resulting in a determination more than 200% too high.

IV. CONCLUSION

The FPD will continue to be one of the most popular detectors for the analysis of sulfur compounds in environmental samples due to its high selectivity, ease of operation, and low maintenance requirement. However, the use of alternative detectors will gain in popularity, particularly for applications involving complex sample matrices. The SCD 350 demonstrated high sensitivity, linearity, and

Table 8. SCD 350 Detection of Sulfur Compounds in Complex Matrices

Matrix components	(ppmv)	Compound	Found[a] (ppmv)	Theory (ppmv)[b]	% Difference
Aromatic hydrocarbons					
Benzene	10.1	H_2S	4.91	5.05	-3
Toluene	9.32	CH_3SH	5.42	5.40	+1
p-Xylene	10.8	CH_3SCH_3	4.75	4.49	+6
o-Xylene	11.0	CH_3SSCH_3	5.81	5.30	+10
Halocarbons					
Dichloromethane	0.49	H_2S	5.10	5.05	+1
Chloroform	0.17	CH_3SH	5.40	5.40	0
1,2-Dichloroethane	0.25	CH_3SCH_3	4.63	4.49	+3
Trichloroethylene	0.10	CH_3SSCH_3	5.56	5.30	+5
Carbon tetrachloride	0.05				
Tetrachloroethylene	0.05				
1,2-Dibromoethane	0.04				
Trichloroethane	0.10				
Hydrocarbons, CO_2 and CO					
Carbon monoxide	50	H_2S	4.82	5.05	+1
Carbon dioxide	3990	CH_3SH	5.27	5.40	-2
Methane	100	CH_3SCH_3	4.39	4.49	-2
Ethane	25	CH_3SSCH_3	5.27	5.30	-1
Isopentane	10				
Unsaturated hydrocarbons					
Ethylene	9.34	H_2S	4.93	5.05	-2
Acetylene	8.54	CH_3SH	5.27	5.40	-2
Propadiene	9.14	CH_3SCH_3	5.35	4.49	+19
Propylene	9.52	CH_3SSCH_3	5.36	5.30	+1
Methyl acetate	9.00				
Oxygenates					
Acetone	0.97	H_2S	4.89	5.05	-3
Methyl ethyl ketone	0.98	CH_3SH	5.27	5.40	-2
Methyl isobutyl ketone	1.01	CH_3SCH_3	4.42	4.49	-2
n-Butyl acetate	1.01	CH_3SSCH_3	5.28	5.30	0

[a]　Samples consists of sulfur standard:matrix gas = 1.
[b]　Theoretical concentrations calculated relative to sulfur standard:N_2 = 1.

insensitivity to matrix effects; these characteristics are highly desirable for the analysis of environmental air samples. The SCD 350 is more versatile than the SCD 300, which has greater sensitivity to selected compounds with detection limits in the low parts per billion by volume range. The RCD is limited, due to relatively high detection limits for sulfur compounds, but can be used reliably in applications involving parts per million by volume levels of inorganic sulfur gases.

Table 9. FPD Detection of Sulfur Compounds in Complex Matrices

Matrix components	(ppbv)	Compound	Found (ppmv)[a]	Theory (ppmv)[b]	% Difference
Aromatic hydrocarbons					
Benzene	10.1	H_2S	5.07	5.05	0
Toluene	9.32	CH_3SH	5.36	5.40	-1
p-Xylene	10.8	CH_3SCH_3	4.46	4.49	-1
o-Xylene	11.0	CH_3SSCH_3	5.24	5.30	-1
Halocarbons					
Dichloromethane	0.49	H_2S	5.12	5.05	+1
Chloroform	0.17	CH_3SH	5.52	5.40	+2
1,2-Dichloroethane	0.25	CH_3SCH_3	4.41	4.49	-2
Trichloroethylene	0.10	CH_3SSCH_3	5.28	5.30	0
Carbon Tetrachloride	0.05				
Tetrachloroethylene	0.05				
1,2-Dibromoethane	0.04				
Trichloroethane	0.10				
Hydrocarbons, CO and CO_2					
Carbon monoxide	50	H_2S	5.13	5.05	+2
Carbon dioxide	3990	CH_3SH	5.52	5.40	+2
Methane	100	CH_3SCH_3	4.60	4.49	+2
Ethane	25	CH_3SSCH_3	5.50	5.30	+4
Isopentane	10				
Unsaturated hydrocarbons					
Ethylene	9.34	H_2S	5.11	5.05	+1
Acetylene	8.54	CH_3SH	5.40	5.40	0
Propadiene	9.14	CH_3SCH_3	4.50	4.49	0
Propylene	9.52	CH_3SSCH_3	5.37	5.30	+1
Methyl acetate	9.00				
Oxygenates					
Acetone	0.97	H_2S	4.82	5.05	-5
Methyl ethyl ketone	0.98	CH_3SH	5.16	5.40	-4
Methyl isobutyl ketone	1.01	CH_3SCH_3	4.30	4.49	-4
n-Butyl acetate	1.01	CH_3SSCH_3	16.31	5.30	+208

[a] Samples consists of sulfur standard:matrix gas = 1.
[b] Theoretical concentrations calculated relative to sulfur standard:N_2 = 1.

ACKNOWLEDGMENTS

The authors thank Sievers Research, Inc. for the loan of the SCD 300, SCD 350, and RCD, and Mike Shu for his work determining detection limits for the FPD. We are also grateful to Steve Barbosa for many helpful discussions. Finally, we give our thanks to the South Coast Air Quality Management District (SCAQMD) for supporting this project.

REFERENCES

1. Lee, M. L., F. J. Yang, and K. D. Bartle. *Open Tubular Column Gas Chromatography* (New York: John Wiley & Sons, Inc., 1984).
2. Adams, D. F. *Air Pollution*, Vol. 3, 3rd ed., A. C. Stern, Ed, (New York: Academic Press, 1976), Chapter 6.
3. Farwell, S. O., and R. A. Rasmussen. "Limitations of the FPD and ECD in Atmospheric Analysis: A Review," *J. Chromatogr. Sci.* 14:224–234 (1978).
4. Brody, S. S., and J. E. Chaney. "The Application of a Specific Detector for Phosphorus and for Sulfur Compounds — Sensitive to Subnanogram Quantities," *J. Gas Chromatogr.* 4:42–46 (1966).
5. Dressler, M. *Selective Gas Chromatographic Detectors* (New York: Elsevier, 1986), pp. 109–132.
6. Farwell, S. O., and C. J. Barinaga. "Sulfur-Selective Detection with the FPD: Current Enigmas, Usage, and Future Directions," *J. Chromatogr. Sci.* 24:483–494 (1986).
7. Maruyama, M., and M. Kakemoto. "Behavior of Organic Sulfur Compounds in Flame Photometric Detectors," *J. Chromatogr. Sci.* 16:1–7 (1978).
8. Gage, D. R., and S. O. Farwell. "Nonflame, Source-Induced Sulfur Fluorescence Detector for Sulfur-Containing Compounds," *Anal. Chem.* 52:2422–2425 (1980).
9. Patterson, P. L., R. L. Howe, and A. Abu-Shumays. "Dual-Flame Photometric Detector for Sulfur and Phosphorous Compounds in Gas Chromatography Effluents," *Anal. Chem.* 50:339–344 (1978).
10. Patterson, P. L. "Comparison of Quenching Effects in Single- and Dual-Flame Photometric Detectors," *Anal. Chem.* 50:345–348 (1978).
11. Markides, K. E., E. D. Lee, R. Bolick, and M. L. Lee. "Capillary Supercritical Fluid Chromatography with Dual-Flame Photometric Detection," *Anal. Chem.* 58:740–743 (1986).
12. Treybig, D. S., and S. R. Ellebracht. "Vacuum Ultraviolet Plasma Atomic Emission Spectrometer as a Sulfur Specific Detector for Gas Chromatography," *Anal. Chem.* 52:1633–1636 (1980).
13. Genna, J. L., W. D. McAninch, and R. A. Reich. "Atmospheric Microwave-Induced Plasma Detector for the Gas Chromatographic Analysis of Low-Molecular Weight Sulfur Gases," *J. Chromatogr.* 238:103–112 (1982).
14. Ehrlich, B. J., R. C. Hall, R. J. Anderson, and H. G. Cox. "Sulfur Detection in Hydrocarbon Matrices. A Comparison of the Flame Photometric Detector and the 700A Hall Electrolytic Conductivity Detector," *J. Chromatogr. Sci.* 19:245–249 (1981).
15. Gluck, S. "Performance of the Model 700A Hall Electrolytic Conductivity Detector as a Sulfur-Selective Detector," *J. Chromatogr. Sci.* 20:103–108 (1982).
16. Hutte, R. S., R. E. Sievers, and J. W. Birks. "Gas Chromatography Detectors Based on Chemiluminescence," *J. Chromatogr. Sci.* 24:499–505 (1986).
17. Nelson, J. K., R. H. Getty, and J. W. Birks. "Fluorine Induced Chemiluminescence Detector for Reduced Sulfur Compounds," *Anal. Chem.* 55:1767–1770 (1983).
18. Foreman, W. T., C. L. Shellum, J. W. Birks, and R. E. Sievers. "Supercritical Fluid Chromatography with Sulfur Chemiluminescence Detection," *J. Chromatogr.* Preprint.
19. Nyarady, S. A., R. M. Barkley, and R. E. Sievers. "Redox Chemiluminescence Detector: Application to Gas Chromatography," *Anal. Chem.* 57:2074–2079 (1985).

20. Sievers, R. E., S. A. Nyarady, R. L. Shearer, J. J. DeAngelis, R. M. Barkley, and R. S. Hutte. "Selectivity of the Redox Chemiluminescence Detector for Complex Sample Analysis," *J. Chromatogr.* 349:395–403 (1985).

21. Shearer, R. L., D. L. O'Neal, R. Rios, and M. D. Baker. "Analysis of Sulfur Compounds by Capillary Column Gas Chromatography with Sulfur Chemiluminescence Detection," *J. Chromatogr. Sci.* 28:24–28 (1990).

22. Spurlin, S. R., and E. S. Yeung. "On-Line Chemiluminescence Detector for Hydrogen Sulfide and Methyl Mercaptan," *Anal. Chem.* 54:318–320 (1982).

23. Benner, R. L., and D. H. Stedman. "Universal Sulfur Detection by Chemiluminescence," *Anal. Chem.* 61:1268–1271 (1989).

24. Kelly, T. J., J. S. Gaffney, M. F. Phillips, and R. L. Tanner. "Chemiluminescent Detection of Reduced Sulfur Compounds with Ozone," *Anal. Chem.* 55:135–138 (1983).

25. Bruening, W., and F. J. M. Concha. "Improved Gas Chromatographic Ozone Chemiluminescence Detector," *J. Chromatogr.* 142:191–201 (1977).

26. For reasons of legality and confidentiality, source names and locations cannot be revealed.

CHAPTER 9

Measurement of Oxygenated Hydrocarbons and Reduced Sulfur Gases by Full-Scan Gas Chromatography/Mass Spectrometry (GC/MS): EPA Method TO-14

Steven D. Hoyt, Vivian Longacre, and Michael Stroupe

TABLE OF CONTENTS

0-87371-606-0/93/$0.00+$.50

© 1993 by Lewis Publishers

I. INTRODUCTION

There has recently been considerable interest in measuring organic compounds in the ambient air using EPA Method TO-14. This method uses SUMMA canisters and cryogenic preconcentration as described in "The Determination of Volatile Organic Compounds in Ambient Air Using SUMMA™ Passivated Canister Sampling and Gas Chromatographic Analysis."[1] In this method, a full-scan GC/MS procedure is described which uses a Nafion® dryer and whole air cryotrapping, with detection limits in the subparts per billion. The use of the Nafion® dryer has raised some concerns about the loss of oxygenated hydrocarbons such as alcohols, ketones, and aldehydes. Many clients are interested in using TO-14 to analyze indoor air or odor complaint samples containing oxygenated hydrocarbons and reduced sulfur gases with a minimum of modification to the existing method. The objective of this chapter is to describe the performance of EPA Method TO-14 using full-scan GC/MS, to investigate the use of Method TO-14 for oxygenated hydrocarbons and reduced sulfur compounds, and to determine how well the method performs with only minor modifications.

II. SAMPLING

EPA Method TO-14 is based on a whole-air sampling system using specially constructed and passivated stainless steel air sampling canisters. A diagram of the integrated sampler with the 3.2-L canister is shown in Figure 1. The SUMMA™ canisters cannot be used to collect the more reactive reduced sulfur gases such as hydrogen sulfide, but can be used for carbonyl sulfide, oxygenates, carbon disulfide, and other reduced sulfur gases. SUMMA™ canisters can be used for samples containing compounds with concentrations as low as 1 pptv. The canisters are cleaned by connecting them to a two-stage vacuum pump with a liquid nitrogen trap and evacuating them to 10 μm. While under vacuum, the canisters are heated to 100°C in a custom oven designed to prevent excess heat from reaching the valves. The canisters are often capped and sent to the sampling locations under vacuum. Initially, all canisters need to be blanked by filling with zero air and analyzing each canister to verify that all target compounds are less than 0.2 ppbv. After the cleaning procedure has been used and a history of performance established, it is only necessary to blank percentage of the canisters.

Tedlar® bags are often used for the reduced sulfur gases, but many of the bags tested were found to outgas large quantities of carbon disulfide, and smaller

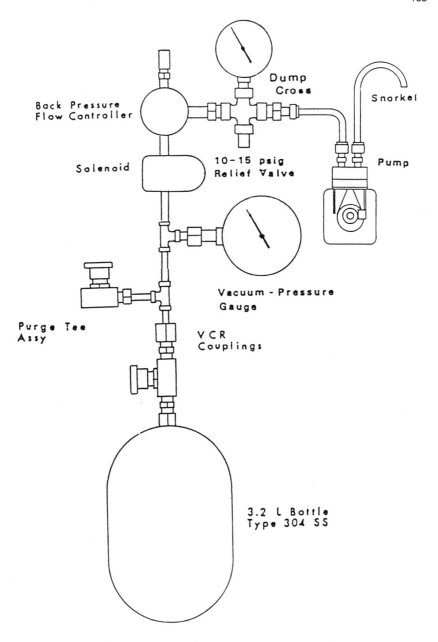

FIGURE 1.   Integrated sampler with 3.2-L canister for collecting whole air samples.

amounts of other sulfur gases. The stability of some reduced sulfur gases and oxygnates is summarized in Table 1. The bags work best in the concentration range of 1 ppbv or larger. The bags are first filled with zero air until taut. The bags

Table 1. Sample Container Suitability Reduced Sulfur and Oxygenate Compounds

Compounds	SUMMA® canister
Hydrogen sulfide	Poor
Carbonyl sulfide	Excellent
Dimethyl sulfide	Good
Carbon disulfide	Good
Methyl mercaptan	Poor
Acetone	Good
2-Butanone	Good
2-Hexanone	Good
4-Methyl-2-pentanone	Good

then sit for 24 hr to check for leaks. If the bags do not leak, they are evacuated until empty using a small pressure/vacuum pump. The bags are refilled with zero air, and this flushing cycle is repeated three times. Two bags from each lot are tested for blank levels of compounds on the GC/MS. The bags are shipped in cardboard boxes to avoid light from contacting the sample.

III. ANALYTICAL METHODOLOGY

The samples are analyzed by using cryotrapping and high resolution GC with full-scan MS as described in EPA Method TO-14. A modification to the method was made by using cryofocusing and an internal standard introduction method. A Nutech Model 8533 cryogenic concentrator with a cryogenic freezeout loop was connected to an HP 5890 GC with a J & W Scientific, 50-m DB-5 fused silica capillary column which is interfaced directly to the source of a HP 5970 MSD mass spectrometer. The data are processed using a HP 300 Pascal computer with probability-based match (PBM) NIST mass spectral library.

The Nutech 8533 concentrator is designed to load ambient air samples in the 50- to 1000-mL range and landfill gas samples, from 1.0 to 20.0 mL. The concentrator has separate pathways, for ambient air and for landfill gas samples to eliminate carryover from high concentration samples. A diagram of the concentrator is shown in Figure 2. The 1000 mL vol is sufficient to achieve a detection limit of 0.5 ppbv for most VOCs in ambient air. Ambient air samples are introduced from the air sampling canister through a Nafion® dryer used to remove water vapor. The sample then goes through a mass flow controller to measure sample volume, and internal standards are added using an internal sample loop in the Nutech. The freezeout loop is held at −150°C by the Nutech and concentrates the air sample. After concentration, the sample is cryofocused onto the beginning of the 50-m fused silica capillary column. When the MS is ready, the cryofocus is heated and the compounds are injected into the GC. The column is temperature

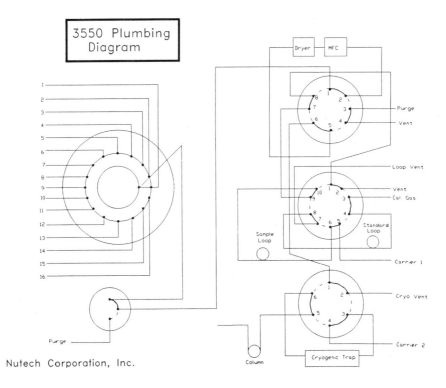

FIGURE 2. Schematic of Nutech 8533 concentrator.

programmed from –40 to 200°C. The MS is operated in the EI mode, and scanned from 33 to 250 amu, using three scans per second. A summary of the analytical conditions is given in Table 2.

For analysis of the reduced sulfur gases, the Nutech concentrator was modified by replacing the nickel lines that came in contact with the sample inside the concentrator with Teflon® lines to reduce the active surfaces. Deactivated fused silica capillary tubing can also be used. The cryogenic trap was left as nickel tubing since it has to be heated for desorption.

The system was evaluated with the standard Nafion® dryer supplied by Nutech.

Ambient air samples were spiked with known concentrations of the reduced sulfur gases carbonyl sulfide, hydrogen sulfide, carbon disulfide, and dimethyl sulfide. These samples were analyzed using the modified Nutech concentrator. Performance of the reduced sulfur gases are shown in Table 3.

Calibration for the VOC compounds is done using a static dilution system, which utilizes the replaceable internal standard loop in the Nutech 8533 concentrator. Commercial NBS traceable gas standards containing between 5 and 10 ppmv of the target compounds are used for calibrations. The standards are

Table 2. Analytical Conditions

Nutech concentrator
  Cryotrap: TO-14 with glass beads
    Trap: −185°C
    Desorb: 100°C
  Cryofocus: direct on capillary column
  Internal standards: 2-mL loop
Hewlett Packard 5890 GC
  Column: J & W Scientific; 50-m, 0.32-mm
  DB-5; 1-µm phase
  Temperature program:
    Initial: −30°C for 2 min
    Program: 12°C
    Final temp: 200°C
    Run time: 25 min for TO-14
Hewlett Packard 5970 MSD
  HP standard conditions and autotune
  Scan: 33–270 amu
  Tune: bromofluorobenzene (BFB)

Table 3. Reduced Sulfur Compounds

Response factors	16.5 ppbv	33 ppbv	66 ppbv
Hydrogen sulfide	4,464	6,418	5,157
Carbonyl sulfide	3,460	2,779	3,706
Dimethyl sulfide	5,632	10,496	6,149
Carbon disulfide	44,324	41,433	36,920

Relative standard deviation (SD)	RSD %	% Recovery
Hydrogen sulfide	19	80
Carbonyl sulfide	14	85
Dimethyl sulfide	36	75
Carbon disulfide	9	90

purchased from Scott-Marrin, Riverside, CA. Several cylinders are needed for calibration since the complete VOC or TO-14 list cannot be blended into a single cylinder. Combinations of these cylinders are analyzed using loop sizes of 0.5–10.0 mL of standard diluted with 1000 mL of humidified zero air. The oxygenated hydrocarbon standards were prepared by injecting 5.0 µL of pure liquid into a Tedlar® bag filled with 50.0 L of zero air.

IV. QUALITY ASSURANCE

For full-scan GC/MS, the following laboratory quality control criteria are used for sample analysis. Samples are analyzed by daily batches, which include the standards, blanks, QC samples, and actual samples. See Figure 3A.

A. Instrument Tuning

Prior to analysis of samples, 50 ng of bromofluorobenzene (BFB) tuning compound is injected through the internal standard loop on the Nutech concentrator. The BFB areas must meet the EPA tuning criteria for BFB. The machine must be retuned at least every 24 hr.

B. Detection Limits

The determination of the detection limits should follow the guidelines in CFR 40, Part 136. For the full scan, the instrument should be capable of producing a complete mass spectrum in which fragmentation ions greater than 20% of the base peak must be observed from a 500-mL sample of a 5-ppbv standard.

C. System Blank

A system blank of the Nutech 8533 concentrator run with 500 mL of humidified zero air must have interfering peaks less than 0.2 ppbv. A dry zero air blank will not check for oxygenate contamination in the Nafion® dryer.

D. Internal Standards

The internal standard recovery for all samples is set to between 80 and 120%. See Figure 3(a).

E. Initial Calibration

An initial calibration curve is prepared for method evaluation at 1.0, 3.0, 5.0, and 10.0 ppbv. The percent RSD of the response factors is less than 10.0% for most compounds. If the daily standard response factors deviate more than 20% from these values, the initial calibration is redone. See Figure 3(b).

F. Daily Calibration

See Figure 3(c).

Figure 3a. Volatile Organic GC/MS Tune and Internal Standard Area Summary

Date: 5/3/90   Instrument: GC/MS 01   File: S05030A3
EPA Method TO-14   HP5890/5970   DB-5 Capillary Column

m/e	Ion Abundance Criteria	Area	% Abundance
50	15.0 to 40.0% of mass 95	14164	31
75	30.0 to 60.0% of mass 95	25776	57
95	Base Peak, 100% relative abundance	45104	100
96	5.0 to 9.0% of mass 95	3685	8
173	Less than 2.0% of mass 174	0	0
174	Greater than 50% of mass 95	33272	74
175	5.0 to 9.0% of mass 174	3117	9
176	95.0 to 101.0% of mass 174	32120	97
177	5.0 to 9.0% of mass 176	2348	7

Internal Standard Area Summary
File: S05030A2

Sample Number	IS 1 Area	RT	IS 2 Area	RT	
12 hr Standard	200819	10.64	1843972	13.04	
Upper Limit	401638		3687944		IS + 100%
Lower Limit	100410		921986		IS − 50%
S05030A3	217925	10.65	1713100	13.07	
S05030A4	216463	10.66	1756221	13.07	
S05030A6	225938	10.66	1759620	13.09	
S05030A8	232210	10.65	1780464	13.07	
S05030A7	174271	10.63	1781202	13.06	
B05030A2	221418	10.72	1707073	13.13	
00397A2	221188	10.65	1742222	13.08	
00398A1	224566	10.64	1785318	13.07	
00399A1	169324	10.68	1775870	13.09	
00400A1	240032	10.65	1761719	13.08	
00400A2	226485	10.66	1762496	13.09	
00401A1	168240	10.67	1779967	13.08	
00402A1	234038	10.63	1765724	13.06	
00403A1	207628	10.64	1688714	13.06	
00404A1	177530	10.68	1699200	13.08	
00405A1	178936	10.62	1812671	13.03	
00406A1	229302	10.59	1752541	13.02	

FIGURE 3a. GC/MS tune and internal standard summary work sheet.

Figure 3b. Initial Calibration Data
EPA Method TO-14: GC/MSD Full Scan   Date: 11/18/88
Lab File: 0   RRF40 = S11188A9   RRF150 = S11188A8
RRF100 = S11188A4   RRF50 = S11188A1   RRF200 =   0

	RRF20	RRF50	RRF100	RRF150	Average	%RSD
Freon 12	1.74	1.60		1.29	1.54	12.2
Chloromethane						NM
Freon 114						NM
Vinyl Chloride	0.47	0.48	0.30	0.39	0.41	18.1
Bromomethane						NM
Chloroethane						NM
Freon 11	1.39	1.15	0.93	0.93	1.10	17.3
1,1-Dichloroethene	0.59	0.63	0.38	0.46	0.51	19.6
Dichloromethane	0.48	0.54	0.28	0.40	0.43	22.3
Trichlorotrifluoroethane						NM
1,1-Dichloroethane						NM
c-1,2-Dichloroethene						NM
Chloroform	0.77	0.76	0.75	0.65	0.73	6.5
1,1,1-Trichloroethane	0.17	0.19	0.21	0.19	0.19	6.7
1,2-Dichloroethane (ion 62)	0.28	0.30	0.32	0.30	0.30	4.6
Benzene	0.91	0.96	0.91	0.84	0.90	4.7
Carbon Tetrachloride	0.25	0.26	0.25	0.25	0.25	2.8
Trichloroethene	0.08	0.08	0.08	0.07	0.08	4.9
1,2-Dichloropropane (ion 63)	0.67	0.71	0.69	0.62	0.67	4.9
t-1,3-Dichloropropene	0.23	0.26	0.24	0.24	0.24	5.0
Toluene	0.53	0.58	0.62	0.53	0.57	6.8
c-1,3-Dichloropropene	0.11	0.13	0.11	0.12	0.12	5.1
1,1,2-Trichloroethane	0.47	0.49	0.40	0.43	0.45	7.4
1,2-Dibromoethane	0.39	0.43	0.37	0.36	0.39	7.2
Tetrachloroethene	0.03	0.04	0.04	0.04	0.04	16.3
Chlorobenzene	0.41	0.46	0.37	0.37	0.40	9.3
Ethylbenzene		0.02	0.03	0.02	0.02	20.9
m,p-Xylene	0.22	0.25	0.20	0.20	0.22	8.1
Styrene	0.80	0.84	0.68	0.68	0.75	9.6
1,1,2,2-Tetrachloroethane						NM
0-Xylene	0.22	0.25	0.20	0.20	0.22	8.1
4-Ethyltoluene						NM
1,3,5-Trimethylbenzene	0.93	0.95	0.79	0.74	0.85	10.3
1,2,4-Trimethylbenzene	0.83	0.90	0.73	0.69	0.79	10.4
m-Dichlorobenzene	0.21	0.39	0.27	0.28	0.29	21.8
Benzyl Chloride	0.22	0.29	0.25	0.26	0.26	10.2
p-Dichlorobenzene	0.21	0.39	0.27	0.28	0.29	21.8
o-Dichlorobenzene	0.21	0.39	0.27	0.28	0.29	21.8
1,2,4-Trichlorobenzene		0.27	0.10	0.19	0.19	37.9
Hexachlorobutadiene			0.06	0.44	0.25	75.2

NM = indicates this compound was not measured.

FIGURE 3b. GC/MS initial calibration data work sheet.

Figure 3c. Daily Continuing Calibration Check
Cal Date 5/3/90
Initial Cal Date: 12/30/89

Compound	Blank (ppbv)	Standard RRF	Standard RRF50	%D	QC Limits
Freon 12	0.00	5.77	6.72	15	25
Chloromethane	0.00	1.99	2.34	17	25
Freon 114	0.00	0.00	0.00	NM	25
Vinyl Chloride	0.00	1.80	2.07	14	25
Bromomethane	0.00	1.75	1.97	12	25
Chloroethane	0.00	3.30	4.55	32	25
Freon 11	0.00	5.13	5.76	11	25
1,1-Dichloroethene	0.00	1.85	1.92	4	25
Dichloromethane	0.64	1.71	1.86	9	25
Trichlorotrifluoroethane	0.00	0.69	0.74	6	25
1,1-Dichlorethane	0.00	4.66	4.89	5	25
c-1,2-Dichloroethene	0.00	7.02	7.36	5	25
Chloroform	0.00	5.65	5.89	4	25
1,1,1-Trichloroethane	0.00	5.30	5.54	4	25
1,2-Dichloroethane (ion 62)	0.00	0.48	0.49	1	25
Benzene	0.00	1.61	1.81	11	25
Carbon Tetrachloride	0.00	0.73	0.78	6	25
Trichloroethene	0.00	0.38	0.41	7	25
1,2-Dichloropropane (ion 63)	0.00	0.45	0.41	9	25
t-1,3-Dichloropropene	0.00	0.49	0.47	5	25
Toluene	0.00	2.04	2.23	9	25
c-1,3-Dichloropropene	0.00	0.38	0.32	18	25
1,1,2-Trichloroethane	0.00	0.64	1.05	49	25
1,2-Dibromoethane	0.00	0.07	0.07	12	25
Tetrachloroethene	0.00	0.47	0.49	4	25
Chlorobenzene	0.00	1.46	1.96	29	25
Ethylbenzene	0.00	0.81	1.04	25	25
m,p-Xylene	0.00	0.81	1.04	25	25
Styrene	0.00	2.11	2.05	3	25
1,1,2,2-Tetrachloroethane	0.00	0.28	0.33	16	25
o-Xylene	0.00	0.81	1.04	25	25
4-Ethyltoluene	0.00	6.30	6.99	10	25
1,3,5-Trimethylbenzene	0.00	5.45	6.10	11	25
1,2,4-Trimehylbenzene	0.00	6.28	7.22	14	25
m-Dichlorobenzene	0.13	1.52	1.94	24	25
Benzyl Chloride	0.00	1.63	1.69	4	25
p-Dichlorobenzene	0.00	1.52	1.94	24	25
o-Dichlorobenzene	0.00	1.52	1.94	24	25
1,2,4-Trichlorobenzene	0.00	4.01	NM	NM	25
Hexachlorobutadiene	0.00	2.46	NM	NM	25

NM - not measured

FIGURE 3c. Daily continuing calibration check work sheet.

Figure 3d. Duplicate Sample/Spike Results

Compound	Sample ppbv	Duplicate ppbv	% RPD	QC Limits
Freon 12	0.42	0.64	42	30
Chloromethane	not detec	not detec		30
Freon 114	not detec	not detec		30
Vinyl Chloride	not detec	not detec		30
Bromomethane	not detec	not detec		30
Chloroethane	not detec	not detec		30
Freon 11	not detec	not detec		30
1,1-Dichloroethene	not detec	not detec		30
Dichloromethane	1.6	1.8	12	30
Trichlorotrifluoroethane	not detec	not detec		30
1,1-Dichlorethane	not detec	not detec		30
c-1,2-Dichloroethene	not detec	not detec		30
Chloroform	not detec	not detec		30
1,1,1-Trichloroethane	0.79	0.73	8	30
1,2-Dichloroethane	not detec	not detec		30
Benzene	1.7	1.6	6	30
Carbon Tetrachloride	not detec	not detec		30
Trichloroethene	not detec	not detec		30
1,2-Dichloropropane (ion 63)	not detec	not detec		30
t-1,3-Dichloropropene	not detec	not detec		30
Toulene	4.8	5	4	30
c-1,3-Dichloropropene	not detec	not detec		30
1,1,2-Trichloroethane	not detec	not detec		30
1,2-Dibromoethane	not detec	not detec		30
Tetrachloroethene	not detec	not detec		30
Chlorobenzene	not detec	not detec		30
Ethylbenzene	not detec	not detec		30
m,p-Xylene	0.52	0.55	6	30
Styrene	not detec	not detec		30
1,1,2,2-Tetrachloroethane	not detec	not detec		30
o-Xylene	not detec	not detec		30
4-Ethyltoluene	1.4	1.3	7	30
1,3,5-Trimethylbenzene	0.27	0.28	4	30
1,2,4-Trimehylbenzene	1.2	1.2	0	30
m-Dichlorobenzene	not detec	not detec		30
Benzyl Chloride	not detec	not detec		30
p-Dichlorobenzene	not detec	not detec		30
o-Dichlorobenzene	not detec	not detec		30
1,2,4-Trichlorobenzene	not detec	not detec		30
Hexachlorobutadiene	not detec	not detec		30

FIGURE 3d. Duplicate/spike analysis results.

G. Duplicate/Spike Analysis

Spiked ambient air samples are prepared by collecting coastal marine air in 25-L SUMMA™ canisters and adding a known quantity of dilute gas standard with a glass syringe. These cylinders are initially calibrated against the primary standards and are then analyzed as spiked air samples for quality control. Duplicate samples are also analyzed. The duplicates can be duplicates collected from collocated samplers or duplicates on a single sample. See Figure 3(d).

V. RESULTS

The application of EPA Method TO-14, and a modified commercial cryogenic concentrator to the analysis of ambient air samples containing oxygenated hydrocarbons and reduced sulfur compounds were evaluated by determining the percent recovery, method detection limit, reproducibility, and the linear range.

The Nutech 8533 cryogenic concentrator was modified as described. The nickel tubing used in the concentrator was replaced with Teflon® tubing in an effort to reduce the number of active sites and improve the recovery of the reduced sulfur gases. This modification would not be necessary for the oxygenated hydrocarbons. The nickel cryogenic trap was not replaced since this would result in a major modification, and the intent of this study was to determine how effective the system could be made without major change. The Nafion® dryer was left in the system so that a 500- to 1000-mL sample could be analyzed. The Nafion® dryer does not effect the recovery of the reduced sulfur compounds, but does effect the recoveries of water soluble compounds such as the oxygenated hydrocarbons. The recoveries were determined by preparing two Tedlar® bags. A high level bag was prepared by spiking about 5.0 µL of pure compound into a 50-L bag and filling the bag with zero air. The concentration in this bag was about 30 ppmv. A 10.0-mL aliquot from this bag was transferred to a 10.0-L Tedlar® bag filled with humidified zero air. The concentration of the second bag was about 30 ppbv. The recovery of the compounds was determined by loading 1.0 mL of the high concentration bag using a sample loop with no dryer, and loading 500 mL of the low concentration humidified bag through the Nafion® dryer in the same manner that an ambient sir sample would be analyzed. Then by comparing the response of the low concentration humidified bag to the high concentration standard bag, the recovery could be determined. The same procedure was used for the reduced sulfur compounds, except a 33-ppmv commercial reduced sulfur standard cylinder was used in place of the high concentration standard bag. As an additional check of the high concentration Tedlar® bag, a 92-ppmv commercial acetone standard was analyzed against the bag to determine the percentage difference between the spiked bag and the commercial standard. The agreement between duplicate analysis was better than 3%. The results for the recovery study are shown in Table 4. A total ion chromatogram (TIC) showing the oxygenated hydrocarbons and internal standards is shown in Figure 4.

Table 4. Recovery Study Reduced Sulfur Compounds, Oxygenated Hydrocarbons

Compound	Average % Recovery
Hydrogen Sulfide	80
Carbonyl sulfide	85
Dimethyl sulfide	75
Carbon disulfide	90
Butanol	0
n-Octanol	5
Acetone	54
2-Butanone	48
4-Methyl-2-pentanone	62
2-Hexanone	33

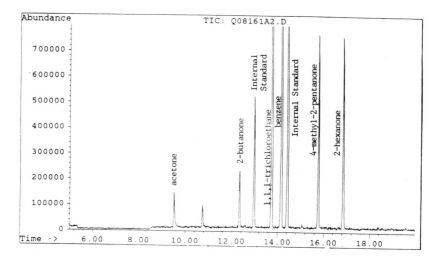

FIGURE 4.   Total ion chromatogram (TIC) of oxygenated hydrocarbons with internal standards added.

The recovery for the reduced sulfur gases was between 75 and 90%. The best recoveries were obtained for carbonyl sulfide and carbon disulfide which are the more stable of the reduced sulfur compounds. The process of replacing the interior lines of the concentrator with Teflon® seemed to remove many of the active sites. The recoveries for the alcohols are almost zero, indicating that these compounds were totally trapped by the Nafion® dryer. The ketones, acetone to 2-hexanone gave recoveries that were low but not unacceptable. The lowest average recovery was for 2-hexanone, which was 33%. The ketones are also trapped by the dryer, but not to the extent of the alcohols. An additional note on this is that when humidified zero air is passed through the system (or other samples) the 2-hexanone is slowly released as carryover in subsequent samples. Extensive flushing of the Nafion® dryer with zero air and a blank analysis are necessary after samples containing ketones to prevent contamination of the following samples.

Table 5. Method Detection Limits

Compound	MDL (ppbv)	CRQL (ppbv)
Hydrogen sulfide	2.0	6.6
Carbonyl sulfide	1.0	3.3
Dimethyl sulfide	1.5	5.0
Acetone	2.0	6.6
2-Butanone	2.5	8.3
4-Methyl-2-Pentanone	2.0	6.6
2-Hexane	2.5	8.3
Freon-12	0.10	0.33
Vinyl chloride	0.10	0.33
Dichloromethane	0.15	0.50
1,1-Dichloroethane	0.15	0.50
Chloroform	0.10	0.33
1,1,1-Trichloroethane	0.10	0.33
Benzene	0.05	0.15
Toluene	0.05	0.15
Perchloroethylene	0.05	0.15

The method detection limit (MDL) for the VOC compounds of interest was determined by analyzing replicate samples near the estimated detection limit and using the procedure of Glasser et al.[2] to determine the MDL, which is based on reproducibility of the measurement process near the detection limit. The results for this process showed a variation in the MDL from 0.15 to 0.50 for the selected VOC compounds. Since the uncertainty of a measurement made at the MDL could be 50%, a limit of quantitation (CRQL) is often used to evaluate the reliability of the reported results. The CRQL (or practical quantitation limit) is defined as 3.3 times the detection limit. Table 5 shows the MDL and the CRQL for selected VOC compounds, as well as some reduced sulfur gases and oxygenated hydrocarbons. The detection limit for the reduced sulfur gases and the oxygenated hydrocarbons was higher than the target VOC compounds, averaging about 2 ppbv. This would be expected since the recoveries of these compounds is less than 100%; there is more variability in the analysis; and, in the case of the oxygenated hydrocarbons, the peaks exhibit tailing which makes integration more difficult. This can be seen in Figure 5. The two alcohols, butanol and octanol, that were tested showed recoveries of almost 0%, and appeared to be totally lost in the Nafion® dryer.

Data for the linearity of the reduced sulfur compounds and oxygenated hydrocarbons for the full scan GC/MS method are shown in Table 6. The GC/MS response factors are calculated as the ratio between the compound response and the internal standard response per ppbv. Normally, ambient air samples will be run at maximum volume (1000 mL); and since the levels of VOCs are usually low in ambient air samples, they rarely exceed the area counts for the upper limit of the calibration curve. The linearity for the oxygenated hydrocarbons compounds was checked from 10 to 100 ppbv, and the reproducibility of the response factors

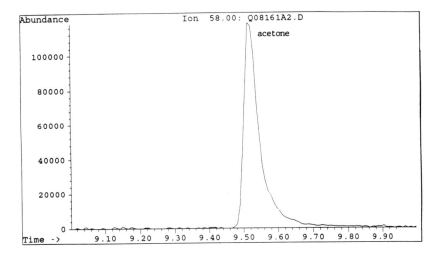

FIGURE 5. Extracted ion chromatogram of ion 58 of acetone peak. This figure shows the peak tailing that is common with oxygenated hydrocarbons.

Table 6. Linearity of Response

Compound	Relative response (RRF)				% RSD
	10 ppbv	20 ppbv	40 ppbv	100 ppbv	RF
Butanol	—	—	—	—	—
n-Octanol	—	—	—	5,635	—
Acetone	46,365	61,855	47,400	70,327	21
2-Hexanone	13,260	18,535	13,515	24,402	30
4-Methyl-2-pentanone	31,840	40,335	26,193	24,088	24
2-Butanone	37,840	39,310	28,145	30,620	17
Hydrogen sulfide	4,464	6,418	5,157	—	19
Carbonyl sulfide	3,460	2,779	3,706	—	14
Dimethyl sulfide	5,632	10,496	6,149	—	36
Carbon disulfide	44,324	41,433	36,920	—	9

Note: The relative response is calculated as the area counts divided by the concentration of the standard in ppbv. RRF = area counts/ppbv.

varied from 17 to 30%. The reduced sulfur gases were tested for linearity and found to have response factors that varied from 20 to 30%.

Method TO-14 has been used for several years to analyze various types of ambient air samples. In this case, the target list includes the EPA 625 target VOC compounds and the hazardous substance list (HSL) compounds which are mostly oxygenated hydrocarbons and carbon disulfide (a reduced sulfur gas) as well as the newer EPA CLP target compound list (TCL) containing the ketones.

VI. CONCLUSIONS

EPA Proposed Method TO-14 using an automated cryogenic concentrator with full-scan GC/MS can be used to provide a survey of the volatile organic compounds present in air samples. By performing some simple modifications to the Nutech automated cryogenic concentrator in order to deactivate most of the surfaces that come in contact with the sample, many of the oxygenated hydrocarbons (excluding alcohols) and reduced sulfur gases can also be detected using this method. This expands the versatility of the TO-14 method for use in odor surveys and for indoor air pollution work which may involve polar oxygenated compounds. While this procedure gave good recoveries and quantitation, it is not meant to be used as the primary analytical method for either category of these compounds. A comprehensive quality assurance program is needed to ensure that the instrumentation meets its initial performance specifications for the more reactive reduced sulfur gases and oxygenates.

REFERENCES

1. EPA. "The Determination of Volatile Organic Compounds in Ambient Air Using SUMMA™ Passivated Canister Sampling and Gas Chromatographic Analysis," U.S. EPA/EMSL., Research Triangle Park, NC (May, 1988).
2. Glasser, J. A. M., D. L. Forest, G. D. McKee, S. A. Quave, and W. L. Budde. *Environ. Sci. Technol.* 15: 1426 (1981).

CHAPTER 10

Sampling of Atmospheric Carbonyl Compounds for Determination by Liquid Chromatography after 2,4-Dinitrophenylhydrazine Labeling

Appathurai Vairavamurthy, James M. Roberts, and Leonard Newman

TABLE OF CONTENTS

0-87371-606-0/93/$0.00+$.50

I. INTRODUCTION

Determination of carbonyl compounds, aldehydes (RCHO), and ketones (R_1COR_2), in the ambient atmosphere is receiving increasing attention because of the critical role these compounds play as pollutants and as key participants in tropospheric photochemistry. Carbonyls are involved in photochemical reactions as products of the oxidation of hydrocarbons, precursors of oxidants including ozone and peroxycarboxylic nitric anhydrides ([PAN]-type compounds), and as sources of free radicals and organic aerosols. Formation of carbonyls in the atmosphere proceeds through a series of free-radical reactions. It is initiated by the formation of a carbon-centered radical (R•), usually through reaction of hydroxyl radical (OH•) with a hydrocarbon, although photolysis of labile compounds (such as another carbonyl) or reaction with nitrate radical (NO_3•) is also possible. Reactions of OH•, O_3, or NO_3• with alkenes proceed through addition to the double bond, forming a carbon-centered radical on the adjacent carbon. The photochemical mechanisms of carbonyl production in the atmosphere are dealt with in several recent reviews.[1-4] In addition to the in situ photochemical generation, a number of carbonyls are emitted directly in auto exhaust, and by a variety of both anthropogenic and biogenic sources.[5,6] There is a potential for increased carbonyl emissions resulting from changes in fuel technology such as the use of methanol, ethanol, etc. as gasoline substitutes.[7,8]

Correct understanding and assessment of the role of carbonyls in tropospheric chemistry require the accurate and precise measurement of these compounds along with their parent and product compounds. However, measurement of carbonyl compounds in the ambient atmosphere poses challenging problems because of their trace concentrations (sub- or low parts per billion volume in clean air[9] to higher parts per billion volume in urban and polluted air[10]) and interferences arising from atmospheric copollutants[11] (e.g., ozone). In the 1970s, chromatographic techniques in conjunction with chemical derivatization methods paved the way for sensitive and selective determination of carbonyls in ambient air. Although many chromatographic methods have been proposed, derivatization with 2,4-dinitrophenylhydrazine (DNPH) coupled to liquid chromatographic separation has received widespread acceptance.[12]

Chromatographic methods for ambient carbonyl measurements, such as the DNPH-liquid chromatography (LC), involve two separate operational steps: (1) integrated collection of target carbonyls, and (2) chromatographic analysis of the collected sample. Since chromatography allows simultaneous separation of individual species, interference problems arising from similar compounds are greatly minimized. However, air sampling remains the most critical step affecting the accuracy and precision of the measurements. The goal of integrated sampling is to concentrate the sample in order to improve the sensitivity of the method. A classical method used for sampling and preconcentration of airborne organics is cryogenic collection. Since many other components present in air are also concentrated along with the target molecules, the concentration effect may accelerate many reactions which are kinetically not significant in the ambient air such as the reaction of ozone with alkenes. In DNPH methods for carbonyls, this problem has been alleviated, at least partially, by simultaneous derivatization and collection, which also improves collection efficiency. This selective enrichment has usually been achieved by sampling with reagent-loaded, solid-phase cartridges, or impingers charged with reagent solution.

In spite of the numerous studies concerned with integrated air sampling, especially with the DNPH method, several questions regarding interferences and sampling artifacts have not yet been adequately addressed. The major concerns with air sampling of carbonyls which can affect the accuracy of the method are: (1) incomplete collection of carbonyls, (2) loss of carbonyls by physical processes such as adsorption or chemical reaction with ambient compounds such as SO_2 and O_3, (3) generation of carbonyls as sampling artifacts, (4) formation of various interfering compounds, and (5) variable blanks resulting from contamination of the reagent and sampling instrument. Here we discuss the different techniques used for time-integrated collection of carbonyls in the DNPH-based liquid chromatographic methods, emphasizing the principles, advantages, and limitations. These techniques illustrate the complexity and variability involved in time-integrated sampling of atmospheric carbonyls.

II. 2,4-DINITROPHENYLHYDRAZINE (DNPH) DERIVATIZATION AND LIQUID CHROMATOGRAPHY

The acid-catalyzed derivatization of DNPH proceeds by nucleophilic addition to the carbonyl followed by 1,2-elimination of water to form the 2,4-dinitrophenylhydrazone (Figure 1). Although gas chromatography (GC) can be used for separation and determination of DNP hydrazones,[13-15] the GC methods

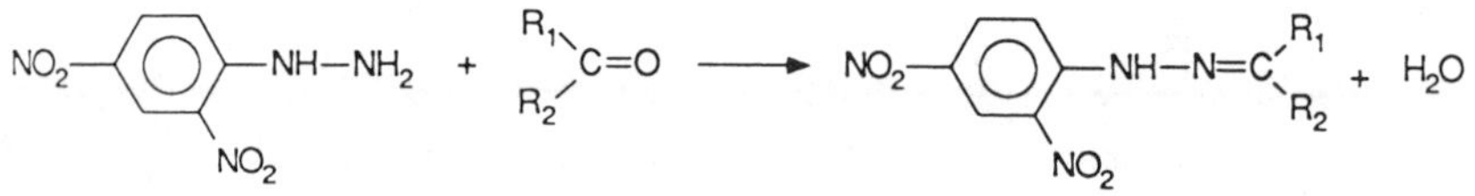

FIGURE 1. Reaction of carbonyls with 2,4-dinitrophenylhydrazine to form hydrazone derivatives.

have not found widespread acceptance because of the low volatility of the derivatives and the formation of double peaks (due to *syn-* and *anti-* isomers) by some derivatives, which may hamper identification and quantitation of compounds in complex samples. In contrast to GC methods, liquid chromatographic separation of hydrazones combined with UV absorption detection has become the most popular method for determination of carbonyls in air samples.[16-27]

Usually, separation of hydrazones has been accomplished with a reversed-phase C_{18} column (4.6-mm i.d. x 150-mm length) using either isocratic or gradient elution and a water-acetonitrile solvent combination (Figure 2). A major problem has been coelution or poor resolution of certain compound combinations (e.g., acrolein, acetone, propionaldehyde, and furfural; iso-butyraldehyde, n-butyraldehyde, and 2-butanone; and iso-valeraldehyde and 2-methylbutyraldehyde). Smith, et al.[28] showed that the use of a ternary gradient mobile phase results in good separation of the C_3 carbonyls, acrolein, and acetone, as well as butanal and the isomers of 2-butanone (Figure 3). Elevated column temperatures (e.g., 60°C) provided adequate separation of acrolein, propionaldehyde, and furfural but not for other combinations.[29]

In methods focused specifically on HCHO, a variety of wavelengths have been used for the detection of HCHO-DNP hydrazone. In a recent study, Grömping and Cammann[30] recommended 345 nm for formaldehyde, based on the UV spectrum of the derivative which shows two peaks, a small peak at 250 nm and a larger one at 345 nm. The detector wavelength used for simultaneous analysis of many carbonyls (usually in the 360–375-nm range) reflects a compromise, because the absorption maxima of the different hydrazone derivatives vary significantly (Table 1). In some recent studies, the use of a diode-array detector allowed the full spectra to be stored and processed later, thus aiding in the identification of the compounds.[25,31] In a few studies, mass spectromeric detection was used for confirmation of identification made by LC, determination of compounds in unresolved chromatographic peaks, and characterization of unidentified peaks.[32,33]

III. SAMPLING PROCEDURES

A. Impinger Sampling

Impinger sampling of carbonyls involves two mechanisms: (1) physical dissolution, and (2) formation of less volatile hydrazones by derivatization. The derivatization reaction will not be quantitative within the very short residence time (order of seconds) of air in the sampling solution and, therefore, dissolution plays

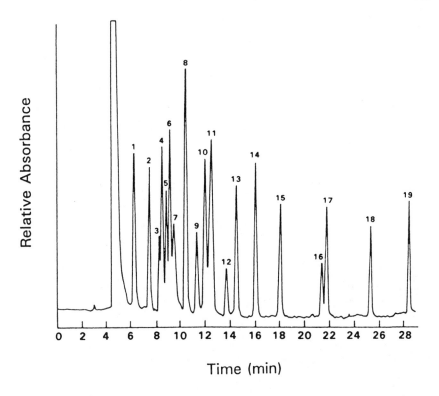

FIGURE 2. Chromatogram of various DNPH derivatives separated on a Zorbax-ODS column; mobile phase: acetonitrile-water (67:33) at 0.7 mL/min for 8 min, then 1.0 mL/min and gradient to acetonitrile-water (90:10) over 17 min, then gradient to 100% acetonitrile over 3 min. Peaks: 1, formaldehyde; 2, acetaldehyde; 3, furfural; 4, acrolein; 5, acetone; 6, propanal; 7, salicylaldehyde; 8, crotanaldehyde; 9, butanal; 10, glyoxal; 11, benzaldehyde; 12, glutaraldehyde; 13, pentanal; 14, p-tolualdehyde; 15, hexanal; 16, 3-heptanone; 17, heptanal; 18, octanal; 19, nonanal. (From Lipari, F., and S. J. Swarin. *J. Chromatogr.* 247: 297–306 (1982). With permission.)

an important role in controlling carbonyl collection. The dissolved carbonyls will subsequently undergo derivatization. Since organic solvents are better than aqueous solution for dissolution of carbonyls, they result in increased collection efficiency. Furthermore, the reduced surface tension of the organic solvent enhances mixing of the air stream with the liquid reagent during collection.

The first application of microimpinger sampling for determination of carbonyls by the DNPH-LC method was that of Kuwata et al.[16] who used DNPH reagent (5 mM) in 2 N HCl as sampling solution. Two bubblers (each containing 10 mL) in series were found adequate for quantitative trapping of carbonyls in air mixtures. For LC analysis, the sampling solutions were combined, extracted with

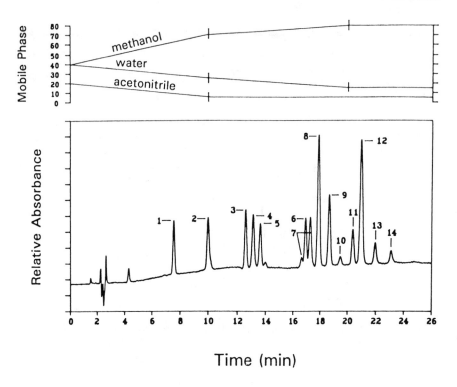

Time (min)

FIGURE 3.   (a) Ternary gradient mobile-phase composition. (b) Standard chromatogram. Peaks: 1, formaldehyde; 2, acetaldehyde; 3, acrolein; 4,
acetone; 5, propanal; 6, butanal; 7, *anti-* and *syn*-2-butanone; 8,
cyclopentanone (internal standard); 9, benzaldehyde; 10, glyoxal,
11, pentanal; 12, cyclohexanone (internal standard); 13, p-
tolualdehyde; 14, methyl glyoxal. (From Smith, D. F., T. E. Kleindienst,
and E. E. Hudgens. *J. Chromatogr.* 483: 431–436 (1989). With
permission.)

chloroform, evaporated to dryness, and the residue reconstituted in 2 mL acetonitrile
for injection into the LC. Later, investigators introduced several modifications in
the preparation of the sampling solution: (1) using a different acid, (2) trapping
carbonyls with an organic solvent (e.g., acetonitrile) compatible with LC analysis,
and (3) trapping carbonyls in a two-phase, aqueous-organic system. The main
reasons for these modifications were: (1) to improve collection efficiency, (2) to
reduce the volume of collecting solution, and (3) to minimize sample handling
steps between collection and LC analysis.

In the modified method by Kuntz et al.,[17] the impinger solution used was a 1.25
mM DNPH in acetonitrile, acidified with concentrated H_2SO_4 (0.2 mL/L). In this
case, quantitative collection was claimed with a 4-mL solution at a flow rate of
0.5 liters per minute (Lpm) for 1 hr. The use of acetonitrile also allows direct
injection of the sample into the LC system. In some studies, perchloric acid was

Table 1. Absorption Maxima of Carbonyl DNP Hydrazones

Carbonyl compound	λ-max (nm)	Carbonyl compound	λ-max (nm)
DNPH reagent	357[a]	Acrolein	373,[a] 367[b]
Formaldehyde	353,[a] 350,[b] 345[c]	Crotanaldehyde	378[d]
Acetaldehyde	363,[a] 360[b]	Glyoxylate	355,[a] 351[b]
Propanal	365[a]	Pyruvate	369,[a] 351[b]
n-Butanal	363[a]	Acetoacetate	375[b]
Isobutanal	363[a]	Acetone	367[a]
n-Pentanal	363[a]	Methylethylketone	367[a]
n-Hexanal	363,[a] 350[b]	Hydroxyacetone	360[b]
n-Heptanal	359[a]	Dihydroxyacetone	367[b]
Benzaldehyde	385[a]	2-Methylcyclohexanone	371[a]
Hydroxybenzaldehyde	393[b]	5-Hydroxy-2-pentanone	369[a]
Glyoxal	437[a]	Methylglyoxal	427[a]

[a] In 55:45 CH_3CN, H_2O medium.[25]
[b] In 60:40 CH_3CN, H_2O (pH 2.6) medium.[31]
[c] In 75:25 CH_3OH, H_2O medium.[30]
[d] In 65:35 CH_3CN, H_2O medium.[29]

used instead of sulfuric acid to acidify the DNPH solution.[23,24] De Bortoli et al.[35] observed an increase in the rate of derivatization for ketones, when phosphoric acid was used in place of perchloric acid. The use of hydrochloric acid in acetonitrile produces a white precipitate (DNPH hydrochloride); a similar effect has not been found with other acids used. Tanner and Meng[22] used DNPH in acetonitrile as impinger solution to determine formaldehyde, acetaldehyde, and benzaldehyde in ambient air with a detection level of <1 ppbv. They cooled the microimpinger to ice temperature, which further enhanced collection efficiency by the dissolution mechanism.

Grosjean used a two-phase system containing 10 mL of an aqueous, acidic (2 N HCl) solution of DNPH and 10 mL of a 9:1 by volume mixture of cyclohexane and isooctane as opposed to using a relatively polar organic solvent as acetonitrile.[10,36] A major advantage of using a two-phase system is that the derivatization reaction is accelerated because of the in situ organic phase extraction of hydrazones, which shifts equilibrium toward hydrazone formation. Grosjean claimed that the two-phase system was required to obtain quantitative recovery of aliphatic and aromatic carbonyls other than formaldehyde. In contrast, work by Van Langenhove et al.,[21] who compared derivatization in a one-phase system with that in a two-phase system, indicated no advantage of using a two-phase system for C_2-C_9 carbonyls. However, higher carbonyls showed a decreasing conversion due to their hydrophobicity.

B. Sampling with DNPH-Coated Solid Sorbents

Although DNPH-based impinger techniques have been used in many studies to determine atmospheric carbonyls, they are cumbersome and not well suited to large field studies or to sampling at remote locations when samples have to be stored and transported to a central laboratory for analysis. The solid sorbent technique results in much higher sensitivity than the impinger method because the derivatives are usually preconcentrated to a high degree in the sample. For these reasons, the DNPH-coated solid sorbents are a convenient alternative to impinger sampling and have recently been increasingly used.[24-27] A number of solid sorbents, both commercial and laboratory made, have been used for this purpose. The solid sorbents include glass beads, glass fiber filters, silica gel, Chromosorb P, Florisil, Carbopack B, XAD-2, and C_{18} (silica). Several solid sorbents including silica gel, Florisil, and C_{18} (silica) are now commercially available as prepacked cartridges or syringe columns with polypropylene or polyethylene casings (e.g., Sep-Pak brand cartridges manufactured by Waters Associates, Milford, MA), which have several advantages including convenience of use, reproducibility, and low blanks. Recently, inert sampling devices made with only glass and Teflon® parts have also been introduced (Inert Columns, Burdick and Jackson, Muskegon, MI).

1. Glass Beads and Glass Fiber Filters

Grosjean and Fung examined DNPH-coated glass beads (20 mesh size) packed in glass tubes (100-mm length x 6-mm i.d.) for carbonyl sampling.[18,36] The glass beads were coated with DNPH by immersing in a DNPH reagent (a saturated, acidic solution with added polyethylene glycol to increase viscosity) and evaporating the reagent to obtain a film around the bead. Hydrazones were extracted with a mixture of hexane and methylene chloride (7:3v/v), which was then washed with water to remove excess DNPH and acid. The extract was then evaporated and reconstituted in methanol prior to injection into the LC. The collection efficiency was found to be highly variable, especially affected by humidity variations. Carbonyl collection using DNPH-coated glass fiber filters (organic- and binder-free) were also affected by humidity variations.[37,38] As in the case with glass beads, the preparation of DNPH-coated glass fiber filters was cumbersome, and they performed poorly at Hi-Vol sampling rates.[22]

In a recent study, de Andrade and Tanner[39] reported that bisulfite-coated cellulose filters can be used for ambient air sampling of formaldehyde at high volume flow rates. The hydroeyalkanesulfonate formed is then extracted and treated with base to regenerate formaldehyde, which is then determined by DNPH derivatization and LC.

2. Silica Gel Cartridge

DNPH-coated silica gel was used initially by Beasley et al.[40] for sampling formaldehyde in air. Later, Tejada[24] simplified the silica gel technique by using commercial Sep-Pak silica gel cartridges (contains ca. 0.7 g of silica gel) and also examined the technique for sampling other carbonyls. In this study by Tejada, the DNPH cartridges were prepared by passing a DNPH solution acidified with HCl through a prewashed cartridge, which produced DNPH loading of ca. 1.9 mg per cartridge. Typical blank concentrations, when a DNPH cartridge was eluted with

5-mL acetonitrile, ranged 0.1–0.3, 0.05–0.1, and 0.1–0.25 nmol/mL for formadehyde, acetaldehyde, and acetone, respectively. The hydrazone derivatives formed during sampling were eluted with acetonitrile and analyzed by LC.

The silica gel cartridge technique was compared with the DNPH-acetonitrile impinger technique for sampling carbonyls in ambient air and in diluted automotive exhaust emissions. Results indicated a discrepancy in the two methods with respect to olefinic aldehydes such as acrolein and crotanaldehyde, but stable species including formaldehyde, acetaldehyde, propionaldehyde, benzaldehyde, and acetone correlated very well. The acrolein derivative degraded partially on the cartridge and formed an unknown product. For stable carbonyls, the sample integrity on cartridge was maintained for over a month under refrigerated storage. The cartridge technique was found to provide adequate preconcentration for sampling carbonyls at sub- to low parts per billion volume level in ambient air. However, recent work by Arnts and Tejada[11] showed a dramatic negative interference by ozone in the determination of formaldehyde and put in question the validity of silica gel sampling technique unless a carbonyl-passive ozone scrubber is employed. The effect of ozone on sampling with DNPH-coated solid sorbents is discussed in Section V.B.

3. Florisil Cartridge

Florisil is the brand name of purified magnesium silicate, manufactured by Floridian Company. Commercial, prepacked Florisil cartridges of the make Thermosorb/F (Thermoelectron Corporation, Waltham, MA) coated with DNPH were used by Lipari and Swarin[41] for determination of formaldehyde in ambient air and in diluted automotive exhaust emissions. The cartridges are constructed of polyethylene tubing (2.0-cm length x 1.5-cm o.d.) and contain about 1.2 g of dry sorbent. These cartridges allow high sampling rates up to 4.0 L/min.

In the method by Lipari and Swarin, the sorbent was coated by filling the cartridge with a DNPH solution in methylene chloride, without any acid, which resulted in a 3-mg loading. The formaldehyde hydrazone was extracted with acetonitrile and analyzed by HPLC. The hydrazone was found to be stable on cartridges kept for more than 3 weeks at 21°C, as long as the end caps were properly installed. The detection limit, which was ca. 1 ppbv in 100-L air, was limited by the blank level of 0.5 ppbv in 100-L air. Excellent agreement between the cartridge and the DNPH/acetonitrile impinger sampling methods was obtained for formaldehyde. However, no comparative results were presented for the other carbonyls (such as acetaldehyde, acetone, acrolein, etc.) that are known to be present in these sample types. Interferences from NO_2 (550 ppbv), SO_2 (100 ppbv), and humidity were examined and found to have no effect. However, the effect of ozone was not studied.

4. C_{18} Cartridge (Octadecylsilane-Bonded Silica)

As opposed to silica gel and Florisil, which are polar sorbents, C_{18} provides non-polar, hydrophobic, and relatively inert surface characteristics. Because of these surface properties, C_{18} sorbents easily retain relatively nonpolar organic compounds by hydrophobic interactions. The adsorbed molecules can be eluted quantitatively from the sorbent with organic solvents. Due to these advantages, C_{18} sorbents have been used successfully to enrich and clean up trace organic compounds in many environmental and biological applications involving aqueous samples. Recently, there has been an increasing interest in the use of C_{18} cartridges to sample organics in air, especially carbonyls in conjunction with DNPH.

The use of a DNPH-impregnated C_{18} cartridge for sampling carbonyls in air was first introduced by Kuwata et al.,[20] who used the Sep-Pak C_{18} cartridges (Waters Associates, Milford, MA). The study focused on aldehydes, with no results on ketones presented. The cartridge was coated with 1.0- to 1.2-mg DNPH by passing through a 2-mL acetonitrile solution containing 0.2% DNPH and 1% phosphoric acid. Blank levels for formaldehyde and acetaldehyde were in the range 0.2–0.5 ppbv for 100-L air sample. The collection efficiency was found to be >95% on the first cartridge, when two were used in series, for 100 L of sample at 0.7–1.2 L/min of sampling rate. This study highlighted the simplicity and usefulness of the method, but did not address questions regarding possible interferences and sampling artifacts.

Tejada[24] noted some shortcomings in the C_{18} cartridge method: (1) significant acetone contamination from the cartridge and (2) formation of carbonyls with molecular weight greater than hexanal and their increase with storage of the cartridge. However, no systematic study was undertaken to reevaluate Kuwata's method. Recently, Druzik et al.[25] essentially followed the sampling procedure of Kuwata et al. in their method using UV absorption diode array detection of the hydrazones following LC separation. This study did not report any major problem in the method and agreed with that of Kuwata et al.[16] Furthermore, based on indirect evidence, Druzik et al.[25] noted that co-pollutants including ozone do not interfere in the sampling with C_{18} cartridges.

The most recent work on the C_{18} sampling technique has been that of Zhou and Mopper.[27] Important modifications were proposed in this study regarding reagent purification and cartridge preparation in order to reduce blank levels for clean, marine air applications. An aqueous DNPH solution (ca. 0.15 mM) was used for cartridge loading, instead of an acetonitrile solution used by previous workers. The use of aqueous reagent allowed effective removal of hydrazone blanks by solvent extraction. With two cartridges in series and a flow rate of 0.7 L/min, greater than 96% collection efficiency was obtained for all compounds of interest, except acetone (92%). The study claimed detection limits in the 0.01- to 0.02-ppbv range for most carbonyl compounds for a 100-L air sample.

In agreement with other studies, humidity was found by Zhou and Mopper[27] to have no effect on the collection efficiency. The authors claimed no interference by ozone at a level of ca. 50 ppbv, based on a comparison study with and without stripping of ozone by a KI solution. Exposure of cartridges to sunlight was found to cause significant production of carbonyls and was eliminated by wrapping the cartridges in aluminum foil during sampling and storage.

5. XAD-2, Carbopack™ B, and Chromosorb P

In contrast to silica gel, Florisil, and C_{18}, these sorbents are not commercially available as prepacked cartridges; thus they have to be laboratory-packed, which is a marked disadvantage in large field studies. XAD-2 is a styrene-divinylbenzene polymer and requires tedious procedures for cleanup. Andersson et al.[42,43] used DNPH-coated XAD-2 for sampling formaldehyde, acrolein, and glutaraldehyde. The sorbent appeared to contribute high carbonyl blanks, especially acetaldehyde. XAD-2 has not been used in any recent study for sampling carbonyls in air.

Carbopack B™, a graphitized carbon black of specific surface area (ca. 100 m^2/g) and supplied by Supelco (Bellefonte, PA), was used by Ciccioli et al.[44] as the solid sorbent with $DNPH/H_3PO_4$ to sample carbonyls in air. The hydrazones formed during sampling were eluted with acetonitrile (5–10 mL) and determined by LC and UV detection. Carbopack B™ cartridges were made by packing 20–40 mg of material in glass tubes (0.5-cm i.d. x 5-cm length); the sorbent was held in place by a 100 mesh, stainless steel screen at the trap inlet and by a glass wool plug at the outlet. The cartridges, despite their very small size, allowed quantitative collection of formaldehyde, acetaldehyde, acrolein, propionaldehyde, and acetone; and made possible their determination at sub-parts per billion volume levels when about 100- to 200-L air was sampled.

Recently, Chromosorb P coated with DNPH has been used by Grömping and Cammann[30] for determination of formaldehyde in air. Chromosorb P is diatomaceous silica which is acid and base washed to remove both inorganic and organic contaminants, and was found suitable for coating with acids. For formaldehyde, the collection efficiency on the Chromosorb cartridge was >95%, and the data agreed well with the standard impinger technique. The study did not report on collection efficiency for other carbonyls or problems related to interferences from other airborne pollutants, including ozone.

6. Kinetics of DNPH Derivatization in Aqueous Medium and on Solid Phase

As discussed previously, in impinger sampling both derivatization and physical dissolution aid in the initial trapping of airborne carbonyls into the collecting solution. Subsequently, derivatization proceeds to completion. Therefore, an understanding of the effects of variables on derivatization is important in order to optimize sample collection and analysis in impinger-based methods. In solutions, the derivatization yield depends on many variables including reaction pH, reagent concentration, and temperature. Some of these questions have been addressed with formaldehyde. Although other low molecular weight aldehydes are expected to behave similarly, the results may not be valid for ketones.

The effect of pH on reaction yield was studied for formaldehyde and acetaldehyde.[45] For formaldehyde, a smooth relationship was observed over the pH range of 1.7–7.0, with maximum around pH 4 (Figure 4). The reaction yields did not change significantly over the 3- to 5-pH range. These results were remarkable

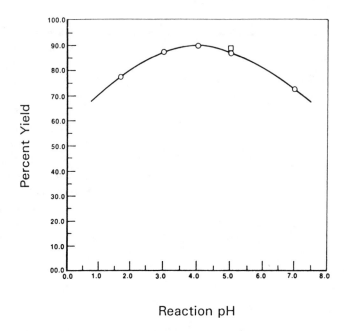

Reaction pH

FIGURE 4.   Effect of reaction pH on percent yield of HCHO-DNP hydrazone.
Symbols: circles, phosphate buffer; square, acetate buffer. (From
Bicking, M. K. L., and W. Marcus Cooke. *J. Chromatogr.* 455:310–
315 (1988). With permission.)

because most studies used very low pH (<3) for derivatization. In contrast, for
acetaldehyde, the maximum yield was observed at pH 1.7 and the yield was nearly
constant between pH 3 and 5. Protonation of the carbonyl group at low pH
promotes the nucleophilic addition, but concurrently reduces the amount of
unprotonated DNPH, which is the reactive nucleophile. Because of these compet-
ing effects, the rate passes through a maximum at a characteristic pH.

Tuss et al.[46] studied the effect of temperature on reaction yield at pH 3 for
formaldehyde. Their results are summarized in Figure 5. At 25°C the reaction was
nearly complete after 20 min. Similar results were obtained by Lowe et al.[47]
However, results obtained by Cofer and Edahl[48] indicated that longer derivatization
times (ca. 2 hr) were required for completion at pH 2. The reaction yield was also
dependent on the molar ratio of reagent (DNPH) to carbonyls. The data by Tuss
et al.[46] showed that a DNPH molar ratio in excess of 40 was required for
quantitative derivatization. An interesting observation was that when HCHO-
DNP hydrazone was added to a DNPH solution, the added hydrazone dissociated
to variable extent, forming HCHO and DNPH, if the DNPH was present in less
than 40-*m* ratio. However, the added hydrazone was fully recovered at DNPH
molar ratios >40.

In contrast to liquid-phase derivatization, the mechanism of carbonyl trapping
with DNPH-coated solid sorbent is not well understood. The derivatization can

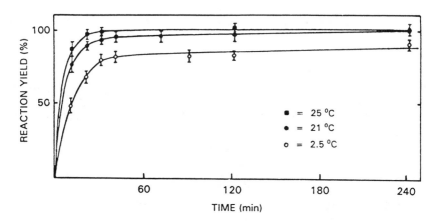

FIGURE 5.   Reaction yield of DNPH derivation of formaldehyde as function of
temperature and time (corrected for incomplete extraction recovery).
(From Tuss, H., V. Neitzert, W. Seiler, and R. Neeb. *Fresenius Z.
Anal. Chem.* 312: 613–617 (1982). With permission.)

take place in a liquid-phase film or as a gas-solid phase reaction. In analogy to
liquid-phase derivatization, probably both derivatization and dissolution are involved in initial trapping, followed by time-dependent derivatization of dissolved
carbonyls. Past studies, which documented collection efficiency using two cartridges in series, assumed complete derivatization immediately. Because of the
high degree of DNPH enrichment on cartridges, the derivatization may proceed
faster on solid sorbents than in liquid medium. Furthermore, the reduced water
activity on cartridges may facilitate equilibrium toward hydrazone formation.

C. Miscellaneous Sampling Techniques

1. Cryogenic Collection

This sampling is based on the principle that soluble species (which include
carbonyls) are collected along with condensed or solidified water vapor and CO_2
which are present in the sample air. Since the collection solution is derived
entirely from the small amount of condensable water present in air, cryogenic
collection results in very high air: water ratios. This approach was used by Neitzert
and Seiler[49] and Tuss et al.[46] to sample clean air for formaldehyde determination
by DNPH derivatization. For preconcentration, air was passed (1 Lpm) through a
glass trap (200 mL) cooled with liquid nitrogen, thereby separating HCHO from
atmospheric nitrogen and oxygen and fixing it in the ice and CO_2 matrix. After
sampling, the cooling trap was warmed up and the DNPH solution added to the
sample at a temperature of 5°C for the derivatization of formaldehyde. The
derivative (DNP hydrazone) was then extracted with carbon tetrachloride and the
extract used for analysis by high-performance liquid chromatography (HPLC)
or GC.

Tests performed by using two traps in series indicated that the collection efficiency in the first trap was >95% at a sampling rate of 1 Lpm. Excellent agreement was observed between this sampling technique and the impinger technique suggesting no production or destruction of formaldehyde in the cold trap. No loss of HCHO was observed during the warm-up step because of the ease with which HCHO dissolves in liquid water. However, this step may lead to losses for higher molecular weight carbonyls. The collection devices are not simple to construct and, operate, and especially, great care must be taken to keep all collection surfaces clean to promote uniform wetting.

2. High Volume, Rotating Cylinder Sampling

The high volume, high efficiency rotating flask sampler was used by Lowe et al.[47,50] as the device to strip formaldehyde from air at high flow rates (ca. 40 Lpm), into DNPH solution acidified with sulfuric acid. The HCHO-DNP hydrazone was determined by HPLC. The sampler was a pyrex cylinder (ca. 9-cm i.d. x 24-cm length) packed with Raschig rings (1 L of 4 x 4 mm) and held in place by two glass sieve plates. Tight packing of the rings is required to avoid channeling effects that would affect sampling efficiency. About 40–100-mL DNPH solution (0.3 mM) was added. During sampling, the cylinder was rotated at an optimum speed of ca. 30 rpm, ensuring that all Raschig rings were wet with the DNPH solution, thereby facilitating efficient scrubbing of carbonyls. The collection efficiency was found to be >80% with 40-mL DNPH solution and at an air flow rate of 40-Lpm. About 1000-L air were processed for each sample. The method was found to be useful for formaldehyde determination at low mixing ratios of the order of 0.1 ppbv.

3. Nebulization/Reflux Concentration

In common with impinger sampling, nebulization/reflux concentration is based on the extraction of target molecules present in air into a liquid scrubber, but presents a marked improvement in extraction efficiency. The scrubbing solution is dispersed into fine drops (a mist), generating large interfacial surface area, which promotes extremely effective mass transfer between the gas and liquid phases.[51,52] The collection process is described as consisting of sorption onto the droplet surface, rather than dissolution. A method based on this technique has been used by Cofer and Edahl[48] in conjunction with DNPH derivatization, for collection and determination of formaldehyde in air.

In the nebulizer method during sampling, air is drawn through a commercially available (DeVilbiss 40) glass-nebulizing nozzle at ca. 7.5 Lpm, aspirating the DNPH solution from the reservoir into the airstream, where the solution is atomized by impaction into small droplets, forming an air/droplet mist (Figure 6). The DNPH solution (4–6 mL total) was aspirated at a rate of ca. 2 mL/min. The upward drawn air/droplet mist impinges on a Teflon® filter (Zeflour, 1-µm diameter) which traps the solution droplets while allowing the scrubbed sample air to pass out of the collector. The trapped solution droplets containing the

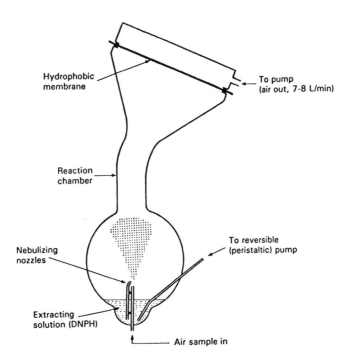

FIGURE 6. Schematic diagram of the nebulization/reflux concentrator. (From Cofer, W. R., III, and R. A. Edahl, Jr. *Atmos. Environ.* 20(5): 979–984 (1986).

scrubbed compounds coalesce into larger droplets, which subsequently roll back down the collector into the reservoir to be recycled. After 15–20-min scrubbing runs, followed by a 1-min rinse with 3- to 4-mL fresh DNPH solution using ultrapure nitrogen, the solutions were withdrawn, mixed, and prepared for HPLC analysis.

The nebulization/reflux concentration technique presents a marked advantage over other liquid scrubbing methods because of the highly efficient collection mechanism. Since maximum flow (V_g) through the minimum volume (V_x) — achievable without causing loss of trapped analytes — translates into the most concentrated solution, the extraction efficiency can be represented as V_g/V_x. For polar gases, this ratio is ca. 300 for impingers but with nebulizer collectors a value of >3000 can be achieved.[52] A problem encountered in these devices is the evaporation of the collecting solution when sampling low humidity air, because of the high gas:liquid ratio. This problem can be minimized by placing the collector in an ice bath. Further studies are required in order to use the nebulizer collector for sampling and analysis of carbonyls other than formaldehyde. Furthermore, modifying and adapting the nebulizer/collector design for high-repetition analysis is highly desirable.

IV. REAGENT BLANKS

In DNPH-based methods, the limits of detection (LOD) for different carbonyl compounds are limited either by the analytical detection limit (i.e., the lowest quantifiable limit) or by the blank level. For the most common carbonyl compounds (namely, formaldehyde, acetaldehyde, and acetone), the blank levels determine the detection limit. Therefore, reducing the blank to the lowest possible level is necessary to achieve the lowest detection limit, especially for clean, marine air applications. Several different sources including the reagent, water, chemicals, solvents, and apparatus used for sampling and subsequent sample preparatory steps can contribute to the carbonyl concentrations appearing in the blank. Ambient air contact with prepared reagents, gradual leaching from plastics as well as formation by unidentified mechanisms during storage increase blank levels. It is difficult to eliminate carbonyl blanks completely, but they can be minimized to allow parts per thousand level detection. Important ways to achieve lowest blanks are: (1) to use highest purity DNPH (usually recrystallized twice in acetonitrile), (2) to purify the reagent solution thoroughly, (3) to avoid air and light contact with prepared reagent, and (4) to use highest purity solvents (sometimes distilled with DNPH).

Several solvents including hexane, chloroform, and carbon tetrachloride have been used to purify aqueous DNPH reagent by extracting hydrazones. The relative polarity of the solvent is an important factor that determines extraction efficiency. Hexane is non-polar and, therefore, less efficient than chloroform and carbon tetrachloride in extracting polar hydrazones. Chloroform being more polar than carbon tetrachloride removes appreciable amounts of the reagent (DNPH) along with hydrazones.[46] Thus, chloroform is less suitable than carbon tetrachloride. Tuss et al.[46] studied the extraction efficiency of hydrazones with carbon tetrachloride, and their results (Table 2) suggest that quantitative extraction can be achieved by extracting three times using a 10:1 ratio of DNPH solution to carbon tetrachloride. Once purified, the aqueous DNPH solution can be maintained for at least 2 weeks in glass containers, if air contact is avoided. This can be achieved by purging the headspace with highest purity nitrogen, which will also allow dispensing the reagent.

Recently, Zhou and Mopper[27] used CCl_4 to purify the DNPH reagent in their C_{18} cartridge-based technique. Three successive extractions of the reagent (500 mL) with 5- to 10-mL CCl_4 resulted in lower blanks of ca. 0.3 nmol for formaldehyde and acetone, and undetectable for other carbonyls per cartridge with ca. 0.9-mg loaded DNPH. These blanks corresponded to 0.07 ppbv for formaldehyde and acetone and less than 0.02 ppbv for other carbonyl compounds for a 100-L sample. This was a significant improvement over previous studies.

Although extremely low blanks can be achieved in the home laboratory with freshly prepared cartridges, it is much more difficult to guarantee that low blanks are maintained during shipment and storage of these cartridges at field sites. Thus, assignment of proper blank values is critically important in field sampling studies using cartridges. It was observed that when DNPH-coated C_{18} cartridges were

Table 2. Extraction Efficiency[a]

CCl$_4$ (mL)	HCHO-DNPH (ng)		Recovery after first extract (%)	Recovery after three extracts (%)
	Added	Found		
50	400	363		
		375	94	95
		384		
5	400	303		
		292	76	92
		318		
1	400	173		
		217	34	56
		136		

Source: Tuss et al.[46]

[a] A 50-mL DNPH solution containing 400-ng added HCHO-DNPH was used for extraction.

stored at ambient temperature, blank values for formaldehyde and acetone increased with time at ca. 0.5 nmol/day, probably due to leaching from cartridges.[27] For this reason, cartridges were prepared in the field within 2 hr prior to use by Zhou and Mopper.[27] For similar reasons, the cartridge should also be extracted immediately after sampling, rather than storing the hydrazones on the cartridge itself. The hydrazones extracted with acetonitrile appear to maintain the sample integrity for a considerable period of time (at least up to 2 weeks).[27] The blank problems associated with reagent-coated C$_{18}$ cartridges on storage have been well studied, while the relative behavior of different types of cartridges such as silica gel and Florisil are not well understood.

Beasley et al.[40] recommended that Bakelite bottle caps should be avoided in any utensils used in the determination of formaldehyde. Bakelite is a polymer prepared from formaldehyde and phenol and may contain enough free formaldehyde to cause a low-level background. New glass bottles require conditioning with DNPH prior to use because of adsorption loss on new glass surfaces. Polyethylene bottles do not exhibit this negative interaction, but appear to result in an increase of the formaldehyde signal with time. Lowe et al.[47] explained this HCHO contamination as originating from the polyethylene, but another possible source could be diffusion through the plastic.

V. INTERFERENCES

In air samples, potential interfering compounds for the determination of carbonyl compounds include ozone, nitrogen dioxide, and sulfur dioxide. Ordinarily, NO$_2$ does not interfere with determination of carbonyls in ambient air. For example, Lipari and Swarin[41] studied the effect of nitrogen dioxide on HCHO sampling using Florisil cartridges coated with DNPH and found no interference at concentrations as high as 550 ppbv and NO_2/HCHO ratios of 7:1.

A. SO$_2$ Interference

The interference by sulfur dioxide is due to the formation of carbonyl-bisulfite addition compound (hydroxyalkane sulfonic acid) which can reduce the recovery of the carbonyl compound.[53] It appears that SO_2 has no effect on sampling gas-phase HCHO with DNPH containing impingers or cartridges. For example, at gas-phase concentrations of 1000-ppbv SO_2 and 92-ppbv HCHO, no interference was observed for HCHO collection on DNPH-coated Florisil cartridges.[41] However, SO_2 could potentially affect the DNPH derivatization in atmospheric water depending on the physicochemical conditions influencing the formation-dissociation of hydroxymethane sulfonic acid. It was shown that the yield of HCHO-DNP hydrazone was reduced when derivatization reaction was carried out by addition of HCHO to a DNPH solution in the presence of sulfite and at near neutral pH. However, under acidic conditions (pH less than 3), the derivatization of HCHO was complete despite added sulfite corresponding to a SO_2 mixing ratio in air of 90 ppbv, suggesting that DNPH reaction with HCHO dominated over the formation of hydroxymethanesulfonate under these conditions.[47] Similarly, in collecting HCHO as hydroxymethanesulfonate on bisulfite-coated filters, it was found that S(IV) interfered with DNPH derivative formation unless removed by acidification and heating.[39] In contrast, if hydroxymethanesulfonate is already formed, the bound HCHO does not react with DNPH. To determine this bound HCHO, hydroxymethanesulfonate must be dissociated first at high pH (ca. pH 13) prior to DNPH derivatization.[54]

B. Effect of Ozone

Ozone is one of the most abundant reactive gases in air, hence could potentially cause sampling artifacts. The effect of ozone on DNPH-based methods for carbonyls can be threefold: (1) formation of carbonyls as artifacts from reaction with sampling substrates, (2) degradation of DNP hydrazones, and (3) formation of other interfering compounds. The reagent (DNPH) itself reacts with ozone. For example, a DNPH solution rapidly became colorless when high concentrations of ozone in air (0.1%) were passed through, but the reaction products were not identified.[47] In contrast, the reactions of ozone with hydrazine, monomethylhydrazine, and dimethylhydrazine have been studied, and products including hydrogen peroxide and formaldehyde have been shown to form under simulated atmospheric conditions.[55] Formation of formaldehyde from the reaction of ozone with DNPH has not been studied, but cannot be excluded.

In a recent study by Arnts and Tejada,[11] the reaction of ozone with HCHO-DNP hydrazone was identified as a potential problem when DNPH-coated silica cartridges were used for formaldehyde sampling. In this study, synthetic mixtures of humidified air containing formaldehyde (20–140 ppbv) and ozone (0–770 ppbv) were sampled. The loss of HCHO-DNP hydrazone increased markedly with increase in ozone concentration; at 25 ppbv of HCHO and 120 ppbv of O_3, about

Table 3. Formaldehyde Recovery as a Function of Ozone Using the DNPH-Coated Silica Gel Cartridge Technique

Added ozone, ppbv	Ratios of HCHO[a] with/without ozone	
	Average response	SD
0[b]	1.0	0.063
120	0.63	0.084
300	0.39	0.053
500	0.27	0.074
770	0.15	0.077

Source: Arnts and Tejada[11]

[a] Formaldehyde concentrations of 20, 40, and 140 ppb were used in the experiment.

[b] Ozone concentration in the ambient background ranged from 0 to 20 ppbv with no ozone addition.

48% of HCHO was lost (Table 3). Also noticed on silica cartridges were concurrent large losses of DNPH. In contrast to the silica cartridges, impingers charged with DNPH acetonitrile solutions did not show any loss of HCHO-DNP hydrazone, but DNPH was markedly reduced. It was concluded that the silica cartridge exhibited such large reductions in formaldehyde response because the DNPH derivative, which is largely formed at the front of the cartridge and immobilized, was being destroyed by O_3. In the case of impingers, the HCHO-DNP hydrazone is protected by the DNPH, which is always present in excess and well dispersed. However, a matter of concern in the impinger technique is that the products formed from the DNPH-O_3 reaction can interfere with resolution of the formaldehyde peak in HPLC separation. Recently, this problem was addressed by Smith et al.[28] who used a ternary gradient mobile phase (refer to Figure 2) to obtain good chromatographic separation of the formaldehyde peak from interfering artifact peaks (Figure 7). In contrast to the silica cartridges, C_{18} cartridges (Sep-Pak brand by Waters Associates, Milford, MA) exhibited no loss of the HCHO derivative up to 120 ppbv of O_3. In this case, it was reasoned that C_{18} substrate itself reacted with O_3, thereby preventing attack on formaldehyde-DNP hydrazone.

The mechanisms of ozone initiated reactions in the above cases are not clearly understood. Atkinson and Carter[56] suggested that a chain of free-radical reactions can be initiated when ozone reacts with hydrazines, either by addition to a nitrogen, or abstraction of a hydrogen from a weak N-H bond. Arnts and Tejada[11] pointed out that under the acidic conditions of the DNPH reaction, O_3 addition to the protonated nitrogen is restricted, and hydrogen abstraction could be the preferred route. The following pathway has been proposed by Atkinson and Carter[56] for hydrogen abstraction:

$$RHNNH_2 + O_3 \rightarrow RNNH_2 \cdot \text{ (or } RHNNH \cdot) + O_2 + OH \cdot$$
$$RNNH_2 \cdot \text{ (or } RHNNH \cdot) + O_2 \rightarrow RN = NH + HO_2 \cdot$$
$$RN = NH + O_3 \text{ (or } OH \cdot) \rightarrow RN = N \cdot + OH \cdot + O_2 \text{ } (+H_2O)$$
$$RN = N \cdot \rightarrow R \cdot + N_2$$

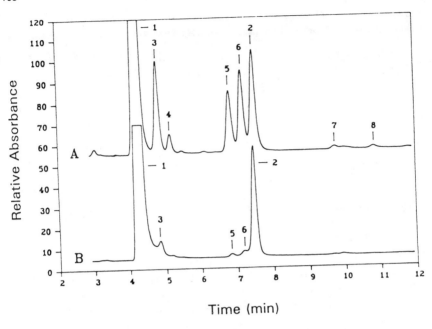

FIGURE 7.   Ternary gradient separation of HCHO-DNP hydrazone from ozone-
DNPH reaction artifacts at constant HCHO concentration. (A) High-
level ozone (514 ppbv); (B) low-level ozone (16 ppbv). Peaks: 1,
DNPH reagent; 2, formaldehyde; 3–8, ozone-DNPH reaction arti-
facts. (From Smith, D. F., T. E. Kleindienst, and E. E. Hudgens. *J.
Chromatogr.* 483: 431–436 (1989). With permission.)

Arnts and Tejada suggested that when DNPH-coated C_{18} substrate is used for
sampling, the radicals generated by O_3 attack can be scavenged by the C_{18}, thus
limiting further attack on DNPH or the hydrazones. In recent studies by
Vairavamurthy[57] using O-(pentafluorobenzyl)hydroxylamine (PFBOA) for carbonyl derivatization, it has been observed that when O_3 was passed through
PFBOA-coated C_{18} cartridges several carbonyl compounds were generated (Figure 8). An increased production was observed with inert glass columns packed
with C_{18} material as against polypropylene cartridges containing C_{18}, suggesting
that O_3 was partially destroyed by the polypropylene cartridge before it impinged
on the C_{18} material.

From this discussion, it is clear that ozone is a serious interference in methods
using both C_{18} and silica gel cartridges coated with DNPH for sampling carbonyls
in the ambient air. The ozone effect is exhibited in different ways in these
cartridges; ozone causes the production of artifact carbonyls in reagent-coated C_{18}
cartridges, while the predominant effect in silica gel cartridges is destruction of
hydrazones. Preliminary studies using PFBOA indicate that the effect of ozone on
a reagent-coated Florisil cartridge is similar to that of a C_{18} cartridge;[57] however,

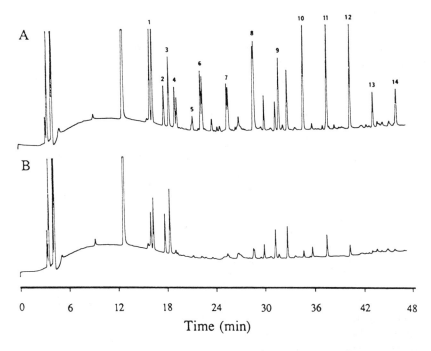

FIGURE 8. Ozone effect on C_{18} cartridges loaded with PFBOA. (A). Carbonyls generated by passing 10-L zero air containing 100 ppbv through a C_{18} cartridge (900 mg, Burdick & Jackson brand) loaded with 3 mL of 5 mM PFBOA in pH 3.2 buffer. 1, acetaldehyde; 2, internal standard (2-bromochlorobenzene); 3, acetone; 4, propanal; 5, 2-butanone; 6, n-butanal; 7, n-pentanal; 8, n-hexanal; 9, n-heptanal; 10, n-octanal; 11, n-nonanal; 12, n-decanal; 13, n-undecanal; 14, n-dodecanal. (B) Ozone removed with a CuO (5 g) cartridge attached in front of the C_{18} cartridge.

the artifact carbonyls are formed at much lower concentrations than those obtained with C_{18} cartridge. It is thus clear that prior removal of ozone from the airstream is critical when reagent-coated solid-phase cartridges are used for sampling of carbonyls in ambient air. Arnts and Tejada[11] reported that they have obtained encouraging results in a preliminary study using potassium iodide-coated copper tubing inlet to remove O_3 prior to collection with a DNPH-coated silica gel cartridge. Ozone can also be removed by using a CuO cartridge in front of the sampling cartridge. Studies with gas-phase standards indicate that concentrations of carbonyl compounds are not affected by using the CuO cartridge.[57] Gas-phase titration of O_3 with nitric oxide (NO), as used by Tanner et al.[58] in the determination of hydrogen peroxide in the ambient atmosphere, is also a potential technique to overcome ozone interference.

VI. CONCLUSION

DNPH labeling followed by liquid chromatographic separation and UV detection is currently the most popular chromatographic technique used for the determination of atmospheric carbonyls. Among the variety of sampling techniques that have been used with this derivatization, carbonyl collection with DNPH-coated solid-phase cartridges such as C_{18} has been preferred in recent studies because of convenience and other logistic reasons. Although the DNPH method has been in widespread use, it is surprising that some important analytical problems (e.g., ozone interference) have not been resolved yet. Because of the increasing demand for time-series measurements in field studies, an automated method for continuous sampling and analysis of carbonyls is very much required. The DNPH method employing a commonly used sampling technique (e.g., cartridge sampling) may not lend itself for this purpose due to the lengthy collection times required to achieve subparts per billion volume detection limits. Because of this inherent problem, the DNPH-LC method is also unsuitable for studies in which short time resolution is required or for sampling from aircraft, unless a suitably designed high volume sampler (e.g., the nebulization/reflux concentrator) is used. It appears that new analytical approaches are required in the development of a suitable field method with real-time capabilities for carbonyl measurements in the ambient air. However, when problems associated with integrated sampling are resolved, particularly ozone interference, the DNPH-LC method employing cartridge sampling will be a useful batch method for the measurement of atmospheric carbonyl compounds.

ACKNOWLEDGMENTS

This work was conducted under Contract No. DE-AC02-76CH00016 with the U.S. Department of Energy under the Atmospheric Chemistry Program within the Office of Health and Environmental Research and supported in part by an appointment of Appathurai Vairavamurthy to the U.S. Department of Energy Alexander Hollaender Postdoctoral Fellowship Program administered by Oak Ridge Associated Universities.

REFERENCES

1. Carlier, P., H. Hannachi, and G. Mouvier. "The Chemistry of Carbonyl Compounds in the Atmosphere," *Atmos. Environ.* 20(11):2079–2099 (1986).
2. Seinfeld, J.H. *Atmospheric Chemistry and Physics of Air Pollution*, 2nd ed. (New York: John Wiley & Sons, Inc., 1986), pp. 111–193.
3. Finlayson-Pitts, B.J., and J.N. Pitts, Jr. *Atmospheric Chemistry* (New York: John Wiley & Sons, Inc., 1986), p. 1098.

4. Warneck, P. *Chemistry of the Natural Atmosphere* (New York: Academic Press Inc., 1988), p.753.

5. Levaggi, D.A., and M. Feldstein. "The Collection and Analysis of Low Molecular Weight Carbonyl Compounds from Source Effluents," *J. Air Pollut. Control Assoc.* 19:43–45 (1969).

6. Seizinger, D.E., and B. Dimitrides. "Oxygenates in Exhaust from Simple Hydrocarbon Fuels," *J. Air Pollut. Control Assoc.* 22(1):47–51 (1972).

7. Tanner, R.L., A.H. Miguel, J.B. de Andrade, J.S. Gaffney, and G.E. Streit. "Atmospheric Chemistry of Aldehydes: Enhanced Peroxyacetyl Nitrate Formation from Ethanol-Fueled Vehicular Emissions," *Environ. Sci. Technol.* 22(9):1026–1034 (1988).

8. Williams, R.L., F. Lipari, and R.A. Potter. "Formaldehyde, Methanol and Hydrocarbon Emissions from Methanol-fueled Cars," *J. Air Waste Manage. Assoc.* 40:747–756 (1990).

9. Lowe, D.C., and U. Schmidt. "Formaldehyde (HCHO) Measurements in the Nonurban Atmosphere," *J. Geophys. Res.* 88(C15):10844–10858 (1983).

10. Grosjean, D. "Formaldehyde and Other Carbonyls in Los Angeles Ambient Air," *Environ Sci. Technol.* 16(5):254–262 (1982).

11. Arnts, R.R., and S.B. Tejada. "2,4-Dinitrophenylhydrazine-Coated Silica Gel Cartridge Method for Determination of Formaldehyde in Air: Identification of an Ozone Interference," *Environ. Sci. Technol.* 23(11):1428–1430 (1989).

12. Intersociety Committee. "Determination of C_1-C_5 Aldehydes in Ambient Air and Source Emissions as 2,4-Dinitrophenylhydrazones by HPLC," in *Methods of Air Sampling and Analysis*, 3rd ed., J.P. Lodge, Jr., Ed. (Chelsea, MI: Lewis Publishers, Inc., 1989), pp. 293–295.

13. Kalio, H., R.R. Linko, and J. Kaitaranta. "Gas-Liquid Chromatographic Analysis of 2,4-Dinitrophenylhydrazones of Carbonyl Compounds," *J. Chromatogr.* 65:355–360 (1972).

14. Hoshika, Y., and Y. Takata. "Gas Chromatographic Separation of Carbonyl Compounds as Their 2,4-Dinitrophenylhydrazones Using Glass Capillary Columns," *J. Chromatogr.* 120:379–389 (1976).

15. Johnson, L., B. Josefsson, and P. Marstorp. "Determination of Carbonyl Compounds in Automobile Exhausts and Atmospheric Samples," *Int. J. Environ. Anal. Chem.* 9:7–26 (1981).

16. Kuwata, K., M. Uebori, and Y. Yamasaki. "Determination of Aliphatic and Aromatic Aldehydes in Polluted Air as Their 2,4-Dinitrophenylhydrazones by High Performance Liquid Chromatography," *J. Chromatogr. Sci.* 17:264–268 (1979).

17. Kuntz, R., W. Laumeman, G. Namie, and L.A. Hull. "Rapid Determination of Aldehydes in Air Analysis," *Anal. Lett.* 13:1409–1415 (1980).

18. Fung, K., and D. Grosjean. "Determination of Nanogram Amounts of Carbonyls as 2,4-Dinitrophenylhydrazones by High-Performance Liquid Chromatography," *Anal. Chem.* 53(2):168–171 (1981).

19. Maskarinec, M.P., D.L. Manning, and P. Oldham. "Determination of Vapor-Phase Carbonyls by High-Pressure Liquid Chromatography," *J. Liq. Chromatogr.* 4(1):31–39 (1981).

20. Kuwata, K., M. Uebori, H. Yamasaki, and Y. Kuge. "Determination of Aliphatic Aldehydes in Air by Liquid Chromatography," *Anal. Chem.* 55(12):2013–2016 (1983).

21. Van Langenhove, H.R., M. Van Acker, and Niceas M. Schamp. "Quantitative Determination of Carbonyl Compounds in Rendering Emissions by Reversed Phase High Performance Liquid Chromatography of the 2,4-Dinitrophenylhydrazones," *Analyst* 108:329–334 (1983).

22. Tanner, R.L., and Z. Meng. "Seasonal Variations in Ambient Atmospheric levels of Formaldehyde and Acetaldehyde," *Environ. Sci. Technol.* 18(9):723–726 (1984).

23. Götze, H.-J., and S. Harke. "Determination of Aldehydes and Ketones in Natural Gas Combustion in the PPBV Range by High-Performance Liquid Chromatography," *Fresenius Z. Anal. Chem.* 335:286–288 (1989).

24. Tejada, S.B. "Evaluation of Silica Gel Cartridges Coated In Situ with Acidified 2,4-Dinitrophenylhydrazine for Sampling Aldehydes and Ketones in Air," *Int. J. Environ. Anal. Chem.* 26:167–185 (1986).

25. Druzik, C.M., D. Grosjean, A. Van Neste, and S.S. Parmar. "Sampling of Atmospheric Carbonyls with Small DNPH-Coated C18 Cartridges and Liquid Chromatography Analysis with Diode Array Detection," *Int. J. Environ. Anal. Chem.* 38:495–512 (1990).

26. Fung, K., and B. Wright. "Measurement of Formaldehyde and Acetaldehyde Using 2,4-Dinitrophenylhydrazine-Impregnated Cartridges During the Carbonaceous Species Methods Comparison Study," *Aerosol Sci. Technol.* 12:44–48 (1990).

27. Zhou, X., and K. Mopper. "Measurement of Sub-Parts-per-Billion Levels of Carbonyl Compounds in Marine Air by a Simple Cartridge Trapping Procedure Followed by Liquid Chromatography," *Environ. Sci. Technol.* 24:1482–1485 (1990).

28. Smith, D.F., T. E. Kleindienst, and E.E. Hudgens. "Improved High-Performance Liquid Chromatographic Method for Artifact-Free Measurements of Aldehydes in the Presence of Ozone Using 2,4-Dinitrophenylhydrazine," *J. Chromatogr.* 483:431–436 (1989).

29. Puputti, E., and P. Lehtonen. "High-Performance Liquid Chromatographic Separation and Diode-Array Spectroscopic Identification of Dinitrophenylhydrazone Derivatives of Carbonyl Compounds from Whiskies," *J. Chromatogr.* 353:163–168 (1986).

30. Grömping, A., and K. Cammann. "Some New Aspects of a HPLC-Method for the Determination of Traces of Formaldehyde in Air," *Fresenius Z. Anal. Chem.* 335:796–801 (1989).

31. Kieber, R., and K. Mopper. "Determination of Picomolar Concentrations of Carbonyl Compounds in Natural Waters, Including Seawater, by Liquid Chromatography," *Environ. Sci. Technol.* 24(10):1477–1481 (1990).

32. Grosjean, D. "Chemical Ionization Mass Spectra of 2,4-Dinitrophenylhydrazones of Carbonyl and Hydroxycarbonyl Atmospheric Pollutants," *Anal. Chem.* 55:2436–2439 (1983).

33. Olson, K.L., and S.J. Swarin. "Determination of Aldehydes and Ketones by Derivatization and Liquid Chromatography-Mass Spectrometry," *J. Chromatogr.* 333:337–347 (1985).

34. Lipari, F., and S.J. Swarin. "Determination of Formaldehyde and Other Aldehydes in Automobile Exhaust with an Improved 2,4-Dinitrophenylhydrazine Method," *J. Chromatogr.* 247:297–306 (1982).

35. De Bortoli, M., H. Knoppel, E. Pecchio, A. Peil, L. Rogora, H. Schauenburg, H. Schlitt, and H. Vissers. "Concentations of Selected Organic Pollutants in Indoor and Outdoor Air in Northern Italy," *Environ. Int.* 12:343–350 (1986).

36. Grosjean, D., and K. Fung. "Collection Efficiencies of Cartridges and Microimpingers for Sampling of Aldehydes in Air as 2,4-Dinitrophenylhydrazones," *Anal. Chem.* 54(7):1221–1224 (1982).

37. Levin, J., K. Andersson, R. Lindahl, and Carl-Axel Nilsson. "Determination of Sub-Part-per-Million Levels of Formaldehyde in Air Using Active or Passive Sampling on 2,4-Dinitrophenylhydrazine-Coated Glass Fiber Filters and High-Performance Liquid Chromatography," *Anal. Chem.* 57(6):1032–1035 (1985).

38. Levin, J., R. Lindahl, and K. Andersson. "A Passive Sampler for Formaldehyde in Air Using 2,4-Dinitrophenylhydrazine-Coated Glass Fiber Filters," *Environ. Sci. Technol.* 20(12):1273–1276 (1986).

39. de Andrade, J.B., and R.L. Tanner. "Determination of Formaldehyde by HPLC as the DNPH Derivative Following High Volume Air Sampling onto Bisulfite-Coated Cellulose Filters," *Atmos. Environ.* 26(A):819–825 (1991).

40. Beasley, R.K., C.E. Hoffmann, M.L. Rueppel, and J.W. Worley. "Sampling of Formaldehyde in Air with Coated Solid Sorbent and Determination by High Performance Liquid Chromatography," *Anal. Chem.* 52:1110–1114 (1980).

41. Lipari, F., and S.J. Swarin. "2,4-Dinitrophenylhydrazine-Coated Florisil Sampling Cartridges for the Determination of Formaldehyde in Air," *Environ. Sci. Technol.* 19(1):70–74 (1985).

42. Andersson, G., K. Andersson, C. Nilsson, and J. Levin. "Chemosorption of Form-aldehyde on Amberlite XAD-2 Coated with 2,4-Dinitrophenylhydrazone," *Chemosphere* 10:823–827 (1979).

43. Andersson, G., C. Hallgren, J. Levin, and C. Nilsson. "Solid Chemosorbent for Sampling Sub-ppmv Levels of Acrolein and Glutaraldehyde in Air," *Chemosphere* 10:275–280 (1981).

44. Ciccioli, P., R. Draisci, A. Cecinato, and A. Liberti. "Sampling of Aldehydes and Carbonyl Compounds in Air and their Determination by Liquid Chromatographic Techniques," in *Physico-Chemical Behaviour of Atmospheric Pollutants,* Proceedings of the Fourth European Symposium, 1986, G. Angeletti and G. Restelli, Eds. (Dordrecht: Reidel Publishing Co., 1987), 133–141.

45. Bicking, M.K.L., and W. Marcus Cooke. "Effect of pH on the Reaction of 2,4-Dinitrophenylhydrazine with Formaldehyde and Acetaldehyde," *J. Chromatogr.* 455:310–315 (1988).

46. Tuss, H., V. Neitzert, W. Seiler, and R. Neeb. "Method for Determination of Formaldehyde in Air in the pptv-Range by HPLC after Extraction as 2,4-DinitrophenylHydrazone," *Fresenius Z. Anal. Chem.* 312:613–617 (1982).

47. Lowe, D.C., U. Schmidt, and D.H. Ehhalt, C.G.B. Frischkorn, H.W. Nürnberg. "Determination of Formaldehyde in Clean Air," *Environ. Sci. Technol.* 15(7):819–823 (1981).

48. Cofer, W.R., III, and R.A. Edahl, Jr. "A New Technique for Collection, Concentration and Determination of Gaseous Tropospheric Formaldehyde," *Atmos. Environ.* 20(5):979–984 (1986).

49. Neitzert, V., and W. Seiler. "Measurement of Formaldehyde in Clean Air," *Geophys. Res. Lett.* 8(1):79–82 (1981).

50. Lowe, D.C., U. Schmidt, and D.H. Ehhalt. "A New Technique for Measuring Tropospheric Formaldehyde [CH_2O]," *Geophys. Res. Lett.* 7(10):825–828 (1980).

51. Vecera, Z., and J. Janak. "Continuous Aerodispersive Enrichment Unit for Trace Determination of Pollutants in Air," *Anal. Chem.* 59:1494–1498 (1987).

52. Cofer W.R., III, V.G. Collins, and R.W. Talbot. "Improved Aqueous Scrubber for Collection of Soluble Atmospheric Trace Gases," *Environ. Sci. Technol.* 19:557–560 (1985).

53. Kok, G.L., S.N. Gitlin, and A.L. Lazrus. "Kinetics of the Formation and Decomposition of Hydroxymethanesulfonate," *J. Geophys. Res.* 91(D2):2801–2804 (1986).

54. Ang, C.C., F. Lipari, and S.J. Swarin. "Determination of Hydroxymethanesulfonate in Wet Deposition Samples," *Environ. Sci. Technol.* 21(1):102–105 (1987).

55. Tuazon, E.C., W.P.L. Carter, A.M. Winer, and J.N. Pitts, Jr. "Reactions of Hydrazines with Ozone Under Simulated Atmospheric Conditions," *Environ. Sci. Technol.* 15(7):823–828 (1981).

56. Atkinson, R. and W.P.L. Carter. "Kinetics and Mechanisms of the Gas-Phase Reactions of Ozone with Organic Compounds Under Atmospheric Conditions," *Chem. Rev.* 84:437–470 (1984).

57. Vairavamurthy, A. Unpublished results (1991).

58. Tanner, R.L., G.Y. Markovits, E.M. Ferreri, and T.J. Kelly. "Sampling and Analysis of Gas-Phase Hydrogen Peroxide Following Removal of Ozone by Gas-Phase Reaction with Nitric Oxide," *Anal. Chem.* 58:1857–1865 (1986).

Aerosol Sampling and Analysis Developments

CHAPTER 11

Methods of Analysis for Complex Organic Aerosol Mixtures from Urban Emission Sources of Particulate Carbon

Monica A. Mazurek, Lynn M. Hildemann, Glen R. Cass, Bernd R. T. Simoneit, and Wolfgang F. Rogge

TABLE OF CONTENTS

I. INTRODUCTION

Study of the sources of organic aerosols found in urban atmospheres is of key importance because organic particulate matter comprises typically 30% of the fine aerosol mass (i.e., nominal particle diameter <2.1 μm).[1] These airborne carbonaceous particles contribute to visibility reduction[2,3] and have complex chemical compositions that include carcinogenic and mutagenic organic compounds.[4-10] In the past, a few of the organic aerosol compounds present in urban atmospheres have been traced back to their origin, but the vast proportion of the aerosol material remains to be assigned to its source. This chapter describes the experimental methods that have been developed for: (1) the collection of fine organic aerosols from combustion sources; (2) the construction of chemical mass balances based on the mass of fine organic aerosol emitted from major urban sources of particulate carbon; and (3) the generation of discrete emission source chemical profiles derived from chromatographic characteristics of the organic aerosol components.

II. EXPERIMENTAL APPROACH

A. Source Sampling

A dilution stack sampler was designed and field-evaluated for collection of fine organic aerosols from combustion sources.[11,12] The sampler simulates the cooling and dilution processes that occur in the plume downwind of a combustion source, so that the organic compounds which condense onto preexisting particles under ambient conditions are collected in the sampler as particulate matter. To ensure collection of a representative aerosol sample, the following design goals were established for the sampler: (1) choose dimensions of the sampler to minimize the loss of particles and the condensation of supersaturated vapors onto wall surfaces; (2) simulate atmospheric dispersion processes by ensuring that the emissions are highly diluted and cooled to ambient temperature before sample collection; and (3) include a residence time in the sampler sufficient to allow complete condensation of the supersaturated vapors to occur.

Rigorous checks of sampler performance were conducted for quality control/quality assurance (QA/QC) purposes. To minimize sample contamination and artifact collection, the following strategies were adopted: (1) exclude use of rubber, plastics, greases, and oils as sampler components to avoid offgassing of organic materials into the sampled airstream (use Teflon® for all gaskets and O-ring seals); (2) construct the sampler entirely from stainless steel to facilitate thorough cleaning between field experiments; (3) preclean the dilution air using activated carbon and HEPA filtration; and (4) store filters containing the collected organic aerosol in annealed glass containers. Between field tests, the entire sampling system was cleaned before a new source type was measured. Large system pieces were vapor-degreased with tetrachloroethylene, and subsequently

Table 1. Urban Sources of Fine Carbonaceous Aerosol

Anthropogenic sources
    Oil-fired boiler: no. 2 fuel oil
    Fireplace: natural wood, synthetic log
    Vehicles: catalyst and noncatalyst cars, diesel trucks
    Home appliances: natural gas
    Meat cooking: charbroiling (extra-lean and regular meat), frying
    Road dust
    Brake dust
    Tire dust
    Cigarettes
    Roofing tar pot
Biogenic sources
    Vegetative detritus: dead leaves, green leaves (cultivated and native plant species composites)

heated (70°C for 4 hr) to volatilize possible remaining organic impurities. All open ends were wrapped in clean aluminum foil to prevent recontamination. Small sampler parts were immersed in an ultrasonic cleaner and were cleaned sequentially with high purity (glass-distilled) methanol and hexane for 5 min in each solvent. The small pieces were covered with aluminum foil immediately after cleaning to protect the interior surfaces during storage and transport. Before collection of a sample from a new source type, the clean assembled sampler was evaluated for leaks. A system blank was then collected by passing precleaned dilution air through the system for the same length of time that would be required for collection of an actual source sample. Filtered samples of the dilution air were analyzed for artifacts by the same analytical methods used for the actual source samples and included measurement by high-resolution gas chromatography with flame ionization detection (HRGC-FID) and high-resolution gas chromatography/mass spectrometry (HRGC/MS). Further details of sampler design and sample acquisition procedures are presented elsewhere.[11,12]

Urban emission sources were selected on the basis of previous mass emission inventories of particulate carbon compiled for metropolitan Los Angeles, sufficient to account for close to 80% of the organic aerosol emissions in that air basin.[13,14] The urban emission sources of particulate carbon examined in this study are summarized in Table 1. Details of the quantitative mass emission characteristics corresponding to each of the source types are available.[12]

B. Analysis

Several considerations are important to the overall design of the emission source characterization study. First, quantitative links between the total fine aerosol mass and the carbonaceous subfractions must be developed. This is a critical requirement for relating the mass of potential molecular tracers to the total fine particle mass emission rate of a given source type. Second, the analytical protocol must be applicable to a wide range of aerosol mass loadings and chemical compositions. Here the objective is to generate analytical methodologies which are dynamic with respect to sample size, thus placing fewer constraints on field sampling operations to recognize the fact that source emission rates, and hence sample sizes, often will not be known until after the source test has been conducted. An additional benefit of a broad measurement capability is the ability to assess in a quantitative fashion as many carbonaceous compounds as possible that are associated with a given emission source. A third design consideration is the incorporation of QA/QC procedures which adequately evaluate laboratory protocol, as well as sampling background (including dynamic and static system blanks). Without close scrutiny of laboratory and sampling blanks at the molecular level (i.e., individual organic compound analysis by mass spectrometry), it is not possible to identify definitively the carbonaceous chemical components that are present as fine aerosol emissions. Fourth, the precision of the analytical protocol must be within acceptable limits. As a final design consideration, the analytical method must generate data which are suitable for computer manipulation. This last requirement is intended to facilitate quantitative comparisons of the different chemical attributes that characterize the sampled emission sources of fine particulate carbon.

The analytical objective of this study is to construct a mass balance for each emission source by examining the relationships between total aerosol mass and certain carbonaceous fractions:

Total mass : total carbon (organic + elemental) : elutable organics : molecular tracer

An array of analytical methods is required for these mass determinations. The total fine aerosol mass is collected on preweighed Teflon® filters. The total mass of fine aerosol is determined gravimetrically under conditions of constant temperature and humidity using a microbalance. Total carbon (TC) analysis is performed by a combined pyrolysis/combustion measurement technique.[15] The method provides a quantitative measurement of the mass concentrations of elemental carbon (EC) and organic carbon (OC) present on a sampled quartz microfiber filter and uses laser transmittance to correct for the conversion of OC to EC during the initial pyrolysis step. Quantitation of total solvent-soluble, elutable organics (i.e., lipids having 6 to 40 carbon atoms) is achieved by HRGC-FID analysis which uses both a surrogate standard (i.e., internal recovery standard) and a suite of n-alkane external standards.[16-18] Individual molecular tracers present in the total extracts from the source aerosol filters are identified and quantitated by HRGC/MS analysis.[18-20] In this chapter we summarize the results of the first three mass balance resolution steps listed (i.e., TC, OC, and total elutable organics) for the source samples, the analytical blanks, and the sampling system blanks. An example of the fourth step, molecular characterization of the source samples, is given by Rogge et al.[20] The mass emission characteristics of the investigated sources and their chemical composition are reported and discussed at length elsewhere.[11,12]

Micro-methods have been developed for the quantitative recovery of extractable organic matter in the atmospheric fine aerosol fraction.[16-18] The analytical protocol is designed to monitor losses associated with volatilization, incomplete extraction, or instrumental bias. To provide sufficient organic mass for the HRGC-FID and HRGC/MS analyses (i.e., minimum of 300-μg OC per filter composite for a single-source type), up to 18 separate parallel filters were configured into the source sampling device to produce the samples necessary for the above mass determinations. The organics are extracted from the filters by ultrasonic agitation using successive additions of hexane (two volume additions) and benzene/isopropanol (three volume additions). The serial extracts are filtered and then combined. The total extracts are reduced to volumes of 150–2500 μL. The neutral fraction of the organics (neutral elutable organics) is defined operationally as that fraction which elutes from the bonded phase (DB-1701) of the analytical column and is detected by the FID of the HRGC without further derivatization. An aliquot of the total extract is derivatized by addition of diazomethane.[16,18] This step converts reactive organic acids to the respective methyl ester or methyl ether analogs. Injection of this derivatized fraction onto the HRGC column produces chromatographic data for the acid plus neutral (acid + neutral) fraction (total acid + neutral elutable organics). The mass of the acid fraction (acidic elutable organics) of the solvent-soluble organics is determined by difference. Quantitation of the total extracts is accomplished by computerized HRGC-FID analyses that incorporate the combined application of: (1) area counts relative to a coinjection standard (1-phenyldodecane); (2) relative response factor for the perdeuterated surrogate standard (n-$C_{24}D_{50}$); (3) recovery of the perdeuterated surrogate standard for each source sample extract; and (4) relative response factors for a suite of n-alkane external standards (17 n-alkane homologues from n-$C_{10}H_{22}$ to n-$C_{36}H_{74}$).[16-18]

III. RESULTS

The mass determination per filter for the various emission source samples, along with averages of the laboratory blanks and system blanks, is presented in Table 2. The mass loadings of the source samples on a per filter basis range over three orders of magnitude, from 13 μg to 20 mg. Diverse compositions of the source samples are evident based on the relative proportions of the TC, OC, total organics (i.e., OC mass multiplied by a factor of 1.2 provides a lower mass estimate of the hydrogen, oxygen, nitrogen, etc. atoms that are bound or bonded with operationally-determined atomic organic carbon),[1,21] and total elutable organics measured.

A. Quality Control/Quality Assurance

Source sampling system blanks and laboratory blank filter composites were generated routinely throughout the source testing and analytical workup

Table 2. Mass Determinations of Carbonaceous Materials from Urban Sources of Fine Particulate Carbon

Sample	Combined filters[a]	Fine mass μg/filter	Total carbon μg/filter	Organic carbon μg/filter	Total organics[b] μg/filter	Total elutable organics[c] μg/filter	Ratio of elutable organics to total organics (%)
Blanks							
Blank filters (avg of ≥7 analyses)	4–16	<3[d]	<2[e]	<2[e]	<2[e]	8.9 ± 4.5	—
Source sampler system blanks (avg of ≥7 analyses)	15	<3	10.3 ± 2.0	9.5 ± 1.6	11.4 ± 1.9	7.0 ± 1.7	49 ± 17
Anthropogenic[f]							
Boiler, no. 2 fuel oil (Experiment 2)	18	1,230	323	44	52	26	50
Boiler, no. 2 fuel oil (Experiment 5, residence chamber)	9	2,330	408	56	67	29	43
Boiler, no. 2 fuel oil (Experiment 5, tunnel)	9	2,340	388	59	70	38	54
Automobiles, catalyst-equipped	45	30	23	17	21	25	119
Automobiles, noncatalyst	45	83	60	52	63	87	138
Trucks, heavy-duty diesel	21	197	147	63	76	66	87
Roofing tar pot (Experiment 1)	2	19,700	11,900	11,900	14,300	24,500	171[g]
Roofing tar pot (Experiment 2)	2	10,600	6,350	6,350	7,650	11,800	154[g]
Tire wear (Experiment 1)	4	2,290	1,180	825	990	419	42
Tire wear (Experiment 2)	1	9,630	4,940	3,470	4,160	1,970	47
Fireplace, oak	15	1,750	873	827	987	439	44
Fireplace, pine	15	2,850	1,460	1,370	1,650	841	51
Fireplace, synthetic log	15	947	659	543	651	497	76
Cigarette smoke (Experiment 1)	3.75	2,500	1,500	1,490	1,790	1,420	79
Cigarette smoke (Experiment 2)	3.75	2,470	1,490	1,470	1,770	1,520	86

Burgers, extra-lean beef, charbroiled	15	68	41	41	49	15	31
Burgers, regular beef, charbroiled	15	381	218	217	261	65	25
Burgers, mixed beef, fried	30	13	7	7	9	3	33
Brake wear	4	8,080	1,080	—[h]	—[h]	14	—[h]
Natural gas home appliances	39	17	15	14	17	14	82
Paved road dust (Experiment 1)	4	3,350	488	453	543	144	27
Paved road dust (Experiment 2)	4	3,880	565	523	628	111	18
Biogenic[f]						
Vegetative detritus, green leaves (Experiment 1)	4	595	180	176	211	36	17
Vegetative detritus, green leaves (Experiment 2)	4	360	109	106	128	21	16
Vegetative detritus, dead leaves (Experiment 1)	4	1,038	340	328	393	53	13
Vegetative detritus, dead leaves (Experiment 2)	4	378	140	136	163	20	12

a Value represents the number of 47-mm diam quartz fiber filters that were grouped for the elutable organics mass determinations. Fine mass, total carbon, organic carbon, and organics values were determined for samples collected on single 47-mm diam Teflon® or quartz microfiber filters.[12]

b Calculated as 1.2 times the mass of organic carbon present.[1,21]

c Quantified as the sum of all area counts eluting between 16 and 80 min. Values corrected for contributions of artifacts and surrogate standard additions.

d Below detection limit of gravimetric measurement method.

e Below detection limit of pyrolysis/combustion method.

f Mass determinations for anthropogenic and biogenic sources have been blank-corrected on a per sample basis for the elutable organics and on a per filter basis for the other measurements.

g Underestimation of organic carbon by pyrolysis/combustion method.

h Quantity uncertain due to interference with organic carbon measurement.

procedures adopted for this study. All blanks were evaluated by analytical procedures identical to those used for the collected source filter samples. Mass loadings of the principal carbon-containing fractions for the laboratory blank filter composites (each containing 4, 8, or 16 of the 47-mm quartz microfiber filters) and the dynamic source sampler system blanks are listed in Table 2. Both types of blanks contain total masses which are below the detection limit of the gravimetric measurement method (i.e., <5 μg/filter total weighable mass). Total carbon, organic carbon, and total organics masses are <2 μg/filter for the laboratory blank analyses and 10–11 μg/filter for the dynamic system blanks, while elutable organics mass loadings for these blanks average 7 and 9 μg/filter, respectively. The three independent measurements used to quantitate mass confirm the upper bound mass estimates for the blank samples as 9 μg/filter for the laboratory analytical blanks and 11 μg/filter for the dynamic sampling blanks. The standard deviation of the blanks is also given in Table 2, and is seen to be 2 μg/filter or less for the source system sampler blanks, and 4.5 μg/filter for the total elutable organics in the laboratory blanks. These data on the variability of the blanks can be used to assess the detection limits and quantitation limits of the procedures used according to the method of Keith et al.[22]

The replicate analyses of blank samples via HRGC-FID also enabled identification of and correction for contaminant peaks. Using HRGC/MS analyses, spurious compound peaks seen in the blank analyses were identified and determined to be solvent artifacts. Consequently, during HRGC-FID analyses of the source samples, the retention times of the principal contaminant peaks were monitored routinely; and when they were found, the respective resolved areas were subtracted from the total integrated area.

An assessment of the QA/QC achieved for the entire sampling and analytical methodology used for the carbonaceous aerosol emissions study can be obtained from the observed total mass loadings of the source sampler system blanks (Table 2). These blank samples represent the best measure of the combined residual artifacts associated with the sampling and analytical protocols. The reproducibility with which the organic compound distribution from a single source can be characterized using these techniques is evaluated by Hildemann et al.[17] Briefly, mass distributions of replicate pairs as filter composites were analyzed for six separate source types. Using the percent of total eluted mass falling between the elution points of adjacent n-alkanes as the common unit of measure, it was estimated that the standard deviation of a single step within the mass histogram plot (e.g., Figure 1b) was equal to 0.66% of the total eluted organic mass comprising the whole sample. This degree of agreement between sample replicates for the final mass distributions, as represented by the histogram plots, represents the reliability of sampling, analytical, and chromatographic data reduction procedures used in the present study of carbonaceous source emissions.

B. Preliminary Test Filters

Test aerosol samples were evaluated and consisted of authentic fine ambient aerosol samples and replicate source samples. Following the procedure developed

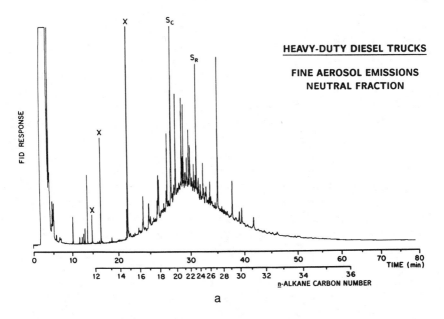

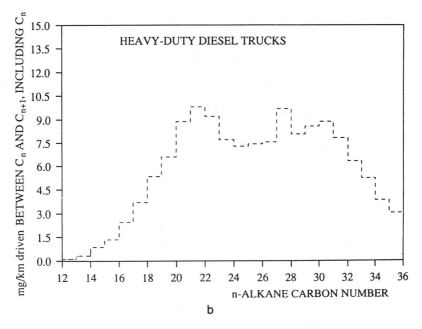

FIGURE 1. Organic mass distributions obtained from heavy-duty diesel truck
fine particulate emissions. X denotes a known solvent artifact; S_C is
the coinjection standard (1-phenyldodecane); and S_R is the recovery
standard (n-$C_{24}D_{50}$). (a) HRGC plot of the neutral elutable organics,
and (b) the computed mass distribution of neutral elutable organics
plotted vs normal alkane carbon number.

by Mazurek,[16-18] n-$C_{24}D_{50}$ was used as an internal standard to facilitate quantitation of the organics present. The amount of n-$C_{24}D_{50}$ to be added to the sample filter as a surrogate standard before sample extraction was determined from the total mass of organic carbon (OC) contained on a filter (collected in parallel) that was analyzed by pyrolysis/combustion; the ratio of OC mass to the mass of added n-$C_{24}D_{50}$ was chosen to be 150 µg OC:1 µg n-$C_{24}D_{50}$.

Results of the test samples provided performance evaluations of the analytical protocol. These analytical procedures were conducted to eliminate blind analyses of the critical source samples. First, the efficiencies of the extraction procedures could be checked using the perdeuterated surrogate standard. The recoveries of n-$C_{24}D_{50}$ obtained for the source sample filter composites ranged typically (within ± 1 SD) between 59 and 93%. Second, by knowing the organic carbon concentration by combustion plus the likely extraction efficiency prior to extraction, the mass of organic material eluting through the HRGC analytical column could be estimated in advance. This estimate provided the opportunity to determine the appropriate number of source sample filters to be extracted, and/or the appropriate degree to which the extracted material should be concentrated for measurement by HRGC-FID. Based on these preliminary extractions of the test filters, it was determined that a minimum of 300 µg of OC for a single-source type provided a quantity of total elutable organics sufficient for mass distribution measurement by HRGC-FID and for molecular tracer identification by HRGC/MS. However, if available, a target quantity of 800-µg OC per source type was adopted for the extraction of an actual source composite. This higher OC mass level for a given source type facilitated quantitation and identification by HRGC/MS of single molecular species present in a complex mixture of total elutable organics. Last, the amount of surrogate standard to be added before extraction could be adjusted based on the nature of the HRGC trace for each source, ensuring that the mass of elutable organics present could be quantitated accurately.

HRGC-FID plots of the total compound distributions (n-$C_{10}H_{22}$ to n-$C_{36}H_{74}$) for the preliminary test filters of the sources sampled showed very different elution profiles. Although the original surrogate spike ratio of 150 µg OC:1 µg n-$C_{24}D_{50}$ was, in general, a good estimate for most source composites, slight adjustments were necessary for a few of the filter composites. The adjustment in the mass of surrogate standard was scaled such that the n-$C_{24}D_{50}$ peak observed in a HRGC-FID plot of total elutable organics for a given source type was twice the height of the tallest resolved peak within the mixture. Although not critical in the case of HRGC/MS analyses of the source extract mixtures, this scale factor was necessary for HRGC-FID quantitation of the extract mixtures.

C. Mass Quantitation

Mass determinations of the carbonaceous materials from urban sources of fine aerosol are given in Table 2. All the data shown are blank-corrected. It is possible to compare the total organics (determined by pyrolysis/combustion) and total elutable organics (determined by HRGC-FID) mass values for the source samples because the results of both analyses are blank-corrected, and the elutable organics values are corrected for the mass contributions of extraction artifacts and standards.

The chemical nature of urban organic aerosol emissions varies greatly from source to source, as shown by the ratios of total elutable organics mass to total organics mass (Table 2). These percentages range from <20 to 100%. Differences between the mass determinations obtained by combustion to those obtained by HRGC-FID can be attributed to several factors. Most importantly, a portion of the total organics present from some sources may not be solvent soluble, or may not elute through the HRGC analytical column. One would expect that a fraction of the organic material in vegetation fragments and in tire dust, for example, will be insoluble or of such high molecular weight that it will not pass through the HRGC analytical column. When this fraction is significant, then the ratio of total elutable organics to total organics is much less than 100%. Second, the mass of total organics is calculated based on the organic carbon mass present, multiplying by a factor of 1.2 to account for the associations of hydrogen, oxygen, nitrogen, etc. with the carbon atoms.[1,21] While this factor is considered appropriate for ambient samples taken in urban areas (based on a survey of the elemental compositions of organic particulate matter sampled in urban atmospheres), it may vary from source to source. Hence, the values calculated for the total organics mass may be somewhat high or low for certain source emissions that have unusual organic composition relative to an average urban aerosol composition. For example, in the case of the roofing tar pot emissions, the ratio of >150% obtained indicates that the total organics mass (by combustion) has been underestimated significantly for this one source test.

D. Elutable Organics Mass Distributions

Besides mass determinations, the total elutable organics fraction determined by HRGC can be characterized by subdividing the HRGC-FID trace into discrete mass segments. An external standard mixture containing 17 normal alkane homologues that range between n-$C_{10}H_{22}$ and n-$C_{36}H_{74}$ is used to quantitate the mass corresponding to each portion of the complex organic mixture eluting between C_n and C_{n+1}. The HRGC-FID relative response factor corresponding to the C_n normal alkane (RRF-C_n) is used for quantitation of the total integrated area eluting between C_n and C_{n+1}. Consistent response factors were obtained for multiple injections of the n-alkane standard mixture throughout the total time period needed to analyze all standard and source samples by HRGC-FID. Relative response factors for the n-$C_{12}H_{26}$ to n-$C_{18}H_{38}$ homologues ranged typically from 0.9 to 1.3 (average value of 1.1). Higher homologues from n-$C_{20}H_{42}$ to n-$C_{34}H_{70}$ showed a systematic increase in RRF with increasing carbon number with RRF values of 1.2 for the smaller homologues up to 2.1 for the larger homologues. RRF values for $C_{36}H_{74}$ were in the range of 2.2 to 3.3. Use of the RRF-C_n was adopted instead of an average RRF determined for a C_n and C_{n+1} homologue pair to maintain a consistent mathematical conversion over the entire range of mass elution (beyond n-$C_{20}H_{42}$, RRFs for even-numbered homologues only were obtained since odd-numbered homologues were not present in the standard mixture). By using the RRF-C_n instead of an average RRF for two adjacent n-alkane homologues, the underestimation of the total mass eluting between two alkane standards with high carbon numbers $>n$-$C_{24}H_{50}$ is <3% for the worst-case example.

A HRGC-FID elution pattern and its computed mass distribution histogram can be utilized as "fingerprint" profiles for a single-source emission type. As an example, Figure 1 shows the HRGC-FID plot for the underivatized extract (neutral elutable organics) of heavy-duty diesel truck fine aerosol (Figure 1a) and the computed organic mass distribution (Figure 1b) of the neutral elutable organics.

Generation of the computed organic mass distribution profile has the advantage of producing a simplified, quantitative fingerprint for each sample of interest. First, because the mass of the species that elutes between the C_n and the C_{n+1} n-alkanes is adjusted for relative response of the HRGC-FID, the computed profile shown in Figure 1b, for example, reflects a more accurate distribution of species concentrations within the complex mixture. Chromatographic bias against the higher molecular weight components which occurs routinely for HRGC-FID plots has been accounted for in the computed elutable organic mass distribution. This adjustment is apparent when comparing the distributions corresponding to $>n$-$C_{24}H_{50}$ for Figures 1a and 1b. A second advantage to the computed profile is that the data on the hundreds of unknown peaks present in the original chromatogram are compressed into a well-defined smaller number of compound groups (i.e., histogram area segments) that can be tracked by computer models for the atmospheric transport of aerosol emissions. Although hundreds of individual resolved compound peaks can be detected in a HRGC-FID profile, these raw data (i.e., retention time vs area counts) are too numerous to be handled easily by computer-based mathematical models. Formulation of the computed organic mass distribution profiles allows a quantitative representation of urban emission sources. Hence, this approach provides a database on organic aerosol source emission characteristics that can be related to the composition of ambient aerosol samples that have been processed in a similar fashion (see References 20 and 23).

IV. SUMMARY AND CONCLUSIONS

An analytical procedure has been developed which provides quantitative links between the total fine aerosol mass and the carbonaceous subfractions that are associated with urban sources of fine aerosol. The method is applicable to a variety of emission sources, where both organic carbon mass loadings and the organic chemical compositions are highly diverse. Emission source samples having organic carbon mass loadings of 7–12,000 μg per filter are analyzed, and individual organic compounds in the range of C_6 to C_{40} are characterized further via HRGC-FID and HRGC/MS. Low levels of analytical and sampling-derived artifacts are observed. Highly reproducible computed mass distribution profiles (i.e., source fingerprints) are obtained for the total elutable organics fraction. These features of the analytical protocol permit a very accurate approach to describing the key chemical attributes of organic aerosols that are emitted to urban atmospheres from major emission sources. This approach, as applied to fine aerosol emissions, will assist in determining the origin of organic particulate matter present in urban atmospheres via atmospheric transport modeling techniques.

ACKNOWLEDGMENTS

This work was supported by the U.S. Environmental Protection Agency (U.S. EPA) grant number R-813277-01-0, and by gifts to the Environmental Quality Laboratory at the California Institute of Technology. Partial support was provided also by the U.S. Department of Energy under the Atmospheric Chemistry Program within the Office of Health and Environmental Research under Contract No. DE-AC02-76CH00016. This chapter has not been subject to the U.S. EPA peer and policy review, and hence does not necessarily reflect the views of the U.S. EPA. Mention of trade names or commercial products does not constitute U.S. EPA endorsement or recommendation for use.

REFERENCES

1. Gray, H. A., G. R. Cass, J. J., Huntzicker, E. K. Heyerdahl, and J. A. Rau. "Characteristics of Atmospheric Organic and Elemental Carbon Particle Concentrations in Los Angeles," *Environ. Sci. Technol.* 20:580–589 (1986).
2. Larson, S. M., G. R. Cass, and H. A. Gray. "Atmospheric Carbon Particles and the Los Angeles Visibility Problem," *Aerosol Sci. Technol.* 10:118–130 (1989).
3. Pratsinis, S., T. Novakov, E. C. Ellis, and S. K. Friedlander. "The Carbon Containing Component of the Los Angeles Aerosol: Source Apportionment and Contributions to the Visibility Budget," *J. Air Pollut. Control Assoc.* 34:643–649 (1984).
4. Alfheim, I., G. Lofroth, and M. Moller. "Bioassays of Extracts of Ambient Particulate Matter," *Environ. Health Perspect.* 47:227–238 (1983).
5. IARC Working Group. "An Evaluation of Chemicals and Industrial Processes Associated with Cancer in Humans Based on Human and Animal Data," *Cancer Res.* 40:1–12 (1980).
6. Gibson, T. L. "Nitro Derivatives of Polynuclear Aromatic Hydrocarbons in Airborne and Source Particulate Matter," *Atmos. Environ.* 16:2037–2040 (1985).
7. Lioy, P. J., and J. M. Daisey, "Airborne Toxic Elements and Organic Substances," *Environ. Sci. Technol.* 20:8–14 (1986).
8. Pierson, W. R., R. A. Gorse, Jr., A. C. Szkariat, W. W. Brachaczek, S. M. Japar, and F. S.-C. Lee. "Mutagenicity and Chemical Characteristics of Carbonaceous Particulate Matter from Vehicles on the Road," *Environ. Sci. Technol.* 17: 31–44 (1983).
9. Pitts, J. N. "Formation and Fate of Gaseous and Particulate Mutagens and Carcinogens in Real and Simulated Atmospheres," *Environ. Health Perspec.* 47:115–140 (1983).
10. Schuetzle, D. "Sampling of Vehicle Emissions for Chemical Analysis and Biological Testing," *Environ. Health Perspect.* 47:65–80 (1983).

11. Hildemann, L. M., G. R. Cass, and G. R. Markowski. "A Dilution Stack Sampler for Collection of Organic Aerosol Emissions: Design, Characterization and Field Tests," *Aerosol Sci. Technol.* 10:193–204 (1989).

12. Hildemann, L. M., Markowski, G. R., and Cass G. R. "Chemical Composition of Emissions from Urban Sources of Fine Organic Aerosol," *Environ. Sci. Technol.* 25:744–759 (1991).

13. Gray, H. A. "Control of Atmospheric Fine Primary Carbon Particle Concentrations," Environmental Quality Laboratory Report #23, California Institute of Technology, Pasadena, CA (1986).

14. Cass, G. R., P. M. Boone, and E. S. Macias. "Emissions and Air Quality Relationships for Atmospheric Carbon Particles in Los Angeles," in *Particulate Carbon Atmospheric Life Cycle*, G. T. Wolff and R. L. Klimisch, Eds. (New York: Plenum Press, 1982), pp. 207–240.

15. Johnson, R. L., J. J. Shah, R. A. Cary, and J. J. Huntzicker. "An Automated Thermal-Optical Method for the Analysis of Carbonaceous Aerosol," in *ACS Symp. Series No. 167, Atmospheric Aerosol: Source/Air Quality Relationships*, E. S. Macias and P. K. Hopke, Eds. (Washington, DC: American Chemical Society, 1981), 223–233.

16. Mazurek, M. A., B. R. T. Simoneit, G. R. Cass, and H. A. Gray. "Quantitative High-Resolution Gas Chromatography and High-Resolution Gas Chromatography/Mass Spectrometry Analyses of Carbonaceous Fine Aerosol Particles," *Int. J. Environ. Anal. Chem.* 29:119–139 (1987).

17. Hildemann, L. M., Mazurek, M. A., Cass, G. R., and Simoneit, B. R. T. "Quantitative Characterization of Urban Sources of Organic Aerosol by High-Resolution Gas Chromatography," *Environ. Sci. Technol.* 25:1311–1325 (1991).

18. Mazurek, M. A., G. R. Cass, and B. R. T. Simoneit. "Biological Input to Visibility-Reducing Aerosol Particles in the Remote Arid Southwestern United States," *Environ. Sci. Technol.* 25:684–694 (1991).

19. Mazurek, M. A., G. R. Cass, and B. R. T. Simoneit. "Interpretation of High-Resolution Gas Chromatography/Mass Spectrometry Data Acquired from Atmospheric Organic Aerosol Samples," *Aerosol Sci. Technol.* 10:408–420 (1989).

20. Rogge, W. F., Hildemann, L. M., Mazurek, M. A., Cass, G. R., and Simoneit, B. R. T. "Sources of Fine Organic Aerosol. I. Charbroilers and Meat Cooking Operations," *Environ. Sci. Technol.* 25:1112–1125 (1991).

21. Gray, H. A., G. R. Cass, J. J. Huntzicker, E. K. Heyerdahl, and J. A. Rau. "Elemental and Organic Carbon Particle Concentrations: A Long-Term Perspective," *Sci. Total Environ.* 36:17–25 (1984).

22. Keith, L. H., W. Crummett, J. Deegan, Jr., R. A. Libby, J. K. Taylor, and G. Wentler. "Principles of Environmental Analysis," *Anal. Chem.* 55:2210–2218 (1983).

23. Hildemann, L. M. "A Study of the Origin of Atmospheric Organic Aerosols," PhD Thesis, California Institute of Technology, Pasadena, CA (1990).

CHAPTER 12

Vapor Adsorption Artifact in the Sampling of Organic Aerosol

Stephen R. McDow and James J. Huntzicker

TABLE OF CONTENTS

0-87371-606-0/93/$0.00+$.50

© 1993 by Lewis Publishers

I. INTRODUCTION

Organic carbon represents a significant fraction of both urban and rural ambient particulate matter.[1] Artifact problems associated with the sampling of the aerosol organic fraction include adsorption of organic vapors on both collection filters and collected particulate matter, volatilization of collected organic particulate matter, and chemical reaction of collected particulate species with reactive atmospheric gases during sampling. Consequently, apparent concentrations of either particulate organic compounds, classes of compounds, or total particulate organic carbon (POC) have been observed to vary with sampling conditions such as face velocity,[2,3] sampling period duration,[4,5] and filter type.[6,7]

Loss from filters due to volatilization of organic particulate matter during sampling has received considerable attention as a possible explanation for the variation of apparent concentrations of organic particulate matter with sampling duration,[4,5] and flow rate.[8] This concern originates in part from results of early experiments which showed that particulate polycyclic aromatic hydrocarbons were lost from filters on exposure to a stream of nitrogen or purified laboratory air.[9-11] Similar volatilization loss from ambient samples has been observed for n-alkanes and carboxylic acids.[11] Volatilization loss of polycyclic aromatic hydrocarbons during sampling has received a theoretical treatment based on equilibrium vapor pressure considerations.[12]

More recently, adsorption of organic vapors on particle collection filters has also emerged as an important sampling artifact problem.[13-18] The adsorption artifact is significant at typical urban concentrations and especially important for low organic aerosol concentrations.[2] We have reported that the adsorption artifact is face velocity dependent and that it is the most likely reason for variations of apparent POC concentrations with face velocity.[2] These observations led us to investigate the role of adsorption artifact in variations of apparent POC concentrations with sampling duration and filter type.

II. EXPERIMENTAL

A. Sampling Procedures

Ambient samples were collected on the roof of the central fire station in downtown Portland, OR at various times from January through July 1985. With only a few exceptions, samples were collected for 24 hr. A tubular sampling manifold containing six sampling ports was constructed for simultaneous collection of six ambient aerosol samples.[2,19] The sampler is described in Figure 1. Each

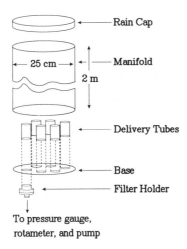

FIGURE 1. Six-port filter sampler.

of the six sampling ports consisted of the following components in series: (1) a 1-μm impactor, (2) a 60 cm in length and 4.45 cm in diameter aluminum delivery tube, (3) a 47-mm filter holder, (4) a regulating valve, and (5) a rotameter with attached pressure gauge. Ambient air was drawn through the six sampling ports at a volumetric flow rate of 8.9 ± 0.1 liters per minute (Lpm) with three carbon vane pumps. Variability between samples was generally less than 5% (1 σ).[2]

Both 47-mm quartz fiber filter (Pallflex QAOT) and Teflon® supported Teflon® membrane (Membrana Zefluor, 2-μm pore size) filters were used. Four different filter arrangements were used for the ambient sampling experiments. Different face velocities were obtained by reducing the filter area with annular masks.[19] For example, at the 8.9 Lpm flow rate, a nominal face velocity of 4 cm/sec was achieved by restricting the exposed filter area to 3.75 cm².

To study apparent POC variations with sampling duration, samples in three sampling ports were collected for 48 hr at face velocities of 15, 40, and 80 cm/ sec. In the other three sampling ports, samples were also collected during the same 48 hr period at the same face velocity, but with filters changed after 24 hr (Figure 2a). The sum of POC collected on two filters sampled for 24 hr was directly compared to POC simultaneously collected during the entire 48 hr for each of the three face velocities employed. In one experiment, sampling durations of 3 and 6 hr were used instead of 24 and 48 hr. Both front and backup filters were analyzed.

The variation of apparent POC concentration between glass and quartz fiber filters was investigated by sampling with quartz fiber filters in three sampling ports and with glass fiber filters in the other three sampling ports at a face velocity of 10 cm/sec (Figure 2b). In additional experiments, three glass fiber filters and three quartz fiber filters were placed behind Teflon® membrane filters at three different face velocities (Figure 2c). This arrangement provides an adsorption artifact estimate for the primary quartz and glass fiber filters.[2] To compare quartz

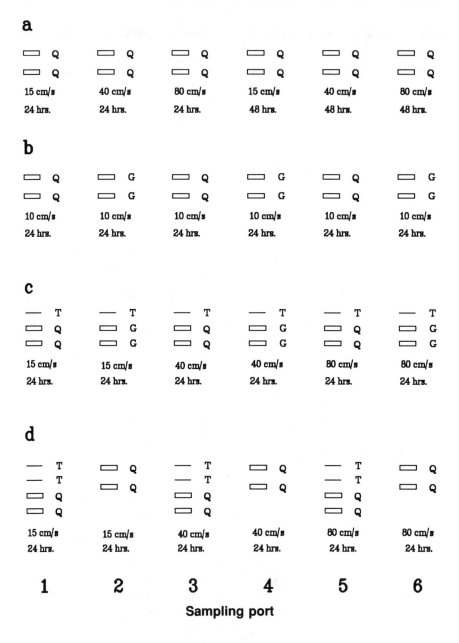

FIGURE 2.   Filter arrangements, filter types, face velocities, and sampling duration for (a)sampling duration comparison, (b) comparison of apparent POC concentration between quartz and glass fiber filters,
(c) comparison of adsorption artifact between quartz and glass fiber
filters, and (d) comparison of apparent POC and adsorption artifact
between quartz fiber and Teflon® membrane filters.

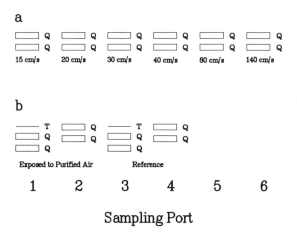

FIGURE 3. Filter arrangements in the investigation of volatilization artifact for (a) exposure to purified air at various face velocities, and (b) filter arrangements for ambient samples used to compare volatilization of adsorbed vapor with volatilization from particle filters.

fiber and Teflon® membrane filters, three sampling ports were equipped with quartz fiber front and backup filters. In the other three sampling ports, two Teflon® membrane filters were placed in series followed by two quartz fiber filters in series (Figure 2d).

Two experiments were conducted in which the variation of volatilization of collected organic carbon with face velocity was investigated. High volume samples were collected for 24 hr on quartz fiber filters at face velocities of 40 cm/sec. 47-mm disks were removed from the high volume filters and placed in appropriate filter holders and pure air from an Aadco pure air generator was drawn across the filters. Two clean filters were placed upstream of the sample filters to further clean the air from the pure air generator. Filters were masked off to allow a range of sampling face velocities for exposure to clean air (Figure 3a). Samples were exposed to purified air for 24 hr. A related experiment was conducted to compare volatilization of collected POC into a clean air stream to volatilization of adsorbed vapor from filters under similar circumstances. For each pair, one sampling port contained two quartz fiber filters in series. In the other sampling port, a quartz fiber filter was placed downstream of a Teflon® membrane filter (Figure 3b). In previous experiments, the quartz fiber filter in this arrangement was a good estimate of adsorption artifact on a quartz fiber filter exposed to the same face velocity.[2] One pair of collected samples was then exposed to the purified air stream for 24 hr at a face velocity of 40 cm/sec. The other pair was used as a reference to determine concentration before exposure to purified air.

B. Analytical Methods

Samples were analyzed for organic and elemental carbon by the Oregon Graduate Center thermal-optical carbon analysis technique.[20,21] A modified method was used for analysis of backup filters.[2] The amount of sample analyzed depended on the amount of unmasked filter available and ranged from 0.25 to 1.5 cm^2. The coefficients of variation for organic, elemental, and total carbon analysis were 4.5, 6.1, and 3.2%, respectively.[2] Before sampling, the quartz fiber filters were baked in air in a muffle furnace for at least 2 hr at 500°C and stored at –10°C within 30 min of the conclusion of sampling. Adsorption during storage was insignificant.

Modifications in the analytical procedure were required for analysis of glass fiber and Teflon® membrane filters. For glass fiber filters, temperatures were only programmed up to 600°C instead of the usual 750°C analysis temperature because the filters melted at higher temperatures. Teflon® membrane filter samples were analyzed at 275°C for total carbon only. Higher temperatures caused significant bleed of the organic filter material.

One ambient sample including a front and a backup filter was quantitatively analyzed by gas chromatography/mass spectrometry (GC/MS). Material in the amount of 5 cm^2 was removed from quartz fiber filters after 24 hr of sampling. Organic compounds were removed from the filters by thermal desorption.[21] Samples were desorbed onto an SE-54 fused silica capillary column of 30-m length and 0.32-mm diameter mounted in a Hewlett Packard (HP)5790A gas chromatograph. The gas chromatograph was interfaced to a Finnigan 4000 mass spectrometer/data system. GC/MS temperature programming was as follows: –80°C during 10-min desorption; –10–250°C at 10°C/min, carrier gas linear velocity 50 cm/sec; and mass spectra — 50–450 amu at 1.0 sec per scan.

III. RESULTS

A. Sample Duration Comparisons

Nine pairs of long- and short-duration samples were obtained from each of the three sampling duration experiments. In all nine samples, apparent POC concentrations were higher for the sum of short-duration samples than for the long-duration samples. A paired t-test indicated a significant difference ($p = 0.05$) in apparent concentrations between the two sampling durations. Apparent concentrations were an average of 16% higher for the short-duration samples. In contrast, a negligible difference was observed for elemental carbon, indicating that general aerosol sampling problems or analytical speciation problems were not responsible for the effect.[2] Results for the samples collected at a face velocity of 15 cm/sec are presented in Figure 4a. These results are consistent with similar experiments carried out in other laboratories in which organic extracts of particulate matter collected on fibrous filters were compared.[4,5]

Adsorbed vapor concentrations estimated from backup filter analysis displayed an even greater variation with sampling duration. On the average, 28% more adsorbed vapor was collected for short than for long sampling durations. A paired t-test indicated that the difference was significant ($p = .05$). With two exceptions, more organic carbon was collected for short than for long durations.

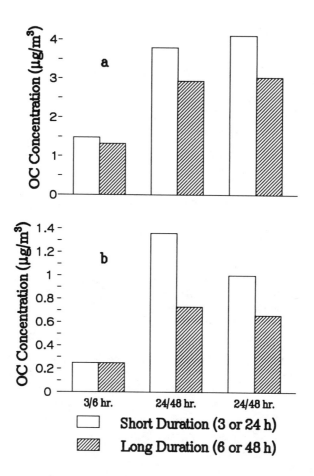

FIGURE 4. Sampling duration effects on (a) apparent POC concentration, and
(b) adsorption artifact.

The exceptions correspond to the two lowest face velocity samples in the short
sampling period experiment (i.e., comparing 3- and 6-hr samples). Results for
samples collected at 15 cm/sec face velocity are presented in Figure 4b.

B. Filter-Type Comparisons

Without blank subtraction, 24–75% more organic carbon was observed on
glass fiber than on quartz fiber filters. In both experiments, the organic carbon
mass collected on glass fiber filters exceeded that on quartz fiber filters by more
than 1.5–µg C per square centimeter. Results are presented in Figure 5. In both
experiments, t-tests indicated a significant difference ($p = .05$) in the amount of
organic carbon collected. Elemental carbon agreed within 0.3-µg C per square
centimeter for both experiments. In both experiments, a substantial portion of the

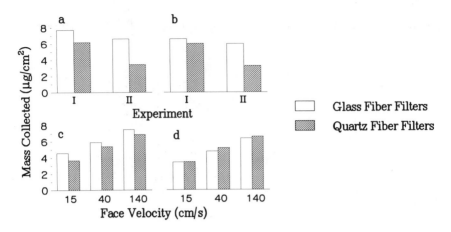

FIGURE 5.   Filter-type effects between glass and quartz fiber filters on apparent
             POC concentration from two experiments (a) before and (b) after
             blank subtraction, and adsorption artifact (c) before and (d) after
             blank subtraction at three face velocities.

difference in organic carbon loading was accounted for by adsorption on glass
fiber filters during storage. The effect of face velocity on apparent POC collected
on glass fiber filters was similar to that for quartz fiber filters reported earlier.[2]

When adsorbed, organic carbon was compared between quartz and glass fiber
filters downstream of Teflon® membrane filters, an average of 14% more organic
carbon was observed on glass than on quartz fiber filters before blank subtraction.
A paired t-test indicated that this difference was significant ($p = 0.05$). However,
after storage blank subtraction, little difference was observed and the average
glass/quartz ratio was 0.96 to ± 0.04. These results are shown in Figure 5b.

Adsorption during storage was further investigated by comparing organic
carbon on both filter types after identical storage conditions at –10°C for various
time periods. The complete databases of storage blanks for each of the filter types
were also compared. All of these comparisons are given in Figure 6. Taken
together, the results from all of the storage experiments indicate that untreated
glass fiber filters adsorb organic vapors more readily than quartz fiber filters, but
it is not clear how much of this is due to adsorption during sampling or during
storage. In contrast, the storage problems associated with quartz fiber filters are
not important compared to adsorption during sampling.

The organic carbon loading on Teflon® membrane and quartz fiber filters are
compared in several ways in Figure 7. An average of 21% more organic carbon
removable at 275°C was observed for samples collected on quartz fiber filters than
for those collected on the Teflon® membrane filters. The amount of carbon
observed on Teflon® backup filters never exceeded 1-µg C per square centimeter.
The relative adsorption artifact estimated for each filter type by the ratio of backup
to primary filters showed 23% as much organic carbon was observed on quartz

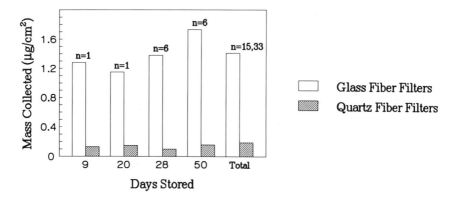

FIGURE 6.   Filter-type effects during storage of quartz and glass fiber filters (n refers to number of blank filters investigated. Total refers to the entire database of blank filters for each filter type).

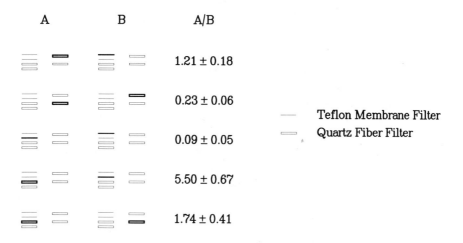

FIGURE 7.   Filter-type effects between quartz fiber and Teflon® membrane filters. A/B represents the ratio of organic carbon mass collected on the bold filters in column A to column B. Filter types are explained in Figure 2d.

fiber backup filters as on primary quartz fiber filters. In contrast, only 9% as much organic carbon was present on Teflon® backup filters as on Teflon® primary filters. An average of more than five times as much organic carbon was observed on quartz fiber filters as on Teflon® membrane filters immediately upstream in the same sampling port. Finally, an indirect comparison of adsorption between quartz and Teflon® filters was obtained by comparing the quartz backup filter downstream of two Teflon® filters to the quartz fiber backup filter directly behind

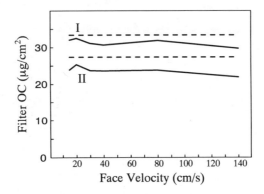

FIGURE 8.   Face velocity effects on volatilization from sample filters into purified air in two experiments. Dotted lines refer to apparent POC collected in ambient sampling. Solid lines refer to organic carbon remaining on filters after exposure to purified air.

a quartz fiber filter at the same face velocity. Downstream of the Teflon® filters 74 ± 41% (1 σ) more organic carbon was observed. These results are consistent with those for similar experiments in which only one Teflon® filter was used[2] and provide further indirect evidence for greater adsorption on quartz fiber than on Teflon® membrane filters.

Direct quantitative comparison of samples collected on quartz fiber and Teflon® membrane filters requires the unlikely assumption that removal temperatures are similar for the two filter types. Consequently, these comparisons must be considered qualitative at best. However, taken together, these direct and indirect comparisons suggest that adsorption artifact effects are substantially greater for quartz fiber filters than for Teflon® membrane filters.

The effect of the reduced analysis temperature of 275°C was compared to the usual analysis temperature of 750°C for quartz fiber filter analysis. The fraction of organic carbon removed from quartz fiber filters after ambient sampling at 275°C averaged 48 and 35% for the two experiments. For 18 backup filters, 63 ± 7% (1 σ) of the total carbon removed at 750°C was removed at 275°C.

C. Volatilization from Sample Filters

Figure 8 shows the POC mass lost per unit area of filter after exposure to purified air. A slight decrease with increasing face velocity was observed in both experiments. However, if results are expressed as mass of carbon lost per unit volume of air drawn across the filter, POC mass lost decreases with face velocity. This is demonstrated in Figure 9. When backup filter loss is considered, mean organic carbon mass agreed within 8% for front filters and 9% for backup filters. Figure 10 shows that if the quartz filters behind Teflon® filters were used as an estimate for the adsorption artifact,[2] the vapor fraction would represent 59 ± 20% of organic carbon volatilized from the particle collection filter.

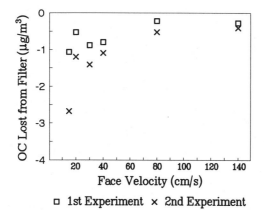

FIGURE 9.   Face velocity effects on volatilization of collected organic carbon per unit volume of purified air.

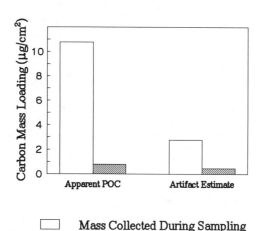

FIGURE 10. Comparison of the volatilization of apparent POC and adsorbed organic carbon into a purified air stream.

D. Composition of Organic Vapor

Table 1 lists compounds identified on front, backup, and blank filters after thermal desorption and gas chromatography/mass spectrometry (GC/MS) analysis. Figure 11 shows the retention time range from about 18 to 25 min for the backup filter. Only n-alkanes from n-tetradecane to n-pentacosane were positively identified on the backup filter. n-Alkanes have been observed previously on backup filters in similar experiments.[23] The unresolved mass of hydrocarbons also displayed aliphatic mass spectral patterns.

Table 1. Compounds Identified on Front, Backup, and Blank Filters after Thermal Desorption and GC/MS Analysis

Compound	Front filter	Backup filter	Blank filter
Tetradecane (c-14)		X	X
Pentadecane (c-15)		X	X
Hexadecane (c-16)	X	X	X
Heptadecane (c-17)	X	X	X
Octadecane (c-19)	X	X	
Nonadecane (c-19)	X	X	
Eicosane (c-20)	X	X	
Heneicosane (c-21)	X	X	
Docosane (c-22)	X	X	
Tricosane (c-23)	X	X	
Tetracosane (c-24)	X	X	
Pentacosane (c-25)	X		

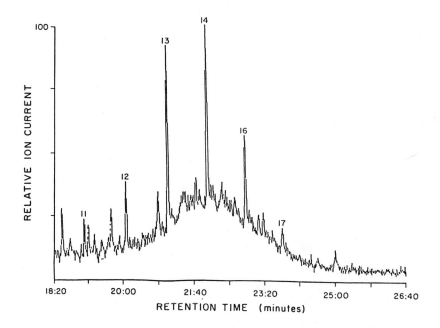

FIGURE 11. Reconstructed ion chromatogram for organic material adsorbed on quartz fiber filter. Numbered peaks are: 11 = nonadecane, 12 = eicosane, 13 = heneicosane, 14 = docosane, 16 = tricosane, 17 = tetracosane.

IV. DISCUSSION

A. Dynamic Adsorption by Quartz Fiber Filters

To interpret differences in apparent POC concentrations and volatilization loss from filter samples into purified air, dynamic adsorption of ambient organic vapors by particle filters is proposed as an alternative explanation to volatilization of collected organic particulate matter. Characteristics of atmospheric organic vapors and quartz fiber filters are favorable for this process. A significant amount of atmospheric organic matter is distributed between the vapor and particulate phases.[11] The composition of this distributed material includes n-alkanes, polycyclic aromatic hydrocarbons, carboxylic acids, and polychlorinated compounds.[11,24] Chromatographically unresolved aliphatic material is probably also distributed between the vapor phase and particulate matter. Moreover, silica is often used as an adsorbent for organic compounds and fibrous filters have large surface areas.[2] Consequently, the quartz fiber filter is likely to be not only an extremely efficient particle collector, but also a moderately efficient sampler of gas-phase semivolatile material as well. Thus, although only a small fraction of the total atmospheric gas-phase organic material is probably retained by the filter, that quantity which is retained is not necessarily negligible compared to POC collected. With that in mind, it would be worthwhile to consider the expected behavior of aerosol sampling filters from a gas-phase sampling perspective and to compare the expected behavior with experimental results.

One of the most important considerations in gas-phase sampling design is the adsorbent breakthrough volume, the sampled volume before a specified fraction of the inlet adsorbate concentration is detected at the sampler outlet. In theory, an adsorption equilibrium is established as the vapor moves through the sampler. For example, in the case of a linear adsorption isotherm, the adsorption equilibrium can be described by:

$$X_a = kX_g \tag{1}$$

where

X_g = vapor phase component of a given atmospheric compound
X_a = filter surface concentration of the same compound at equilibrium

Because analytical results from the thermal carbon analysis method are obtained as mass per unit filter area, atmospheric concentrations are obtained by dividing by the product of face velocity (i.e., flow rate per unit area) and sampling duration. Under conditions of complete breakthrough, an adsorption equilibrium would occur for all samples. The contribution of adsorbed vapor to the apparent particulate concentration would be:

$$\text{“}X_p\text{”} = kX_g/vt \tag{2}$$

where

v = face velocity
t = sampling period
"X_p" = adsorbed vapor fraction of apparent particulate concentration

In this case, the gas-phase contribution to apparent particulate concentration is inversely proportional to both sampling duration and face velocity.

A thorough treatment of the gas-phase sampling efficiency of the quartz fiber filter would require estimates of the breakthrough volume or adsorption equilibrium constants for each important species adsorbed. The observation that mass collected per unit area of filter also varied with respect to both face velocity[2] and sampling period duration indicates nonequilibrium adsorption conditions for at least some of the adsorbed species. In many cases, nonlinearity of the adsorption isotherm might also have to be considered.[25] Breakthrough considerations become more complex if diffusion along the axis flow is significant compared to convective transport (i.e., an extremely low number of theoretical plates),[26] which is probably the case for a thin fibrous filter. However, at high enough flow rates, convective effects become important and boundary layer processes must be considered.[27] This is likely to be the case for the samples collected at the highest face velocities during this study.

Thus, there are more important factors which have not been considered in the simple treatment of Equation 2. However, a more detailed treatment would require additional information about the composition of the adsorbed vapor and a thorough investigation of the dynamic adsorption behavior of individual species identified. However, Equation 2 clearly illustrates that if a significant breakthrough of vapor-phase organic species occurs and the mass of adsorbed organic vapor is not negligible in comparison to collected POC, apparent concentration is expected to decrease with both increasing face velocity and increasing sampling duration.

B. Face Velocity and Sampling Duration Effects

We previously reported the variation of apparent POC with face velocity.[2] In each of the four experiments, a significant decrease of both POC and adsorption artifact estimate was observed. Volatilization of collected POC was considered unlikely to be the cause of this effect. Significant volatilization would be expected because of changing adsorption equilibrium conditions or because of the pressure drop across the filter during sampling. However, changes in atmospheric variables with respect to a set of equilibrium conditions are generally bidirectional and should tend to cancel out over the sampling period, and differences in pressure drop were not sufficient to account for the much greater differences in adsorbed vapor collected.

In that study, a quartz fiber backup filter behind a quartz fiber filter used for particle collection was found to be a relatively poor corrector for adsorption artifact. However, use of a quartz fiber filter mounted downstream of a Teflon® membrane filter operated at the same face velocity removed 90% of the original face velocity dependence and was consequently a much better estimate of adsorbed organic vapor adsorbed on a quartz fiber filter during sampling. It was concluded that the decreasing trend of POC with face velocity was most likely due to organic vapor adsorption during sampling.

Both the previously observed variations of apparent POC concentration with face velocity and the observation of its variation with sampling duration reported here are consistent with the treatment of a quartz fiber filter as an inefficient gas phase sampling device.

C. Volatilization of Adsorbed Organic Vapor

To further pursue the treatment of the filter as a gas-phase sampler, the behavior of adsorbed vapors on exposure to a clean airstream was considered. Figure 8 confirms that volatilization of POC from collected samples is a function of face velocity. Figure 9 indicates that removal is a stronger function of clean air sampling volume at lower face velocities. This is consistent with the dynamic adsorption approach because the breakthrough volume of volatile compounds is more closely approached or is exceeded for more compounds at the higher face velocity. From this experiment, it is not clear whether the organic carbon lost from the filter was associated with collected particulate matter or adsorbed on the filter surface. However, in view of the tremendous increase in surface area experienced by sampled air as it enters the filter[2] and the excellent adsorbent properties of silica for organic vapors, it is likely that more adsorption occurs on the filter surface than occurs on atmospheric particulate matter before the sampled air enters the filter.

This is consistent with the observation described in Figure 10, that adsorbed vapor accounted for the major fraction of organic material which volatilized into the purified air stream. It also demonstrates an important difficulty inherent in correcting for volatilization artifact. In this context, it is interesting to note that members of the homologous n-alkane series found adsorbed on the back filter in Table 1 correspond well both with those distributed between the vapor and particulate phases and those affected by volatilization loss.[11] Moreover, a characteristic broad signal of unresolved material is often observed in chromatograms of extracted urban particulate matter.[28] It is probably composed mainly of cyclic and branched chain aliphatic hydrocarbons. Apparently, the more volatile of those compounds are also retained from the vapor phase by the filter.

V. CONCLUSIONS

The organic vapor adsorption artifact is a serious problem in the sampling of atmospheric organic aerosol. Variations of apparent particulate organic carbon with sampling parameters such as sampling duration, filter type, and face velocity are likely to be caused at least in part by adsorption of organic vapors by filter media. Volatilization from filter samples into purified air streams, a common technique in evaluating the volatilization artifact, appears to include a significant amount of filter adsorbed organic vapors in addition to whatever volatilization from collected organic particulate matter occurs. Moreover, the composition of adsorbed compounds match well with the compounds most susceptible to volatilization. These observations are consistent with the treatment of the particle collection filter as a moderately efficient vapor collector for which incomplete breakthrough occurs. The implications of this study are that the adsorption artifact must be considered in the design of an accurate organic aerosol sampling procedure, and that previous observations considered to be suggestive of volatilization of collected particulate organic matter during sampling are also consistent with an interpretation based on the adsorption artifact. As outlined in previous research, failure to correct for the adsorption artifact can lead to a significant overestimate of particulate organic carbon concentrations, especially at low organic aerosol concentrations.

REFERENCES

1. Shah, J. J., R. L. Johnson, E. K. Heyerdahl, and J. J. Huntzicker. "Carbonaceous Aerosol at Urban and Rural Sites in the United States," *J. Air Pollut. Control Assoc.* 36:254–257 (1986).
2. McDow, S. R., and J. J. Huntzicker. "Vapor Adsorption Artifact in the Sampling of Organic Aerosol: Face Velocity Effects," *Atmos. Environ.* 24A:2563–2571 (1990).
3. Miguel, A. H., and J. B. de Andrade. "Reactivity of Atmospheric Aromatic Hydrocarbons (PAHs) During Aerosol Sampling: Effects of Face Velocities," in *Aerosols: Formation and Reactivity: Proceedings of the Second International Aerosol Conference,* (Oxford, Great Britain: Pergamon Press, Inc., 1986), 519–522.
4. Appel, B. R., E. M. Hoffer, E. L. Kothny, S. M. Wall, M. Haik, and R. L. Knights. "Analysis of Carbonaceous Material in Southern California Aerosol 2," *Environ. Sci. Technol.* 13:98–104 (1979).
5. Schwartz, G. P., J. M. Daisey, and P. J. Lioy. "Effect of Sampling Duration on the Concentration of Particulate Organics Collected on Glass Fiber Filters," *Am. Ind. Hyg. Assoc. J.* 42:258–263 (1981).
6. Lee, F. S.-C., W. R. Pierson, and J. Ezike. "The Problem of PAH Degradation on Several Commonly Used Filter Media," in Polynuclear Aromatic Hydrocarbons: Fourth International Symposium on Analysis, Chemistry, and Biology, (Columbus, OH: Battelle Press, 1980), 543–563.
7. Grosjean, D. "Polycyclic Aromatic Hydrocarbons in Los Angeles Air from Samples Collected on Teflon® , Glass and Quartz Filters," *Atmos. Environ.* 17:2565–2573 (1983).
8. Della Fiorentina, H., F. De Wiest, and J. De Graeve. "Determination par Spectrometrie Infrarouge de la Matiere Organique Non Volatile Associee aux Particules en Suspension dans l'Air II. Facteurs Influencent l'Indice Aliphatique," *Atmos. Environ.* 9:517–522 (1975).

9. Commins, B. T., "Interim Report on the Study of Techniques for Determination of Polycyclic Aromatic Hydrocarbons in Air," *Natl. Cancer Inst. Monogr.* 9:225–233 (1962).

10. Rondia, D. "Sur la Volatilite des Hydrocarbures Polycycliques," *Int. J. Water Air Pollut.* 9:113–121 (1965).

11. Van Vaeck, L., K. Van Cauwenberghe, and J. Janssens. "The Gas-Particle Distribution of Organic Aerosol Constituents: Measurement of the Volatilization Artifact in Hi-Vol Cascade Impactor Sampling," *Atmos. Environ.* 18:915–921 (1984).

12. Pupp, C., J. J. Murray, and R. F. Pottie. "Equilibrium Vapor Concentrations of Some Polycyclic Aromatic Hydrocarbons, As_4O_6 and SeO_2 and the Collection Efficiencies of these Air Pollutants," *Atmos. Environ.* 8:915–925 (1974).

13. Stevens, R. K., T. G. Dzubay, R. W. Shaw, W. A. McClenny, C. W. Lewis, and W.E. Wilson. "Characterization of the Aerosol in the Great Smoky Mountains," *Environ. Sci. Technol.* 14:1491–1498 (1980).

14. Cadle, S. H., P. J. Groblicki, and P. A. Mulawa. "Problems in the Sampling and Analysis of Carbon Particulate," *Atmos. Environ.* 17:593–600 (1983).

15. McDow, S. R., and J. J. Huntzicker. in *Aerosols: Formation and Reactivity: Proceedings of the Second International Aerosol Conference,* (Oxford: Pergamon Press Inc, 1986), 512–514.

16. Ligocki, M. P., and J. F. Pankow. "Measurements of the Gas/Particle Distributions of Organic Compounds," *Environ. Sci. Technol.* 23:75–83 (1989).

17. McMurry, P. H., and X. Q. Zhang. "Size Distributions of Ambient Organic and Elemental Carbon," *Aerosol Sci. Technol.* 10:430–437 (1989).

18. Appel, B. R., W. Cheng, and F. Salaymeh. "Sampling of Carbonaceous Particles in the Atmosphere II," *Atmos. Environ.* 23:2167–2175 (1989).

19. McDow, S. R. "The Effect of Sampling Procedures on Organic Aerosol Measurement," Ph.D. Thesis, Oregon Graduate Institute of Science and Technology (formerly Oregon Graduate Center), Beaverton, OR (1986).

20. Johnson, R. L., J. J. Shah, R. A. Cary, and J. J. Huntzicker. "An Automated Thermal-Optical Method for the Analysis of Carbonaceous Aerosol," in *American Chemical Society Symposium Series No. 167. Atmospheric Aerosol: Source/Air Quality Relationships* (Washington, DC: American Chemical Society, 1982), pp. 223–233.

21. Huntzicker, J. J., R. L. Johnson, J. J. Shah, and R. A. Cary. "Analysis of Organic and Elemental Carbon in Ambient Aerosols by a Thermal-Optical Method," in *Particulate Carbon: Atmospheric Life Cycle* (New York: Plenum Press, 1982), pp.79–88.

22. Ligocki, M. P., and J. F. Pankow. "Assessment of Adsorption/Solvent Extraction with Polyurethane Foam and Adsorption/Thermal Desorption with Tenax-GC for the Collection and Analysis of Ambient Organic Vapors," *Anal. Chem.* 57:1138–1144 (1985).

23. Eichmann, R., P. Neuling, G. Ketseridis, J. Hahn, R. Jaenicke, and C. Junge. "n-Alkane Studies in the Troposphere I. Gas and Particulate Concentrations in Northern Atlantic Air," *Atmos. Environ.* 13:587–599 (1979).

24. Bidleman, T. F., W. N. Billings, and W. T. Foreman. "Vapor-Particle Partitioning of Semivolatile Organic Compounds: Estimates from Field Collections," *Environ. Sci. Technol.* 20:1038–1043 (1986).

25. Grubner, O., and W. A. Burgess. "Calculations of Adsorption Breakthrough Curves in Air Cleaning and Sampling Devices," *Environ. Sci. Technol.* 15:1346–1351 (1981).

26. Senum, G. I. "Theoretical Collection Efficiencies of Adsorbent Samplers," *Environ. Sci. Technol.* 15:1073–1075 (1981).

27. Friedlander, S. K. *Smoke, Dust and Haze* (New York: John Wiley & Sons, Inc. 1977).
28. Simoneit, B. R. T., and M. A. Mazurek. "Organic Matter in the Troposphere II. Natural Background of Biogenic Lipid Matter in Aerosols over the Rural Western United States," *Atmos. Environ.* 16:2139–2159 (1982).

CHAPTER 13

A Sampling System for Reactive Species in the Western United States

Judith C. Chow, John G. Watson, John L. Bowen, Clifton A.
Frazier, Alan W. Gertler, Kochy K. Fung, Dwight Landis,
and Lowell Ashbaugh

TABLE OF CONTENTS

0-87371-606-0/93/$0.00 + $.50

I. INTRODUCTION

The deposition of gaseous and particulate species is one of the most important physical and chemical processes in the troposphere; it affects the atmospheric residence time and distance of travel of numerous atmospheric constituents. The concentration/deposition velocity method of estimating dry deposition fluxes is applied in the California Acid Deposition Monitoring Program (CADMP)[1] to assess deposition fluxes of acidic gases and particles at 10 sites throughout the state of California. This network is acquiring a multi-year database, starting in 1988, at representative locations in California. Seven of these sampling sites represent urban areas (South Coast Air Basin, San Francisco Bay Area, Bakersfield, Santa Barbara, and Sacramento) and three represent forested areas (Sequoia, Yosemite, and Redwood National Parks). A special sampling system was created to meet the needs for monitoring atmospheric concentrations of gaseous sulfur dioxide, nitrogen dioxide, nitric acid, and ammonia along with $PM_{2.5}$ and PM_{10} concentrations of mass, chloride, nitrate, sulfate, sodium, magnesium, potassium, and calcium. This sampling system can be adapted for the measurement of many other pollutants, in addition to those quantified in CADMP.

The objectives of this chapter are: (1) to describe the CADMP sampling system for reactive species in the atmosphere; and (2) to report results of its performance in 2 years of field operations. This detailed description is important for those who will use CADMP data and for researchers who desire to construct specialized aerosol samplers. Several tests have been conducted to compare results from this sampler with those from other samplers and to evaluate individual system components. The results of these tests will be presented in other reports.

II. SYSTEM REQUIREMENTS AND COMPONENT SELECTION

For the CADMP, a gas and particle sampling system was needed which could be operated by part-time technicians at urban and nonurban sites over a period of years. Such a system must be of minimal complexity, maximum durability, and reasonable cost. The particle and gas sampling requirements for the CADMP were: (1) well-defined size fractions ($PM_{2.5}$ and PM_{10}); (2) accurate measurements of acidic species as they would appear in the atmosphere; (3) suitability for application in both urban and nonurban environments; (4) multiple substrates for a variety of analyses; (5) flow rate stability; (6) compatibility with analysis methods; and (7) availability of equipment and ease of operation.

A. Size-Selective Inlets

$PM_{2.5}$ consists of particles which are less than 2.5 μm in aerodynamic diameter, while PM_{10} consists of particles which are less than 10 μm in aerodynamic diameter. These size ranges are operationally defined by inertial separation inlets which transmit 50% of all suspended particles with 2.5- and 10-μm diameters, respectively. Smaller particles penetrate these inlets with higher sampling effectiveness and larger particles penetrate with lower effectiveness. Sampling effectiveness as a function of particle size is determined from wind tunnel tests.[2] Six $PM_{2.5}$ and six PM_{10} inlet designs were evaluated; the inlets chosen were the Bendix $PM_{2.5}$ cyclone[3,4] and the Sierra-Andersen 254 PM_{10} elutriator/impactor[5] inlets. These inlets were selected because they have sharp 50% cut-points at 2.5 and 10 μm, respectively; they are commercially available; and they are well tested. Both inlets operate at 113 liters per minute (Lpm), which allows multiple samples at 10–50 Lpm to be obtained from the sample flow through each inlet.

B. Flow Rate Measurement and Control

Four general methods of sampler air flow control were considered in sampler design: manual volumetric, automatic mass, differential pressure volumetric, and critical orifice volumetric. Differential pressure flow control was selected for the CADMP sampler owing to its reliability as demonstrated in other programs, its low cost, and its ability to make full use of pumping power. Critical orifices induce too large a pressure drop and require additional pumps. Mass flow controllers contain expensive electronics which do not endure hostile environments typical of CADMP sampling sites.

C. Filter Holders

Filter substrates must be protected from contamination prior to, during, and after sampling. The only way to ensure this in the CADMP dry deposition network is to load filters into holders in the laboratory. These holders must: (1) mate to the sampler and to the flow system without leaks; (2) be composed of inert materials (e.g., no metal) which do not absorb acidic gases; (3) allow a uniformly distributed deposit to be collected; (4) have a low pressure drop across the empty holder; (5) accommodate the sizes of commonly available air sampling filters (e.g., 37 or 47 mm); and (6) be reasonably priced.

Open-faced and in-line filter holders composed of stainless steel, polycarbonate, and PFA-molded Teflon® were considered. In-line configurations were rejected because these cause particles to accumulate in the center of the filter just below the entry point. Since sections of the sampled filters are submitted to different chemical analysis methods, a homogeneous sample deposit is essential. Metal filter holders were rejected owing to their cost (several hundred dollars) and their affinity for removal of reactive species. The nitric acid removal characteristics of polycarbonate are unknown, while recent evidence shows that PFA Teflon® is inert to nitric acid removal.

The Savillex 47-mm injection-molded PFA Teflon® open-faced filter holder was selected for CADMP sampling. These filter holders have a tapered extender section (called a receptacle) which can be mated to a sampler plenum with an O-ring in a retainer ring. Several grids can be stacked within the holder to obtain series filtration. The cost is reasonable. The major disadvantages of this holder are nonuniform manufacturing tolerances (diameters can be specified within a 0.01-in. tolerance) and nonuniform porosity of the support grid. A new support grid was designed and manufactured by ATEC, Inc. for this project to reduce flow resistance and to create a homogeneous deposit. The receptacles are individually tested for their dimensions as part of the acceptance criteria.

D. Sampling Surfaces

Certain sampling surfaces absorb or react with gases and particles, thereby preventing their collection on sampling substrates.[6,7] This is especially the case for nitric acid vapor which sticks to nearly everything. In the case of denuders, a surface which *does* absorb nitric acid is desired. In the case of all other sampling components, a surface which *does not* absorb nitric acid is desired. Appel and Povard[8] have tested different materials with respect to their affinity for nitric acid, and they have found that a PFA Teflon®-coated duet or inlet will transmit more than 90% of the nitric acid presented to it after it has been "seasoned" to the nitric acid in a particular environment. These researchers have also determined that annular denuders made of aluminum have an almost infinite capacity for absorbing nitric acid vapor while transmitting $PM_{2.5}$ particles with high efficiency.[9]

For the CADMP sampler, a conical plenum and the Bendix 240 cyclone were coated with PFA Teflon® by washing and sandblasting the metal surface and then spreading it with a mixture of TFE Teflon®, chromic acid, and phosphoric acid. This coating is hardened by baking at 450°C and serves as a primer to which the PFA Teflon® will adhere. The TFE surface and a supply of PFA powder are then given opposite charges and the PFA powder is sprayed at the TFE surface where it is electrostatically deposited. The surface is then heated until the PFA powder melts onto the surface. Prior to use in sampling, these PFA surfaces are washed with a dilute solution of nitric acid which remains in contact with the surface for several hours. They are then washed in distilled water and dried. Laboratory tests have shown that this washing minimizes further absorption of nitric acid by the surface and that the surface does not act as a source of nitric acid when clean air is drawn over it.

E. Substrates

The choice of filter type results from a compromise among the following filter attributes: (1) mechanical stability, (2) chemical stability, (3) particle or gas sampling efficiency, (4) flow resistance, (5) loading capacity, (6) blank values, (7) artifact formation, (8) compatibility with analysis methods, and (9) cost and availability.

Several types of filters were obtained from different vendors and were tested with respect to these attributes. The sampling media which were chosen consist of: (1) Gelman (Ann Arbor, MI) polymethylpentane-ringed 2.0-μm pore size, 47-mm diameter PTFE Teflon-membrane filters (No. R2PI047); (2) Schleicher and Schuell (Keene, NH) 1.2-μm pore size, grade 66, 47-mm diameter nylon membrane filters (No. 00440); (3) Whatman 41 (Maldstone, England) 47-mm diameter cellulose fiber filters (No. 1441047); (4) Pallflex (Puttnam, CT) 47-mm diameter Teflon®-coated glass fiber filters (No. TX40HI20); and (5) Whatman 31ET Chrome 47-mm diameter cellulose fiber chromatography paper.

The Whatman 41 cellulose filters are impregnated with citric acid and potassium carbonate (K_2CO_3) and the Whatman 31ET cellulose filters are impregnated with triethanolamine (TEA). The impregnation solutions consist of: (1) 15% (by weight) K_2CO_3 and 5% glycerol solution (balance being water) for sulfur dioxide (SO_2) sampling; (2) 25% citric acid and 5% glycerol (balance being water) for ammonia (NH_3) sampling; and (3) 25% TEA, 4% ethylene glycol, 25% acetone, and 46% water for nitrogen dioxide (NO_2) sampling.

III. SAMPLER CONFIGURATION

The sampling system is illustrated in Figure 1, which shows the sampler consisting of four separate units: (1) a $PM_{2.5}$ particle module, (2) a PM_{10} particle module, (3) a pump module, and (4) a timer module. Figure 1 is keyed to a list of components in Table 1, which provides a brief description of each component.

Particles and gases are drawn through the Teflon-coated $PM_{2.5}$ and aluminum PM_{10} inlets at 113 Lpm into conical plena. The $PM_{2.5}$ plenum is coated with nitric acid-treated PFA Teflon® and the PM_{10} plenum surface is aluminum. The conical shape of the plena diffuses the airflow and minimizes particle deposition.

The open-faced filter packs are located inside the plena, as shown in Figure 2. These filter packs are connected to a vacuum manifold through differential pressure regulators. Vacuum pumps draw air through these filters when activated by the timer. The timer also switches the flow, via solenoid valves, between two identical sets of filter packs to allow for daytime and nighttime samples of 12-hr duration without operator intervention. Other sample start and stop times can be programmed if needed. A total of 20 Lpm are drawn through two filter packs simultaneously. The additional flow required to maintain the inlet cut points is drawn through two additional sampling ports in each unit. Additional ports are provided for field blanks which are used to evaluate filter loading during passive sampling periods and during filter handling.

Four separate filter packs are used for each 12-hr sample, two in the $PM_{2.5}$ module and two in the PM_{10} module. Figure 3 shows the three-stage filter pack construction and Figure 4 shows how the filter pack is press-fit into the retainer rings in the bottom of the plenum. The filter packs are labeled with a bar code prefix and are also color coded to assure that the proper filter pack is placed in the proper port. The bar code prefixes identify the following four configurations:

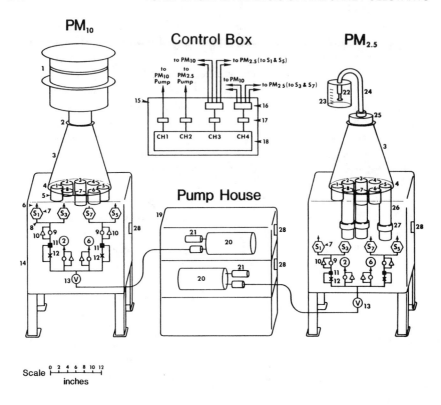

FIGURE 1.   Schematic diagram of CADMP gas/aerosol sampling system.

- A TCK prefix designates a three-stage filter pack in the PM_{10} module containing a Teflon®-membrane filter on the first grid (for PM_{10} particles), followed by a citric acid-impregnated cellulose filter on the second grid (for ammonia gas), and followed by a potassium carbonate-impregnated cellulose filter on a third grid (for sulfur dioxide gas). This filter pack samples from the PM_{10} plenum.
- A GT prefix designates a two-stage filter pack in the PM_{10} module containing a Teflon®-coated glass fiber filter (to remove nitrate particles) on the first grid, followed by a TEA-impregnated cellulose filter (for nitrogen dioxide gas) on the bottom grid. This filter pack samples from the PM_{10} plenum.
- A TN prefix designates a two-stage filter pack in the $PM_{2.5}$ module containing a Teflon®-membrane filter (for $PM_{2.5}$ particles), immediately followed by a nylon filter (for nitric acid gas) on the bottom grid. This filter pack samples from the $PM_{2.5}$ plenum.
- A DN prefix designates a single-stage filter pack in the $PM_{2.5}$ module containing a nylon filter on the bottom grid (for labile and nonlabile $PM_{2.5}$ nitrate). The airstream reaching this filter passes through an annular denuder coated with aluminum oxide to remove gaseous nitric acid prior to reaching the filter pack. This filter pack samples from the $PM_{2.5}$ plenum. The difference between the TN nitrate and the DN nitrate yields nitric acid concentrations by denuder difference.

Figure 5 illustrates the sample flow paths and identifies the analyses which are performed on each substrate. The sampler configuration has been constructed to allow simple modification for: (1) additional sampling ports, (2) sequential switching for up to six samples without operator intervention, (3) up to four simultaneous samples, (4) different filter pack configurations, and (5) adjustable flow rates through filter packs from 10 to 50 Lpm. Several different configurations of this sampler have been used in other monitoring programs.

IV. SAMPLER PERFORMANCE

Sampler performance has been evaluated both operationally, at 10 sites over a 2-year period, and quantitatively by collocated sampling at one site. No major operational failures were found over this period and no major repairs were required. Routine maintenance involved: (1) replacement of worn tubing, (2) changing pump exhaust filters quarterly, (3) annual cleaning of inlets and the sample plena, (4) annual recalibration of flow meters, and (5) biannual replacement of pump vanes.

An audit of flow rates after nearly 2 years of operation showed that flow rate calibrations had not varied from the original calibrations by more than ±10% for most samplers.[10]

The Sacramento monitoring site was equipped with a collocated gas/particle sampling system operated in a manner identical to that of the primary system. The two systems are located within 4 m of each other on the roof of the California Air Resources Board monitoring laboratory.

Figures 6 through 9 show scatter plots, with linear regression statistics, for selected chemical concentrations. Reproducibility is considered to be acceptable when the correlation coefficient (r) exceeds 0.9, the slope is within two or three standard errors of unity, and the intercept is within two or three standard errors of zero. When a few points were identifiable as obvious outliers, the regression statistics were recalculated without these points, and the recalculated values are displayed in parentheses. Most of the comparisons in Figures 6 through 9 meet the criteria for acceptable reproducibility. The exceptions reveal limitations in the measurement process.

All of the PM_{10} measurements are more reproducible than the $PM_{2.5}$ measurements. Potential causes of discrepancies may be one or more of the following: (1) less precise flow rates in the nondenuded $PM_{2.5}$ airstream of one or both sampling systems, (2) lower concentrations of the measured species, (3) different large particle penetration through the $PM_{2.5}$ inlet, and/or (4) instability of the sample deposit.

The flow rate audits[10] showed no biases on the nondenuded $PM_{2.5}$ sampling ports for either the primary or the collocated sampler. If such a bias did exist, it would affect all concentrations on the corresponding sampler in the same way, and the scatter in Figures 6 through 9 would exhibit similar patterns, which it does not. While less precise flow rates may be a cause, the evidence does not support this.

Table 1. Components of CADMP Gas/Particle Sampling System

Code	Part name	Specification and rationale
1	PM_{10} inlet	Sierra-Andersen SA-254 medium-volume PM_{10} inlet, provides a 50% cut point at 10 μm with a flow rate of 113 liters per minute (Lpm); this is the only medium-volume inlet which is commercially available
2	Inlet connection	O-ring seal connects PM_{10} inlet directly to plenum
3	Transitional plenum	Custom constructed of spun aluminum. Inside of $PM_{2.5}$ plenum is coated with PFA Teflon®; conical shape provides for flow expansion to minimize particle loss inside the plenum; the plenum is sealed to the plenum base with 16 clips; a rubber gasket is placed between the plenum and the base
4	Plenum base	Stainless steel with eight 2-in. diameter holes for filter holders and annular denuders; filter holder retainer rings with internal O-rings provide a press-fit for Savillex open-faced filter holders; a sheet of thin PFA Teflon® covers the plenum base in the $PM_{2.5}$ sampler
5	PFA injection molded filter holders	Savillex 47-mm diameter three-stage PFA Teflon® filter holders (Catalog No. 6T-47-6T) with ferrule nut for 3/8-in. tubing and 3-in. long open face receptacle (Catalog No. 4750); this is the only Teflon®/stacked filter holder available within the United States; support grids have been modified to improve the homogeneity of the deposit and to reduce the pressure drop across the holder; open face needed for uniform deposit on filter
6	Tubing	Polypropolene 3/8-in. o.d., since this is downstream of the samples, PFA Teflon® is not necessary
7	2-Way solenoid	ASCO Red-Hat Vacuum Service (Catalog No. 8262C90); maximum valve operating pressure difference is 15 psi with a low vacuum range to 29-in. Hg; valves are mass spectrometer tested; normally closed valves are used
8	Vacuum manifold	Brass pipe "T"s (1/2-in.) manifold connects the solenoid valves to the flow meters; valves switch samples every 12 hr according to timer settings
9	In-line orifice	Custom built, stainless steel; this orifice creates a small pressure drop which is measured by a magnehelic to set flow rates
10	Magnehelic differential pressure gauge	Dwyer low pressure 0–2-in. H_2O differential gauge; used to measure pressure drop across orifice for flow rate adjustment; one magnehelic is attached to every filter and makeup flow channel
11	Differential pressure valve	Conoflow Regulators & Controls, GH21XTXM series fixed differential regulators; maintains a fixed differential pressure across a ball valve downstream from the regulator; one differential pressure valve is attached to every day and night sampling port

Table 1. Components of CADMP Gas/Particle Sampling System (continued)

Code	Part name	Specification and rationale
12	Ball valve	TOYO 3/8-in. stainless steel ball with brass body; this valve is adjusted to set the flow rate at the desired value
13	Vacuum gauge	Valin Craig Christensen, V500, 14 in. L.M., 0–30-in. Hg, 2 1/2 in. dial size; used to test for pump degradation and leaks
14	Cabinet and legs	Custom built, 24 in. high x 24 in. deep x 24 in. wide stainless steel metal cabinet with 12 in. long detachable legs; makes unit stand alone and protects samples from weather
15	Timer control box	Junction box 12 in. high x 6 in. deep x 12 in wide; controls sample start, sample stop, sample switching, and sample duration
16	Elapsed time meter	Cramer elapsed time meter; minute meter with 0.1 min resolution (catalog No. 35F3803) type 10184K/115 A.C., six figures, no-reset, rectangular shape; determines sample duration; one elapsed time meter is associated with each channel day or night
17	Relay	Square D A.C. control relay, DPST-2NO 10 AMP at 277 VAC/5 AMP at 600 VAC, coil 120 V/60 Hz; turns pumps on and opens solenoid valves
18	Timer	Grasslin series 2001; a programmable four-channel micro-processor-based controller with program storage; this timer controls sample start and stop times and channel switching
19	Pump house	Stainless steel cabinet 30 in. high x 12 in. deep x 30 in. wide; protects pumps from exposure
20	Pump	GAST 1022 carbon vane pump, 3/4 HP; pumps pull 25-in. Hg vacuum, applicable for 113 Lpm sampling flow rate
21	Absolute filter	Placed at the exhaust of the carbon vane pump, this filter prevents exhaust recirculation
22	Cyclone separator	Sensidyne (Bendix) model 240 cyclone coated with PFA Teflon® for $PM_{2.5}$ sampling at 113 Lpm; this is the only medium-volume size-selective inlet which is commercially available; well tested in wind tunnels.
23	Inlet stilling chamber	Custom built with bug screen attached to the bottom; coated with PFA Teflon; minimizes large particle entry to cyclone.
24	PFA tubing	Tubing 1/2-in. o.d.; connects cyclone to transitional plenum.
25	Inlet adaptor	Custom built, PFA teflon-coated cone connects cyclone to transitional plenum
26	Annular denuder	Anodized aluminum; outer tube has 2-in. o.d. and 1.834-in. i.d., inner cylinder has 1.75-in. o.d., leaving an annulus which is 0.042" wide; the denuder is blunt at the inlet and has a 30° cone at its base to homogenize the particle concentration by the time it reaches the filter

Table 1. Components of CADMP Gas/Particle Sampling System (continued)

Code	Part name	Specification and rationale
27	Filter retainer rings	Machined aluminum with internal O-ring to allow press-fit of Savillex filter holders into denuders and plenum base; retainer rings are Teflon® coated in $PM_{2.5}$ particle samplers; retainer rings are attached to the plenum base with four cap screws and a rubber sealing gasket to prevent leaks; these cap screws are covered with silicon sealer as the retainer rings are installed to prevent leaks via the tapped holes in the plenum base
28	Fan	Installed in cabinet and pump house for cooling

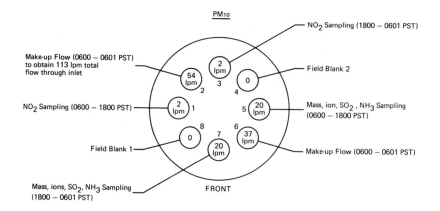

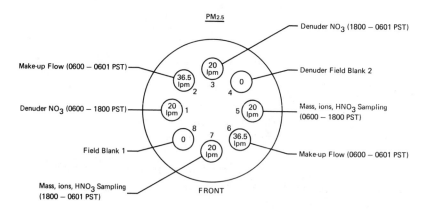

FIGURE 2.   Schematic of the plenum base for the CADMP gas/aerosol sampling system with indicated sampling periods and sampling flow rates.

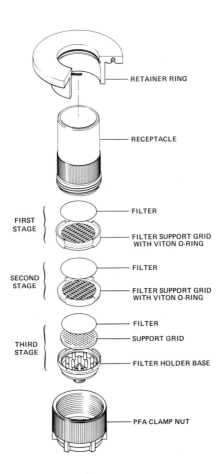

FIGURE 3. Schematic of the CADMP filter holder assembly.

The PM$_{2.5}$ calcium values were close to or less than lower quantifiable limits (LQLs), and this is evident in the comparison of data in Figure 7. This is also the cause of variability at the lower concentrations of PM$_{2.5}$ sodium and magnesium.

Stability of the deposit is probably a major cause of discrepancies between the primary and collocated PM$_{2.5}$ nitrate concentrations, as shown in Figure 8. The first three panels in this figure show: (1) PM$_{10}$ Teflon® nitrate, which is nitrate on a Teflon® filter through an aluminum PM$_{10}$ inlet; (2) PM$_{2.5}$ Teflon® nitrate, which is nitrate on a Teflon® filter through a Teflon®-coated PM$_{2.5}$ inlet; and (3) PM$_{2.5}$ particulate nitrate, which is nitrate measured on a nylon filter after having been drawn through a Teflon®-coated PM$_{2.5}$ inlet and an aluminum nitric acid denuder.

The PM$_{2.5}$ Teflon® nitrate comparison shows much more scatter than any of the other comparisons, yet the PM$_{2.5}$ particulate nitrate and PM$_{10}$ Teflon® nitrate values show good comparisons between primary and collocated measurements.

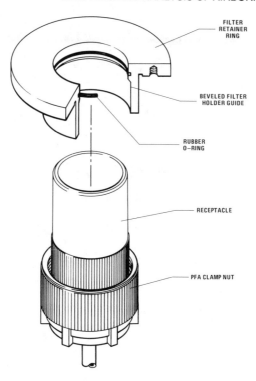

FIGURE 4.   Diagram of the filter retainer ring in the CADMP sampler base and
assembled Savillex filter holder.

These comparisons are consistent with the observations of Stelson et al.,[11] Koutrakis et al.,[12] and others that particulate nitrate dissociates from the Teflon® filters in the primary and collocated samplers in different amounts, and that the deposit on the $PM_{2.5}$ Teflon® prefilter is unstable with respect to nitrate and possibly with other species (such as ammonium) which react with nitrate. On the other hand, the PM_{10} Teflon® nitrate appears to be very stable and is approximately equivalent in concentration to the $PM_{2.5}$ particulate nitrate. The alkaline soil particles present in the coarse particle fraction may serve the same purpose as the $PM_{2.5}$ nylon filter by neutralizing the gaseous nitric acid, and the aluminum PM_{10} inlet may serve the same purpose as the $PM_{2.5}$ denuder in removing nitric acid.

The nitric acid comparison in the last panel of Figure 8 shows four major outliers and a large number of samples near zero. The nitric acid concentration is calculated from three separate measurements. If any one of these measurements is incorrect, or if any one of the samples has been mislabeled as "day" or "night," then the nitric acid value will be affected. The nighttime nitric acid values congregate about the origin and are fairly randomly distributed. These values are close to LQLs and are therefore inherently imprecise. Aside from the four outliers, the daytime values demonstrate a reasonably good reproducibility between the primary and collocated samplers.

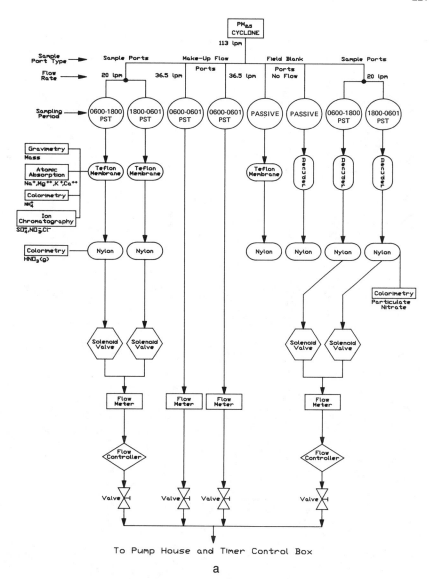

FIGURE 5.   Sampling and measurement flow diagrams of (a) CADMP PM$_{2.5}$ and (b) PM$_{10}$ units.

Figure 9 presents the sulfate (SO$_4^{-2}$), sulfur dioxide (SO$_2$), nitrogen dioxide (NO$_2$), and ammonium (NH$_4^+$) comparisons. The sulfate scatter is similar to that observed for other measurements made on the PM$_{2.5}$ Teflon® filter. The gaseous data are very repeatable, based on these collocated measurements.

These collocated analyses include samples taken between April 16 and September 25, 1989. These samplers are still being operated, and further analysis of

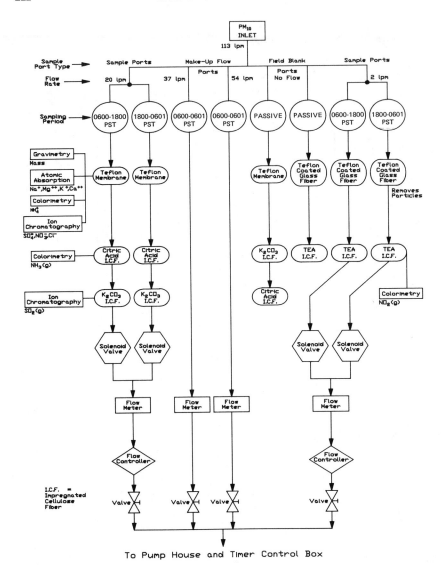

FIGURE 5b.

new data along the lines presented here should be done at 6-month intervals to determine the general applicability of these results. The collocated sampling plots presented in Figures 6 through 9 are similar to those observed for routine collocated sampling of PM_{10} using high-volume samplers, even though the sampling system is more complex and the filter media are more difficult to handle. These data indicate that, for the most part, CADMP gas/particle data are precise, equivalent, and repeatable among samplers of this design. These collocated tests do not demonstrate equivalence of this sampler with other sampler designs. Data for such comparisons are being acquired and will be reported in future publications.

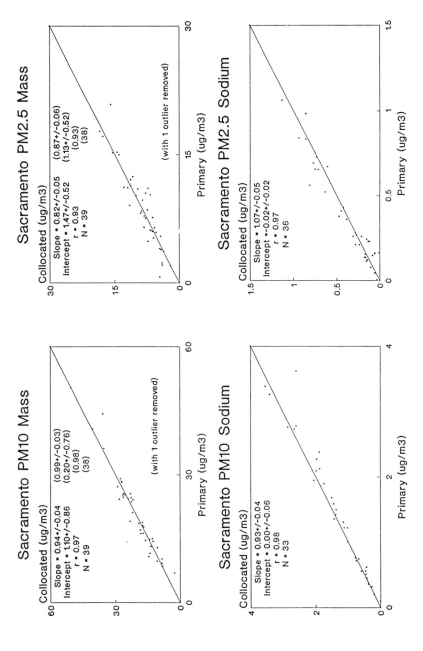

FIGURE 6. Comparisons of collocated measurements of $PM_{2.5}$ and PM_{10} mass and sodium concentrations at Sacramento, CA.

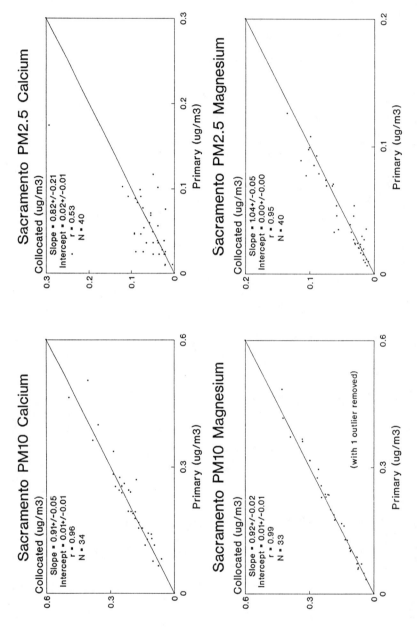

FIGURE 7. Comparisons of collocated measurements of $PM_{2.5}$ and PM_{10} calcium and magnesium concentrations at Sacramento, CA.

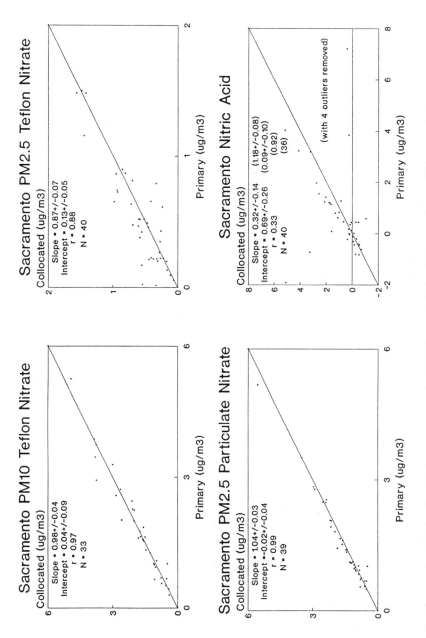

FIGURE 8. Comparisons of collocated measurements of $PM_{2.5}$ and PM_{10} Teflon® nitrate, $PM_{2.5}$ particulate nitrate, and nitric acid concentrations at Sacramento, CA.

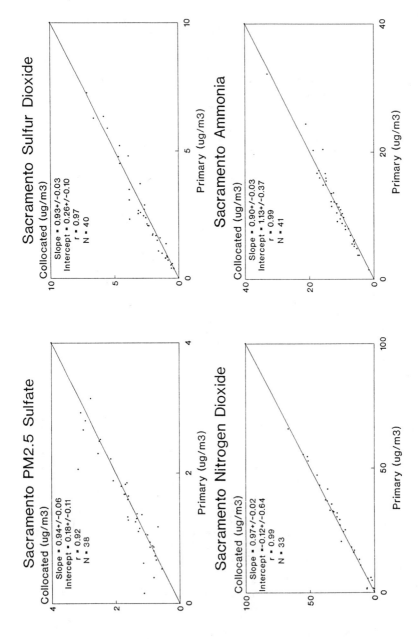

FIGURE 9.   Comparisons of collocated measurements of PM$_{2.5}$ sulfate, sulfur dioxide, nitrogen dioxide, and ammonia concentrations at Sacramento, CA.

V. SUMMARY AND CONCLUSIONS

A sampling system has been designed and deployed for the collection of atmospheric gases and particles on stacked filter media. The system allows sampling through $PM_{2.5}$ and PM_{10} size-selective inlets, adjustable flow rates through filter samples, sample switching for daytime and nighttime sampling, and inert sampling surfaces for the quantification of nitric acid. Eleven of these units have been deployed for multiyear operation in California. These samplers have been operating since 1988 with high data recovery rates and good reliability. Collocated sampling shows species concentrations to be reproducible when they are at levels which exceed lower quantifiable limits. Comparisons with other sampling systems are in progress.

This work was performed under the sponsorship of the California Air Resources Board, Sacramento, CA, under Contract No. A6-076-32.

REFERENCES

1. Watson, J. G., J. C. Chow, R. T. Egami, J. L. Bowen, C. A. Frazier, A. W. Gertler, and K. K. Fung. "Program Plan for California Acid Deposition Monitoring Program," Desert Research Institute Document No. 8868.1F1, prepared for California Air Resources Board, Sacramento, CA (1990).
2. "Ambient Air Monitoring Reference and Equivalent Methods: 40 CFR Part 53." *Fed. Regist.*, 52:24724 (1987).
3. Chan, T., and M. Lippman. "Particle Collection Efficiencies of Sampling Cyclones: An Empirical Theory," *Environ. Sci. Technol.*, 11:377 (1977).
4. Mueller, P. K., G. M. Hidy, J. G. Watson, R. L. Baskett, K. K. Fung, R. C. Henry, T. F. Lavery, and K. K. Warren. "The Sulfate Regional Experiment: Report of Findings, Vol. 1, 2, and 3," Electric Power Research Institute Report EA-1901, Palo Alto, CA (1983).
5. Olin, J. G., and R. R. Bohn. "A New PM_{10} Medium Flow Sampler," paper presented at the 76th Annual Meeting of the Air Pollution Control Association, Atlanta, GA (1983).
6. Hering, S. V., D. R. Lawson, I. Allegrini, A. Febo, C. Perrino, M. Possanzini, J. E. Sickles II, K. G. Anlauf, A. Wiebe, B. R. Appel, et al. "The Nitric Acid Shootout: Field Comparison of Measurement Methods," *Atmos. Environ.* 8:1519 (1988).
7. John, W., S. M. Wall, J. L. Ondo, and H. C. Wang. *Dry Deposition of Acidic Gases and Particles.* (Berkeley, CA: Air and Industrial Hygiene Laboratory, California Department of Health Services, 1986).
8. Appel, B. R., and V. Povard. "Loss of Nitric Acid Within Inlet Devices for Atmospheric Sampling," paper presented at EPA/APCA Symposium on Measurement of Toxic and Related Air Pollutants, U.S. Environmental Protection Agency, Research Triangle Park, NC (1987).
9. John, W., S. M. Wall, and J. L. Ondo. "A New Method for Nitric Acid and Nitrate Aerosol Measurement Using the Dichotomous Sampler," *Atmos. Environ.* 22:627 (1988).

10. Watson, J. G., J. C. Chow, R. T. Egami, J. L. Bowen, C. A. Frazier, A. W. Gertler, D. H. Lowenthal, and K. K. Fung, Measurements of Dry Depository Parameters for the California Acid Deposition Monitoring Program," Desert Research Institute Document No. 8368.2F1, prepared for California Air Resources Board, Contract A6-076-32, Sacramento, CA (1992).

11. Stelson, A. W., S. K. Friedlander, and J. H. Seinfeld. "A Note on the Equilibrium Relationship Between Ammonia and Nitric Acid and Particulate Ammonium Nitrate," *Atmos. Environ.* 13:369 (1979).

12. Koutrakis, P., J. M. Wolfson, J. L. Slater, M. Brauer, and J. D. Spengler. "Evaluation of an Annular Denuder/Filter Pack System to Collect Acidic Aerosols and Gases," *Environ. Sci. Technol.* 22:1463 (1988).

CHAPTER 14

Ice Cores as a Unique Means of Retrieving Records of Chemical Compounds in the Atmosphere: Evidence of an Undocumented Tropical Volcanic Eruption in the Early 19th Century

Jihong Dai, Ellen Mosley-Thompson, and Lonnie G. Thompson

TABLE OF CONTENTS

0-87371-606-0/93/$0.00+$.50

© 1993 by Lewis Publishers

I. INTRODUCTION

The history of the composition of the Earth's atmosphere is critical to understanding the present and forecasting the future role of the dynamic atmosphere in the global climate system. Polar ice sheets of Greenland and Antarctica and certain nonpolar glaciers, through continuous preservation of precipitation, contain proxy records of atmospheric chemistry covering hundreds of thousands of years.[1] Ice cores from carefully selected sites provides a unique means of retrieving these records.[2] For example, air trapped in ice bubbles provides a sample of the history of the atmosphere. Atmospheric components incorporated in the snow/ice lattice contain clues of chemical changes in the atmosphere. Analysis of insoluble particles in the ice and snow can provide information about atmospheric turbidity. However, the extrapolation of ice core results to atmospheric composition records requires caution and development of valid transfer functions.

On a global scale atmospheric chemical composition is affected by volcanism, changes in climate, anthropogenic emissions, as well as other factors. For example, explosive volcanic eruptions emit large amounts of ash and gaseous aerosols to the atmosphere. Most of the gaseous aerosols are sulfur compounds, mainly SO_2, that can have atmospheric residence times from weeks to several years. The effects of these atmospheric changes can be reconstructed from the analyses of ice cores retrieved from sites in strategic locations. This work illustrates the use of ice cores for recovering the record of a large tropical volcanic eruption around 1809 A.D. which significantly altered the global atmospheric sulfur burden for 2 years. This apparently explosive eruption has not been previously recorded in historic chronologies of volcanic eruptions.

In 1985, an ice core 302 m long was recovered from Siple Station in West Antarctica (75°S, 84°W) as part of an extensive glaciologic investigation. In 1989, a core 84 m long was recovered from central Greenland (72°N, 38°W) near the summit of the Greenland Ice Sheet.

II. ICE CORE RETRIEVAL

Recovery of uncontaminated ice cores is the first step in ice core research. The precipitation over the polar areas is the cleanest in the world. Major atmospheric chemical species are present in polar snow and ice in low parts per billion (nanogram per gram of ice) levels.[3,4] The ice core retrieval process consists of the following steps: (1) surveying and selecting an appropriate site; (2) core drilling, recovery, and logging; (3) preliminary analyses if feasible; and (4) packaging for shipment to the laboratory.

The ice cores, cylinders of 8–13 cm in diameter and usually 100 cm in length, are in constant contact with drill, analytical equipment, or packing materials. Efforts are made at each of these steps to prevent serious contamination. In the recovery of the 1985 Siple core, the drill site was selected 2 km upwind from Siple Station so that the chance for contamination from station activities would be minimized. The drilling of the 302-m core was conducted in a 3-m snow trench to avoid contamination by drifting snow and for ease of drill operation.[5]

Similar practices that minimize contamination should be applied to all ice core recovering operations. These include: (1) selecting a drilling site away from stations and camps, (2) pretesting equipment that will come into contact with the snow and ice, (3) using clean laboratory clothing when working and sampling, and (4) precleaning containers and tools.

III. SAMPLE PREPARATION AND ANALYSIS

After the frozen ice cores have been transported to the laboratory, they are prepared and decontaminated for chemical analysis. Stringent contamination control procedures must be applied in the sample preparation process. In this work, a low temperature (–15°C) clean-air room was used to prepare ice samples for ion chromatographic analysis. When handling cores, the analyst wore clean surgical gloves. The cores were sectioned into individual samples (3–5 cm in length) with a precleaned band saw. A minimum amount of snow or ice that came into contact with field gear and containers (even precleaned) used in transportation was removed during sample preparation. Contamination removal consisted of rinsing the ice sample with ultraclean deionized water to melt away exposed ice surfaces. The upper part of a core is firn, or condensed porous snow, that cannot be rinsed without introducing a substantial amount of deionized water. In this case, the exposed outer layers were carefully cut off with the band saw. The decontaminated ice or snow samples were then allowed to melt in precleaned containers. The sample preparation procedures are different for various analyses. For analysis of major chemical species of ice impurities, the decontamination step (removal of exposed ice) is necessary, whereas for oxygen isotope analysis, this rigorous decontamination is not necessary.

Analytical techniques for ice core chemistry include ion chromatography for major ions,[6,7] atomic absorption for metal elements,[8] and acid titrimetry for acidity[9] determinations. For this work all analyses except for oxygen isotope ratio determination were conducted in a Class 100 clean room laboratory. When determining species that may be present in ambient lab air (e.g., trace gases), it is important to melt the samples in capped containers and to determine the analytical background. Three common inorganic anions (chloride, sulfate, and nitrate) were measured with a Dionex 2010i ion chromatograph with an AS4A anion separator column. The eluent was a solution of $NaHCO_3/Na_2CO_3$ (0.0022/0.007 M). The detection limits for Cl^-, NO_3^-, and SO_4^{2-} were ≈ 1 ng/g, with an analytical uncertainty of $\leq 5\%$ at the 50 ng/g level. Working standards were prepared daily from stock solutions and procedural blanks were included in each analytical session for every step in sample preparation beginning with band saw cutting.

IV. DATING OF ICE CORES

Determination of the depth-age relationship for an ice core is essential for meaningful interpretation of proxy data obtained through the analyses. Seasonal (annual) signals in the strata, similar to growth rings in trees, are used to count years from the surface downward.

In polar areas, winter snow is generally isotopically lighter than summer snow. Oxygen isotopic ratios ($\delta^{18}O$) usually provide a reliable method of dating ice cores from areas where annual accumulation of snow exceeds 0.25 m of water equivalent.[10,11] $\delta^{18}O$ for the Siple core is illustrated in Figure 1. The $\delta^{18}O$ oscillations correspond to changes in season and to a large extent reflect seasonal temperature cycles, with less negative (peak) values indicating warmer, summer precipitation.[11] Also shown in Figure 1 are SO_4^{2-} and NO_3^- concentrations. The seasonal variation of SO_4^{2-} is apparent and confirmed when compared to the $\delta^{18}O$ profile. Concentrations of other chemical species have also been found to vary seasonally in polar snow, and the dating technique based on seasonal concentration variations has been referred to as glaciochemical dating.[12] Parameters such as $\delta^{18}O$ and SO_4^{2-} concentration are important tools for dating ice cores. However, the seasonality of a parameter may be different from location to location. For instance, NO_3^- concentrations exhibit a strong seasonal signal in Greenland cores (Figure 2) where they are often used for dating. In contrast, NO_3^- concentrations are much less seasonal in Antarctic cores (Figure 1) and cannot be used as a consistently reliable dating tool. Microparticle concentrations have also been shown to exhibit an excellent seasonal signal in Greenland cores[13] (Figure 2).

The seasonal concentration cycles of some chemical species in snow correspond directly to concentration cycles in atmospheric aerosols.[14] The processes controlling the seasonal variations of the atmospheric species are not very well understood in most cases. It has been proposed that the strong SO_4^{2-} seasonality in Antarctic snow and aerosol results from the available solar radiation controlling the rate of oxidation of reduced atmospheric sulfur species to SO_4^{2-}.[15] Conversely, seasonal characteristics of chemical species concentrations can be used to study the geochemical and atmospheric processes that affect the presence of the airborne chemicals and their relative abundance in the atmosphere.

Using the combination of $\delta^{18}O$ and SO_4^{2-} concentration, the 302-m Siple core was dated to 1417 A.D. with an uncertainty of ± 1 year for the most recent 200 years. The Greenland summit core was dated with the seasonal NO_3^- and insoluble

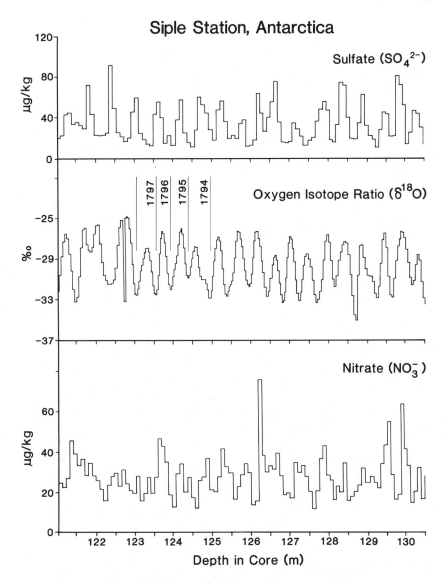

FIGURE 1. Seasonal variations of $\delta^{18}O$ and concentrations of SO_4^{2-} and NO_3^- in the Siple, Antarctica core provide a means for dating the cores. The definition of a "year" in ice cores is different from a calendar year, but covers approximately the same length of time (12 months). Notice the irregular cycles in NO_3^- concentrations relative to SO_4^{2-} and $\delta^{18}O$.

Summit, Central Greenland

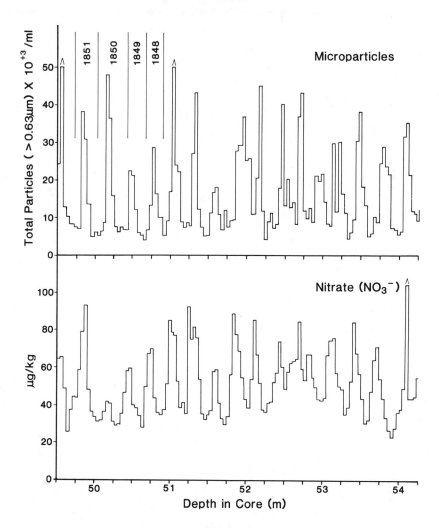

FIGURE 2.   Seasonal variations of microparticle concentrations and NO_3^- in the Greenland summit core provide a means for dating the cores. The seasonality of $\delta^{18}O$ (not shown here) is not as well preserved at this site due to isotopic diffusion. Note that SO_4^{2-} concentration variations are not reliable for dating.

microparticle concentrations (Figure 2). The accuracy of the time scales can be verified with the so-called time stratigraphic "horizons" in cores. The 1783 eruption of the Icelandic volcano, Laki, distributed large amounts of ash and gaseous aerosols to the high northern latitudes. Extremely high SO_4^{2-} or acidity levels associated with

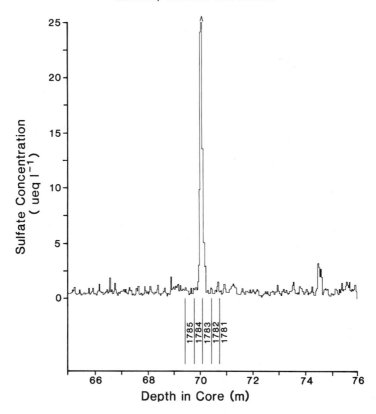

FIGURE 3.   The time stratigraphic "horizon" of the 1783 Laki eruption is evident
in the Greenland summit core. The extremely high SO_4^{2-} concentrations occur in 1783 according to the time scale established using
dust and NO_3^{-} concentrations from the top of the core. In the Siple,
Antarctica core (not shown here) SO_4^{2-} concentrations were not
elevated above background at the depth corresponding to 1783. See
text for discussion.

Laki have been found in all Greenland ice cores.[12,16] It has been demonstrated[17] that
the Laki acidity or SO_4^{2-} levels in Greenland snow are the highest in the last 1000
years. In the Greenland summit core, extremely high concentrations of SO_4^{2-} from
the Laki eruption were found at the depth of 70.2–70.4 m (Figure 3). The time scale
was produced independently using seasonal variations in microparticle and NO_3^{-}
concentrations and the occurrence of Laki in the 1783 horizon confirmed the time
scale. The SO_4^{2-} concentrations measured in this summit core are in agreement with
previously reported Laki SO_4^{2-} concentrations.[16]

V. VOLCANIC ERUPTIONS AND SULFATE IN ICE CORES

Volcanoes are an important source of sulfur compounds in the atmosphere. Worldwide they contribute up to 15% of the total global sulfur emissions to the atmosphere.[18] Compounds such as SO_2 and H_2S are major components of gases emitted from eruptions, and they are oxidized to SO_4^{2-} in the atmosphere.[18] Small and moderate eruptions affect atmospheric chemistry on a regional scale. Large explosive eruptions inject gases directly into the stratosphere, where the chemical species along with fine dust particles are spread globally.[19] The residence time of stratospheric volcanic sulfur can be as long as several years. Stratospheric aerosols consisting largely of sulfuric acid particles[20] can interact with solar radiation and thermal emission from the earth's surface and therefore affect the global radiation balance and initiate changes in the global climate system.[21,22] Recently, the impact of explosive volcanic eruptions on climate has been a subject of intense research.[23,24]

Proxy records of global volcanism are preserved in ice cores as impurity SO_4^{2-} or sulfuric acid concentrations which are elevated above natural background levels, such as in the case of Laki in Greenland cores. Within the depth range 113.3–113.6 m of the Siple core, SO_4^{2-} concentrations of up to 270 ng/g were found vs a background of 10–90 ng/g (Figure 4). This layer of ice was dated as 1816 ± 1 A.D. using $\delta^{18}O$ and SO_4^{2-} seasonal cycles and without reference to horizons. The SO_4^{2-} concentrations at the depth of 60.7–61.0 m of the Greenland summit core are 200–260 ng/g vs a background of 10–60 ng/g (Figure 4). This depth interval was dated as 1816 ± 4, or 1816 ± 0 with respect to the Laki SO_4^{2-} horizon (1783).

Of the known sources of SO_4^{2-} in polar snow and ice, only explosive volcanic eruptions can elevate SO_4^{2-} levels drastically for a period of several years.[3,4] In April, 1815, the Tambora volcano on Sumbawa Island, Indonesia (8°S, 118°E), erupted violently, causing great devastation. The Tambora eruption is one of the largest since the end of the last glaciation (10,000 years ago), and has been regarded as the greatest in historic time.[25] The acidity or SO_4^{2-} signal of Tambora has been found in previous ice cores from both Greenland and Antarctica.[17,26] Comparison with previous ice core data on Tambora and our relatively accurate dates for the elevated SO_4^{2-} concentrations strongly suggest that the Tambora eruption is responsible for the above SO_4^{2-} horizons found in the Siple and Greenland summit cores.

The flux (mass/area) of an atmospheric chemical species to the surface can be calculated from available ice core data, i.e., the concentration of the species in the liquid sample (mass/volume) and the thickness of the firn or ice sample in water equivalent. For example, the SO_4^{2-} flux (mass/area) for an individual sample is calculated by multiplying the SO_4^{2-} concentration (mass/volume) in a small volume of the melted sample by the thickness or length of the sample in water equivalent. Water equivalent depths are obtained by multiplying the core length by the density at that depth.[15] Subsequently, the annual flux can be obtained by adding the individual sample fluxes for all samples in the year. An ice core "year"

Tambora and Unknown Eruptions
Sulfate Concentration
(μeq l^{-1})

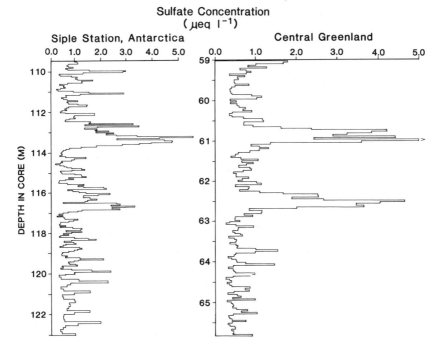

FIGURE 4.   Concentrations of SO_4^{2-} in the Siple and the Greenland summit
cores in early 19th century.

is defined as the depth interval between two adjacent seasonal minima or maxima
(see Figure 1). The annual SO_4^{2-} flux of both cores for 1800 to 1820 is shown in
Figure 5. In both cores, the Tambora eruption is associated with the exceptionally
high fluxes in 1816 and 1817. At Siple, the 1816 flux (10.2 μg/cm^2) is six times
that of the previous year (1.7 μg/cm^2) which is representative of background
annual SO_4^{2-} flux in this area. In the Greenland summit core, the Tambora SO_4^{2-}
flux is similarly higher than the background (Figure 1).

It is significant that the Tambora eruption is recorded in both Siple and
Greenland summit cores. Whether the aerosols of a particular eruption can reach
the polar areas and be recorded in the ice strata depends on many factors, such as
the magnitude of the eruption, chemical composition of the eruption clouds, and
latitude and altitude of the volcano. In Antarctica, due to the lack of large
volcanoes on the continent and in the surrounding oceans, few volcanic eruptions
have left signals sufficiently large to be detected in ice cores. Only extremely large
tropical and southern hemisphere eruptions produce SO_4^{2-} signals detectable
above the background level; and eruptions in the middle and high northern
latitudes ($\geq$20°N) do not produce a detectable signal in Antarctic ice.[26] Similarly,
eruptions south of 20°S latitude are not likely to be recorded in Greenland ice.
Only large explosive eruptions in the low latitudes can significantly perturb

stratospheric sulfur levels in both hemispheres to be recorded in both the Antarctic and Greenland ice sheets.[27] Tambora is found in both the Siple and the Greenland summit cores because of its enormous magnitude and its tropical location. In contrast, the 1783 eruption of the Laki volcano in Iceland (64°N), although strongly recorded in Greenland ice, is not found in any Antarctic cores.

VI. UNKNOWN ERUPTION AROUND 1809 A.D.

Figure 5 also reveals another prominent high SO_4^{2-} flux at 1810 in both cores. Sulfate levels return to their respective background concentrations in both cores after 1811. The striking similarity between this event and the Tambora signal in both flux and concentration profiles (Figure 4) strongly suggests that a volcanic eruption is the most probable cause of this high SO_4^{2-} influx to both polar areas. The suddenness of the concentration or flux increase and short duration of the event are apparent evidence of an earlier injection of volcanic sulfur into the atmosphere and the subsequent removal of the extra sulfur burden from the atmosphere; no other natural source has been demonstrated to cause such changes to the atmospheric sulfur reservoir.[18] The coincidence of the increased flux in both polar cores strongly indicates a single eruption as the source for this global increase in atmospheric sulfur. Figure 5 shows that maximum SO_4^{2-} deposition from the 1815 Tambora eruption occurred in 1816, with reduced but still elevated deposition in the following year (1817). The earlier eruption appears to follow a similar 2–year depositional pattern, suggesting a probable eruption time in 1809 (dating is accurate to ±1 year).

The unknown eruption may be compared with the Tambora eruption just 6 years later. The presence of the SO_4^{2-} signal of the 1809 eruption in both the Greenland and Antarctic cores strongly indicates a low latitude location for this unnamed volcano, probably located between 20°S and 20°N. A more quantitative comparison may be obtained using the calculated fluxes. The net contribution of SO_4^{2-} flux from each eruption is estimated from the total flux corrected for the background. The data are presented in Figure 5 for both cores. The results show that at Siple the net SO_4^{2-} flux from the 1809 eruption is 43% that of Tambora, while in central Greenland it is 58%. Based on these data, the contribution of the 1809 eruption to the global atmospheric sulfur reservoir is estimated to be approximately half that of Tambora. Self et al.[25] estimated the ash and tephra volume of Tambora to be 150 km³. Clausen and Hammer,[16] using Greenland ice cores, calculated the 1816 Tambora eruption to have contributed a total of 145 million tons of sulfur (as SO_2) to the atmosphere. The 1809 eruption thus emitted approximately 73 million tons of SO_2. In comparison, the catastrophic 1883 Krakatau (6°S, 105°E) eruption is estimated to have contributed 19–25 million tons[26] or 36 million tons[16] of SO_2.

If we assume that the ash and tephra volume from the 1809 eruption is also ≈50% that of Tambora, then this particular eruption should undoubtedly be ranked with the most explosive eruptions in historic time. An ash and tephra volume of

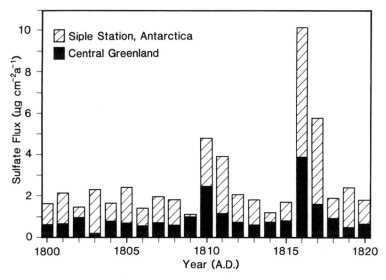

FIGURE 5.   Annual SO_4^{2-} flux derived from cores at Siple in West Antarctica and near the summit in Greenland. The method for flux calculation is described in text. Dating is accurate to ±1 year.

75 km³ would warrant a 5 or 6 on the Volcanic Explosivity Index[28] (VEI); and a VEI of 5 or 6 is suggested for this 1809 eruption. In the most recently compiled comprehensive records of global volcanic eruptions dating back to 8300 B.C. by Simkin et al.,[29] only Tambora is given the highest VEI, 7. No eruption of >3 VEI is recorded between 1800 and 1812 in this volcanic chronology. Only 11 known eruptions over the last 200 years are given VEI ≥5 by Newhall and Self.[28] Examples of these types of large explosive eruptions are the 1883 Krakatau eruption (VEI = 6), the 1835 Cosiguina eruption (VEI = 5), and the 1912 Katmai eruption (VEI = 6). Other more recent, but less violent, eruptions are the 1963 Agung eruption (VEI = 4) and the 1980 Mount St. Helen eruptions (VEI = 3, 5). However, the estimate of magnitude of the 1809 eruption and the assigned VEI must be regarded as tentative, for the eruption volume may not be proportionally reflected in the amount of gaseous sulfur emitted. This represents the general difficulty of estimating volcanic eruptions with ice core data.[26] Only the amount of sulfur in gases emitted by eruptions can be assessed with ice core acidity or SO_4^{2-} measurements. On the other hand, conventional estimates are usually less accurate for gaseous content than for eruption volumes. Therefore, ice core data can be used to complement other methods of indexing volcanic eruptions. In fact, it has been proposed[26] that a glaciological volcanic index be used, based on ice core measurements of explosive volcanic eruptions.

There are several reasons why the 1809 eruption may have escaped earlier detection in Greenland and Antarctic ice cores. High resolution records are not produced for most ice cores because they are not analyzed continuously and in sufficient detail to detect each annual layer. Generally, only selected sections are analyzed by multiple analytical procedures such as $\delta^{18}O$, dust, soluble chemical species, and pH. These selected sections are identified by electrical conductivity measurements as possibly containing unusual chemical signals or are suspected to contain specific "known events" such as Tambora or Krakatau. In fact, Hammer[30] noted elevated acidity around 1810 in the Crête (Greenland) core which he suggested might be due to unnoticed eruptions. Hammer did not investigate this further in his later work. Legrand and Delmas[26] reported two closely spaced (6–8 years) H_2SO_4 or SO_4^{2-} maxima in the early 1800s in several Antarctic ice cores. However, time scale uncertainties arising from the low snow accumulation over East Antarctica (~0.04 m H_2O equivalent) made it impossible to determine which was associated with Tambora. In fact, Tambora was first assigned to the older event, and the younger was assigned to Galunggung (7°S, 108°E, VEI = 5) in 1822; however, subsequent work has led to the correct assignment of Tambora to the younger eruption. No later studies have investigated the 1809 eruption, and no interhemispheric comparisons have been explored.

Our data from well-dated cores from both Antarctica and Greenland lead us to propose a major volcanic eruption in the early 19th century (1809 ± 1). The erupting volcano was probably located in the tropics near the equator. The magnitude of the eruption was quite large, with a contribution of ≈70 million tons of SO_2 to the global atmosphere. The eruption clouds clearly reached the stratosphere where the sulfuric acid aerosols generated by this unknown eruption were distributed globally. The removal of the volcanic sulfur from the atmosphere occurred over the following 2 years.

VII. CONCLUSION

Ice cores obtained from polar areas of the world contain valuable records of the history of the atmosphere because atmospheric chemical constituents are preserved chronologically in accumulated precipitation. Therefore, chemical analyses of ice cores provide unique and valuable information on histories of atmospheric composition. These records can be integrated to reconstruct the paleoclimatic history on a global scale.

Explosive volcanic eruptions leave signals in polar ice sheets by elevating SO_4^{2-} or acid levels above the natural background. Two such volcanic SO_4^{2-} horizons corresponding to two major eruptions were found in ice cores from central Greenland and West Antarctica. The dates for the SO_4^{2-} horizons, correlated in both cores to within 2 years, were determined to be A.D. 1816–1817 and 1810–1811, respectively. The more recent one (1816–1817) is attributed to the eruption of Tambora volcano in April 1815, the greatest historically recorded volcanic eruption. The earlier eruption probably occurred in 1809 but is not found in published volcanic records. The quantity of SO_4^{2-} deposited in Antarctica and Greenland indicates that the eruption was approximately half the magnitude of Tambora. We suggest that it should be assigned a VEI of at least 5 and probably 6 on the scale of explosivity. For comparison, other large low latitude eruptions include Tambora (VEI = 7) and the 1883 Krakatau eruption (VEI = 6). The 1809 eruption contributed a total of 62–84 million tons of SO_2 to the atmosphere and affected the stratospheric sulfur budget on a global scale. The distribution of the SO_4^{2-} from this unknown eruption to both hemispheres indicates a tropical location for the erupting volcano, similar to that of Tambora or Krakatau.

ACKNOWLEDGMENTS

We thank K. Mountain, K. Najmulski, and J. Paskievitch for their participation in the Antarctic and Greenland field programs and the Polar Ice Coring Office of Fairbanks, Alaska (formerly of Lincoln, NE) for conducting the drilling. M. Davis and L. Klein conducted the particulate analyses. Their contribution to this work is gratefully acknowledged. The oxygen isotope measurements were made at the Quaternary Isotope Laboratory of the University of Washington by P.M. Grootes and at the Geophysical Institute of the University of Copenhagen by N. Gundestrup. The diagrams were produced by J. Nagy. This work was supported by NSF grants DPP-8410328 and DPP-8520885. This is contribution 758 of the Byrd Polar Research Center.

REFERENCES

1. Robin, G. de Q., Ed. *The Climatic Record in Polar Ice Sheets.*, (London: Cambridge University Press, 1983).
2. Oeschger, H. "The Contribution of Ice Core Studies to the Understanding of Environmental Processes," in *Greenland Ice Core: Geophysics, Geochemistry and the Environment.* Geophysical Monograph 33, C.C. Langway, Jr., H. Oeschger, and W. Dansgaard, Eds. (American Geophysical Union, 1985), pp. 9–17.
3. Herron, M. M. "Impurity Sources of F^-, Cl^-, NO_3^-, and SO_4^{2-} in Greenland and Antarctic Precipitation," *J. Geophys. Res.* 87(C4):3052–3060 (1982).
4. Legrand, M., and R. J. Delmas. "The Ionic Balance of Antarctic Snow: A 10-Year Detailed Record," *Atmos. Environ.* 18(9):1867–1874 (1984).
5. Mosley-Thompson, E., L. G. Thompson, K. R. Mountain, and J. F. Paskievitch. "Paleoclimatic Ice Core Program at Siple Station," *Antarct. J. U.S.* 21(5):115–117 (1986).
6. Legrand, M., M. de Angelis, and R. J. Delmas. "Ion Chromatographic Determination of Common Ions at Ultratrace Levels in Antarctic Snow and Ice," *Anal. Chim. Acta* 156:181–192 (1984).
7. Saigne, C., S. Kirchner, and M. Legrand. "Ion-Chromatographic Measurements of Ammonium, Fluoride, Acetate, Formate, and Methanesulfonate Ions at Very Low Levels in Antarctic Ice," *Anal. Chim. Acta* 203:11–21 (1987).
8. Boutron, C. F., C. C. Patterson, C. Lorius, V. N. Petrov, and N. I. Barkov. "Atmospheric Lead in Antarctic Ice During the Last Climatic Cycle," *Ann. Glaciol.* 10:5–9 (1988).
9. Legrand, M., A. J. Aristarain, and R. J. Delmas. "Acid Titration of Polar Snow," *Anal. Chem.* 54:1336–1339 (1982).

10. Dansgaard, W., H. B. Clausen, N. Gundestrup, S. J. Johnsen, and C. Rygner. "Dating and Climatic Interpretation of Two Deep Greenland Ice Cores," in *Greenland Ice Core: Geophysics, Geochemistry and the Environment,* Geophysical Monograph 33, C.C. Langway, Jr., H. Oeschger, and W. Dansgaard, Eds. (American Geophysical Union, 1985), pp. 71–76.

11. Dansgaard, W., S. J. Johnsen, H. B. Clausen, and N. Gundestrup. "Stable Isotope Glaciology," *Medd. Grönl.* 197(2) (1973).

12. Herron, M. M. "Glaciochemical Dating Techniques," in *Nuclear and Chemical Dating Techniques,* ACS Symposium Series 176, L.A. Currie, Ed. (Washington, DC: American Chemical Society, 1982).

13. Thompson, L. G., and E. Mosley-Thompson. "Temporal Variability of Microparticle Properties in Polar Ice Sheets," *J. Volcanol. Geotherm. Res.* 11:11–27 (1981).

14. Wagenbach, D., U. Gorlach, K. Moser, and K. O. Munnich. "Coastal Antarctic Aerosol: the Seasonal Pattern of its Chemical Composition and Radionuclide Content," *Tellus* 40B:426–436 (1988).

15. Mosley-Thompson, E., J. Dai, L. G. Thompson, P. M. Grootes, J.K.Arbogast, and J. F. Paskievitch. "Glaciological Studies at Siple Station (Antarctica): Potential Ice Core Paleoclimatic Record," *J. Glaciol.* 37(125), 11–22 (1992).

16. Clausen, H. B., and C. U. Hammer. "The Laki and Tambora Eruptions as Revealed in Greenland Ice Cores from 11 Locations," *Ann. Glaciol.* 10:16–22 (1988).

17. Hammer, C. U., H. B. Clausen, and W. Dansgaard. "Greenland Ice Sheet Evidence of Post-Glacial Volcanism and Its Climatic Impact," *Nature. (London)* 288:230–235 (1980).

18. Cadle, R. D. "A Comparison of Volcanoes with Other Fluxes of Atmospheric Trace Gas Constituents," *Rev. Geophys. Space Phys.* 18:746–752 (1980).

19. Cadle, R. D., C. S. Kiang, and J. F. Louis. "The Global Scale Dispersion of the Eruption Clouds from Major Volcanic Eruptions," *J. Geophys. Res.* 81:3125–3132 (1976).

20. Cadle, R. D. "Composition of the Stratospheric 'Sulfate' Layer," *Eos Trans.* 53:812–820 (1972).

21. Pollack, J. B., O. B. Toon, C. Sagan, A. Summers, B. Baldwin, and W. Van Camp. "Volcanic Explosions and Climatic Change: A Theoretical Assessment," *J. Geophys. Res.* 81:1071–1082 (1976).

22. Chou, M.-D., L. Peng, and A. Arking. "Climate Studies with a Multilayer Energy Balance Model. Part III. Climatic Impact of Stratospheric Aerosols," *J. Atmos. Sci.* 41:759–767 (1984).

23. Bradley, R. S. "The Explosive Volcanic Eruption Signal in Northern Hemisphere Continental Temperature Records," *Clim. Change* 12:221–243 (1988).

24. Mass, C. F., and D. A. Portman. "Major Volcanic Eruptions and Climate: A Critical Evaluation," *J. Clim.* 2:566–593 (1989).

25. Self, S., M. R. Rampino, M. S. Newton, and J. A. Wilff. "Volcanological Study of the Great Tambora Eruption of 1815," *Geology* 12:659–663 (1984).

26. Legrand, M., and R. J. Delmas. "A 220-Year Continuous Record of Volcanic H_2SO_4 in the Antarctic Ice Sheet," *Nature (London)* 327:671–676 (1987).

27. Delmas, R. J., M. Legrand, A. J Aristarain, and F. Zanolini. "Volcanic Deposits in Antarctic Snow and Ice," *J. Geophys. Res.* 90:12,901–12,920 (1985).

28. Newhall, C. G., and S. Self. "The Volcanic Explosivity Index: An Estimate of Explosive Magnitude for Historic Volcanism," *J. Geophys. Res.* 82(C2):1231–1238 (1982).

29. Simkin, T., L. Seibert, L. McClelland, D. Bridge, C. Newhall, and J. Latter. *Volcanoes of the World* (Stroudsburg, PA: Hutchinson Ross, 1981).

30. Hammer, C. U. "Past Volcanism Revealed by Greenland Ice Sheet Impurities," *Nature* (*London*) 270(5637):482–486 (1977).

Optical Remote Sensing

CHAPTER 15

Remote Sensing of Volatile Organic Compounds (VOCs): A Methodology for Evaluating Air Quality Impacts During Remediation of Hazardous Waste Sites

Timothy R. Minnich, Robert L. Scotto,
Margaret R. Leo, and Philip J. Solinski

TABLE OF CONTENTS

0-87371-606-0/93/$0.00+$.50

I. INTRODUCTION

The need for reliable methodologies to assess air quality impacts to on-site workers and downwind populations continues to be a priority for a range of Superfund- and RCRA-related activities. Perhaps the most critical need for these methodologies is associated with excavations at hazardous waste sites or with other remedial activities that offer the potential for large releases of volatile organic compounds (VOCs). Remote optical sensing involving application of open-path spectroscopy as a means to address this need has received much attention in the past few years.[1-8] This chapter provides a detailed methodology for addressing the air monitoring needs, via remote optical sensing, associated with the remediation of hazardous waste sites.

II. BACKGROUND

Air monitoring is conducted during site remediations to achieve two objectives: (1) to ensure that contaminants do not create adverse offsite community health impacts; and (2) to facilitate decisions concerning onsite worker health and safety. For either objective, the concern is typically one of acute health impacts arising from short-term contaminant exposure. Therefore, the need for reliable, real-time instrumentation from which immediate management decisions can be made is essential. Because of the degree of risk to human health, any such monitoring system must be conservative, i.e., must not underestimate the contaminant concentrations and must minimize the chances for a false negative measurement. These requirements represent a significant problem for any system that relies upon the analysis of air samples at a point in space. Because the position of a contaminant plume will continually shift due to changing winds and turbulence, it is very difficult to establish a high degree of confidence that a single point onsite air measurement will always (or ever) represent a plume centerline (maximum) concentration. Remote optical sensing solves this very difficult problem through the determination of a path-integrated concentration.

Having an understanding of the path-integrated concentration is essential before one can work with long-path data, and this concept can be somewhat confusing. Gaseous contaminant concentrations are generally reported in units of mass of contaminant per volume of gas such as micrograms per cubic meter ($\mu g/m^3$), or volume of contaminant per volume of gas such as parts per billion (ppbv) or parts per million (ppmv). Path-integrated concentrations, however, are typically reported in units of parts-per-million-meters (ppm-m). For a remote sensing system, the total contaminant burden is measured within the approximate cylinder defined by the finite cross sections of the light beam at each end and the length of the beam itself. This contaminant burden is then normalized to a pathlength of 1 m. If, for example, a path-integrated concentration of 30 ppm-m is reported, no information concerning the contaminant distribution within the beam can be directly inferred and the instrument response would be identical whether there was a uniform concentration of 30 ppmv over a distance of 1 m, 3 ppmv over a distance of 10 m, 300 ppbv over a distance of 100 m, or 30 ppbv over a distance of 1 km. It is immediately evident that the integrated concentration reported is directly proportional to the total pathlength for a given uniform contaminant concentration. It also follows that for a site from which contaminants are emanating in a plume of narrow width (e.g., 10 m) the same path-integrated concentration will be reported regardless of the pathlength, as long as the narrow plume remains contained within the observing pathlength and there is no upwind contaminant contribution.

Although typically reported in units of ppm-m, path-integrated concentrations are generally converted to units of mg/m^3 times m, or mg/m^2, to facilitate emission rate calculations. The usefulness of this conversion will become apparent later, even though the physical meaning of the mg/m^2 term may not be apparent due to cancellation of the meters units.

There are two types of remote sensing configurations: the unistatic and the bistatic. Most popular today is the unistatic configuration in which a beam in the infrared (IR) or ultraviolet (UV) is propagated across the plume along a pathlength (up to several hundred meters in length) to a retroflector and back to the detector. The bistatic configuration employs the detector and source at opposite ends of the signal. Regardless of the configuration employed, remote sensing offers a cost-effective and very powerful technology to achieve remedial project objectives when used in conjunction with onsite meteorological data.

The Fourier-transform infrared (FTIR) system is frequently the system of choice for making open-path measurements at hazardous waste sites. The Fourier transform is a mathematical process applied to the spectral data. The FTIR spectrometer has a number of significant advantages over other spectral analyzers. For several reasons, it is more efficient at collecting light and analyzing it spectrally. In addition, FTIR spectrometers simultaneously analyze light at all wavelengths within the IR spectrum so that they all are detected at one instant, and none is thrown away. This often is referred to as the "multiplexing advantage."[9]

III. METHODOLOGY

Within classical Gaussian dispersion theory, the general equation for concentration calculated at ground level for a continuously emitting point source is given as follows:[10]

$$X(x, y, 0; H) = \frac{Q}{\pi \sigma_y \sigma_z u} \exp[-1/2(y/\sigma_y)^2] \exp[-1/2(H/\sigma_z)^2] \quad (1)$$

where:

χ = concentration, g/m^3
x = downwind distance to a receptor, m
y = crosswind distance to a receptor, m
z = vertical distance to a receptor, m
H = effective height of emission, m
Q = uniform emission rate, g/s
σ_y = standard deviation of plume concentration distribution in the horizontal direction at the distance of measurement, m
σ_z = standard deviation of plume concentration distribution in the vertical direction at the distance of measurement, m
u = mean wind speed, m/s

This relationship forms the basis for many of the U.S. Environmental Protection Agency (U.S. EPA) atmospheric dispersion models that are currently employed for estimating downwind air quality impact. Examination of this relationship shows that the downwind concentration at a given location increases with increasing source strength, but decreases with increasing wind speed and horizontal and vertical dispersion (as determined via σ_y and σ_z). The standard deviations of the plume concentrations in the horizontal and vertical are, in turn, functions of atmospheric stability and the distance downwind of the source. Nomographs defining σ_y and σ_z as a function of downwind distance for each of six stability classes are frequently used to estimate these parameters. Larger σ_y and σ_z values are associated with unstable atmospheric conditions (greater dispersion) and greater downwind distances.

If one integrates Equation 1 in the y (cross-plume) direction, the resultant representation is a crosswind-integrated concentration instead of a point concentration. Performing this integration with respect to y, from $y = -\infty$ to $y = +\infty$, yields:

$$C(x,0;H) = \frac{2Q}{\sqrt{2\pi}\sigma_z u} \exp[-1/2(H/\sigma_z)^2] \qquad (2)$$

where C = ground-level crosswind-integrated concentration at distance x, g/m^2.

Equation 2 has historically been employed in diffusion experiments to determine vertical dispersion coefficients (standard deviations of the plume concentration in the vertical direction), σ_z, from ground-level data where the source strength, Q, was known and the ground-level crosswind-integrated concentration was determined from a crosswind line or arc of point sampling measurements made at some predetermined downwind distance.

The effective height of emissions, H, is defined as the sum of the actual height of emissions and the buoyancy-induced height increment arising from an elevated effluent temperature. Because most remediations involve sites at ground level and without elevated effluent temperatures, H generally equals zero and Equation 2 reduces to:

$$C(x) = \frac{2Q}{\sqrt{2\pi}\sigma_z u}$$ (3)

Rearranging, Equation 3 may be written as:

$$Q = \frac{\sqrt{2\pi}}{2} C(x)\sigma_z u$$ (4)

Equation 4 is the general emission rate equation for a point source involving path-integrated measurement data. For a measured crosswind-integrated concentration at some specified downwind distance, the emission rate, Q, depends only on σ_z at that distance and on wind speed, u.

To estimate the offsite health impact to downwind communities, reliance upon some type of conservative dispersion model typically offers the only practical alternative. Actual concentrations could be continuously measured at each receptor of concern, but this activity is generally both cost- and labor-prohibitive. All dispersion models rely upon accurate estimates of emission rates. The ability to provide accurate emission rate estimates continually and in real-time is the key to the power of the path-integrated concentration.

As discussed earlier, false negatives must be avoided when estimating short-term exposure during remediations. Therefore, trying to predict concentrations at a precise downwind receptor is neither wise nor practical, because shifting winds may cause higher concentrations. Instead, a much more useful approach is to identify the highest concentration that could exist at the downwind distance of concern if the wind were blowing directly from the source to the receptor. To be conservative, all residents in a defined downwind "sector" would be potentially at risk of being impacted by this worst-case plume-centerline concentration.

The general relationship presented in Equation 1 can be simplified by setting y and H equal to zero to yield the maximum (plume-centerline) concentration:

$$X(x) = \frac{Q}{\pi\sigma_y \sigma_z u}$$ (5)

Equation 5 is the general-case maximum downwind impact equation for a point source. It is noted that this relationship is for a point concentration and is independent of path-integrated measurements. It is frequently employed to estimate the highest concentration to which a downwind receptor could be exposed. Using conditions of worst-case or actual meteorology and atmospheric stability, the σ_y and σ_z values are chosen for the downwind distance of concern. It is a straightforward operation to obtain an emission rate, Q, in the field from Equation 4 and then substitute that value into Equation 5 to obtain the plume-centerline concentration at the downwind distance of concern. Because σ_z is a function of downwind distance, solving Equation 5 for the general case (in which the receptor of concern is farther downwind than the path-integrated measurement) requires first solving Equation 4 for Q. However, for the special case in which the receptor of concern is at the downwind distance of measurement, Equations 4 and 5 can be combined. Beginning with Equation 5, substituting for Q from Equation 4 yields:

$$X(x) = \frac{C(x)}{\sqrt{2\pi}\sigma_y} \qquad (6)$$

Equation 6 is the special-case maximum downwind impact equation for a point source involving path-integrated measurement data. This equation is most likely to be applied to facilitate health and safety decisions involving onsite workers or fenceline receptors. Of note is that a knowledge of Q, σ_z, or u is not required.

IV. APPLICATION

An experienced air dispersion modeler will recognize that a site under remediation may be approximated better as an area source than as a point source. However, by assuming that all VOC emissions are originating from the area being excavated or disturbed and by treating this small area as a point source, one is able to apply the simple relationships in Equations 4, 5, and 6. Methods are available for treating emissions from an area source,[11] but time is of the essence in remedial air monitoring and having only an estimate of downwind impact is generally sufficient to facilitate health-based management decisions. In addition, it is important that any downwind impact estimate is conservative (i.e., overpredicts or errs on the side of safety). Treating an excavation as a point source is always conservative, because it is intuitive that for a given emission rate (mass per time) the maximum downwind concentration will always be greater if all of the mass is assumed to emanate from a single point. With this realization in mind, the following scenario illustrates how remote optical sensing can be used to achieve air monitoring objectives during site remediations.

Assume that a site heavily contaminated with benzene, toluene, ethylbenzene, and xylenes (BTEX) as a result of a major fuel spill is being remediated. Assume also that toxicity data and existing analytical data are such that a decision is made that offsite benzene concentrations are of highest concern. The on-scene coordinator (OSC) has determined that portions of the soil to be removed are saturated with fuel and that vapors released during excavation by the front-end loader offer the potential for adverse health impacts to offsite residents. In addition, there is a high degree of public and media interest in this site, and local residents are demanding that removal operations be conducted in a manner that ensures their continued safety.

An onsite meteorological tower indicates that the wind is blowing steadily from the west at 9:00 A.M. when the crew arrives. The OSC determines that the best downwind location for cross-plume monitoring is 50 m from where excavation activities will begin. Because the weather is expected to be sunny with light winds, the OSC realizes that the pathlength must be long enough to contain the plume when maximum instability is reached during the afternoon and when horizontal dispersion is greatest. Because there are few obstacles at downwind distances greater than 50 m, a decision is made to stay at this downwind location as long as the wind direction remains generally westerly. From existing nomographs,[9] the OSC immediately determines that a pathlength of 120 m will be more than adequate to contain the plume under the most unstable conditions (P-G Stability Class A), while a pathlength of only 60 m is required under current Stability Class C conditions. Because the area is flat and a good line of sight exists, the OSC makes a field decision to employ a pathlength of 200 m at all times as added insurance that the plume remains within the beam despite small variations in the mean wind direction.

The OSC has an FTIR system onsite and would like to conduct baseline monitoring to determine downwind impacts, if any, prior to disturbance of the site. During this monitoring, the wind speed is measured at about 2 m/sec, and the σ_y and σ_z values under Stability Class C at a downwind distance of 50 meters are determined to be 6.58 and 3.95 m, respectively.

Real-time monitoring results indicate that the concentration of BTEX species are too low to be detected. The manufacturer has quoted a detection limit in an otherwise nonpolluted atmosphere of not greater than 10 ppm-m for each species. Applying Equation 4 and the appropriate ppmv conversion factor for benzene (1 ppm-m = 3.25 mg/m^2 at STP), the highest emission rate that could exist is:

$$Q = (1.25) \ (32.5 \ \text{mg/m}^2) \ (3.95\text{m}) \ (2 \ \text{m/sec})$$
$$= 320.9 \ \text{mg/sec}$$

An emission rate of 320.9 mg/sec is not trivial if downwind exposure occurs close to the source. The special-case maximum downwind impact equation (Equation 6) is used to calculate the maximum (detection limit default) concentration to which a worker could be exposed at the downwind distance of measurement (50 m):

$$\chi(50 \ \text{m}) = (10 \ \text{ppmm})/[(2.51) \ (6.58 \ \text{m})]$$
$$= 0.61 \ \text{ppm}$$

It is instructive to note that the same result could have been obtained using the general-case maximum downwind impact equation (Equation 5):

$$\chi(50\text{m}) = (320.9 \ \text{mg/sec})/[(3.14) \ (6.58\text{m}) \ (3.95\text{m}) \ (2 \ \text{m/sec})]$$
$$= 1.97 \ \text{mg/m}^3 \ \text{or} \ 0.61 \ \text{ppm}$$

A comparison of this concentration to a preestablished community action level of 1.0 ppmv for benzene indicates that, under current meteorological conditions, the air quality is well within safe limits with respect to short-term exposure.

By noon the excavation has begun. The measured benzene concentration is about 50 ppm-m, the wind speed has now increased to 4 m/sec, and the Stability Class is still C. The OSC determines, based on the standard deviation of the horizontal wind direction (sigma theta or σ_Θ), that residents within a downwind sector 90° wide centered upon the mean wind direction are potentially at risk. Based on demographic data, he is aware that the nearest residence within this sector is about 0.5 km from the site. Applying Equation 4 and the benzene conversion factor to obtain the emission rate yields:

$$Q = (1.25)\ (162.5\ \text{mg/m}^2)\ (3.95\text{m})\ (4\ \text{m/sec})$$
$$= 3209.4\ \text{mg/sec}$$

Determining σ_y and σ_z at 500 m from the nomographs and substituting the above value for Q into Equation 5 yields:

$$\chi(500\ \text{m}) = (3209.4\ \text{mg/sec})/[(3.14)\ (54.9\ \text{m})\ (32.4\ \text{m})\ (4\ \text{m/sec})]$$
$$= 0.14\ \text{mg/m}^3\ \text{or}\ 0.04\ \text{ppm}$$

Hence, at this time the impacts are still within guidelines and operations can continue.

At about 6 P.M., the OSC observes that the wind speed is now 2 m/sec and that the Stability Class is now F. An emission rate of 3105 mg/sec is determined from Equation 4, some 100 mg/sec less than before. However, application of Equation 5 now yields:

$$\chi(500\ \text{m}) = (3105\ \text{mg/sec})/[(3.14)\ (18.0\ \text{m})\ (8.4\ \text{m})\ (2\ \text{m/sec})]$$
$$= 3.3\ \text{mg/m}^3\ \text{or}\ 1.02\ \text{ppm}$$

Although emissions did not pose an unacceptable health impact during the daytime, as sunset approached and a temperature inversion became established, unacceptable concentrations were calculated to exist as far as 500 m downwind of the site. The OSC, therefore, makes the decision to terminate remedial activities until the next day.

V. CONCLUSION

Remote sensing in combination with onsite meteorological data can be used to achieve the air monitoring objectives associated with the remediation of hazardous waste sites. A series of simple equations are solved to facilitate real-time management decisions concerning worker safety and community exposure to VOCs.

A computerized data acquisition system can easily be configured to accept meteorological data from an onsite tower and, with a minimum of operator input, can be programmed to provide a continual readout of the maximum downwind impact. System design can also include an alarm which is programmed to sound when the maximum concentration at downwind community receptors is calculated to exceed some preestablished action level.

REFERENCES

1. "Application of Long Path Monitors to Superfund Sites (Draft)," PEI Associates, Inc., U.S. EPA Report, Contract No. 68-03-4394, Work Assignment No. 50, PN 3759-50, OAQPS (July 1990).
2. "Remote Sensing Techniques for Measuring Trace Gases in the Ambient Air, Draft Technical Note," Radian Corporation, prepared for A. Pope, U.S. EPA, Non-Criteria Pollutant Programs Branch (November 7, 1989).
3. Stevens, R. K. "Evaluation Program for the OPSIS DOAS," memorandum from R. K. Stevens, Chief, Source Apportionment Research Branch, U.S. EPA Atmospheric Research and Exposure Assessment Laboratory (August 15, 1989).
4. "Remote Sensing of Hydrocarbons and Toxic Pollutants: Workshop Minutes," F. F. Hall, Jr., Ed. prepared for J. L. McElroy, U.S. EPA, EMSL, ORD, Las Vegas, NV (undated).
5. Cline, J. D., G. R. Jersey, L. W. Goodwin, M. N. Crunk, and O. A. Simpson. "Application of a HeNe Laser to Hydrocarbon Leak Detection Over an Oil Field," 1990 EPA/AWMA International Symposium on Measurement of Toxic and Related Air Pollutants, Raleigh, NC (May 1989).
6. Russwurm, G. M., and W. A. McClenny. "A Comparison of FTIR Open Path Data with Method TO-14 Canister Data," 1990 EPA/AWMA International Symposium on Measurement of Toxic and Related Air Pollutants, Raleigh, NC (May 1989).
7. Scotto, R. L., T. R. Minnich, and M. R. Leo. "Emissions Estimation and Dispersion Analysis Using Path-Integrated Air Measurement Data from Hazardous Waste Sites," 83rd Air and Waste Management Association Annual Meeting, Pittsburgh, PA (June 1990).
8. Minnich, T. R., R. L. Scotto, and T. H. Pritchett. "Remote Optical Sensing of VOCs: Application to Superfund Activities," 1990 EPA/AWMA International Symposium on Measurement of Toxic and Related Air Pollutants, Raleigh, NC (May 1990).
9. Minnich, T. R., R. L. Scotto, R. H. Kagann, and O. A. Simpson. "Special Report: Air Monitoring — Optical Remote Sensors Ready to Tackle Superfund, RCRA Emissions Monitoring Tasks," Hazmat World (May 1990).
10. Turner, D. B. "Work Book of Atmospheric Dispersion Estimates," U.S. EPA, Office of Air Programs, Research Triangle Park, NC, OPA Publication No. AP-26 (Revised 1970).
11. Leo, M. R., P. J. Solinski, R. L. Scotto, and T. R. Minnich. "A Detailed Methodology for Estimating Emission Rates From Superfund Sites," HMCRI Superfund 1990 Conference and Exhibition, Washington, DC (November 1990).

CHAPTER 16

The Neutral Gas Laser:
A Tool for Remote Sensing of Chemical
Species by Infrared Absorption

Paul L. Kebabian and Charles E. Kolb

TABLE OF CONTENTS

I. INTRODUCTION

This review discusses the problem of remote measurement of atmospheric trace species by differential absorption of infrared light. We are, in particular, concerned with the use of neutral rare gas lasers as a convenient and practical source of discrete infrared frequencies. In this presentation, what we mean by remote sensing is measurement of the column density:

$$\left(\int_a^b c(x)\,dx \right)$$

where $c(x)$ is the concentration of a given species between two points separated in space. Several such measurements, over different paths can provide some information about the spatial distribution, but we are not dealing with a lidar technique here — the individual measurements do not give any spatial information.

Why use remote sensing? One reason is simply to attain a pathlength that is too large to conveniently realize in an enclosed absorption cell, i.e., over a few hundreds of meters. This long absorption path often is needed to observe species at concentrations in the parts per billion volume range. Also, highly reactive species may be lost by reaction with, or adsorption on, the walls of a sampling system and enclosed cell. Nitric acid vapor is a good example of a species that presents both of these problems.

A second reason is to obtain a representative spatial average of a species being produced by an inherently heterogeneous source. An important example of this is the evolution of methane from a landfill.

The limiting case of a spatially heterogeneous source is a point leak. In this context, remote sensing can be used as a perimeter alarm around, e.g., a chemical plant, to warn of leakage of toxic materials, such as HCN or $COCl_2$.

Finally remote sensing can permit the observation of the concentration of an infrared absorber around an inaccessible or noncooperating source.

Figure 1 illustrates three of the possible configurations for remote sensing. The important thing to note here is that in the bistatic case, or the monostatic case with a retroreflector (such as a cube-corner array), a substantial fraction of the transmitted light ultimately reaches the detector. However, with an incoherent scatter, the typical return signal may be $<10^{-6}$ of that transmitted, even at relatively short ranges.

As a concrete example, consider the case of 100-m range in a bistatic system, at a wavelength of 5 μm. To minimize the size of the optics, the Gaussian laser beam is focused to put the beam waist midway between transmitter and receiver. Then, the beam diameter[1] (measured at $1/e^2$ intensity) at each end is $d = 2(\lambda\ell/\pi)^{1/2}$; when $\lambda = 5$ μm, $\ell = 100$ m, this gives $d = 2.5$ cm. In practice, one would use a receiver diameter somewhat larger than this to avoid the effect of beam wander induced by atmospheric turbulence, but the required size still is small, e.g., 15 cm or less has been used.[2]

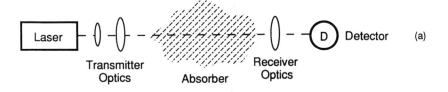

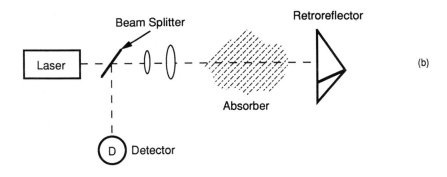

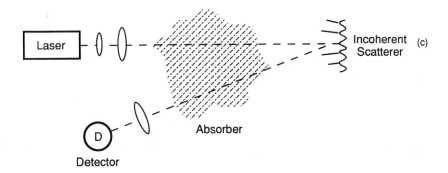

FIGURE 1.   Remote sensing configuration; (a) bistatic, (b) monostatic with a coherent retroreflector, (c) monostatic with an incoherent scatterer. Note that in (b) and (c), the transmitter and receiver are close together, whereas in (a) they are remote. In each case, the transmitted energy is monitored by another detector (not shown).

Now, consider the case of a monostatic system with an incoherent scatterer used at the same range. The scatterer is assumed to be an ideal Lambertian diffuser, and the beam is normally incident onto it. The fraction of the transmitted flux reaching the receiver then is $d^2/4 \ \ell^2$, where d is the receiver diameter. For $d = 15$ cm, $\ell = 100$ m, the fraction collected is 5.6×10^{-7}. Moreover, the signal

Table 1. Summary of Advantages and Disadvantages of Infrared Sources for Remote Sensing

Type	Advantages	Disadvantages
NDIR	Simple, relatively inexpensive	Limited sensitivity, incoherent
FTIR	Versatile	Expensive, incoherent
TDL	Versatile, high resolution	Expensive, limited laser reliability
Neutral gas laser	Simple, relatively inexpensive, good sensitivity	Least versatile

from a real target would be further reduced due to target reflectivity (typically 3%) and off-normal incidence. Thus, although systems with incoherent scatterers have been built,[3-5] they generally are usable only at relatively short ranges.

In general, the infrared absorption by the species to be measured must be distinguished from other factors that can change the received signal, such as optical losses. This is done by comparing transmission (ratio of received to transmitted power) at two or more frequencies, one of which (ideally) is centered on an absorption line of the species of interest. Either laser or incoherent sources can be used for this. Incoherent techniques, such as Fourier-transform infrared (FTIR) and gas filter correlation (nondispersive infrared [NDIR]) spectroscopy, use a continuum of frequencies over a band that includes many of the target's lines. As normally used, a tunable diode laser (TDL) is scanned continuously over a band comparable to a single line of the target. Line-tunable lasers (such as neutral rare gas lasers, or molecular lasers) operate at discrete frequencies, with limited tunability as discussed next.

Each of these IR sources has certain advantages and limits, which are summarized in Table 1. The incoherent sources can only emit a limited flux into a given area, solid angle, and spectral bandwidth, as determined by the source temperature. Thus, in remote sensing use, physically large optics may be needed to collect enough IR flux.

The tunable diode laser, as a result of its tunability, requires elaborate frequency monitoring and control. Moreover, the diodes may show a complex mode structure that can change with time. Most TDLs still must be cooled below 77 K using a closed-cycle cooler, although LN_2 cooling has recently become feasible at the short-wavelength end of the band of interest (e.g., v between 1600 and 2400 cm^{-1}). These factors make TDL spectroscopy the most complicated and expensive of these techniques.

The fixed frequency lasers (molecular and neutral rare gas) depend on an accidental coincidence between a laser line and an absorption line of the target species, and this frequency must be clear of interference from other atmospheric components, such as H_2O and CO_2. The advantage of this type of laser is that the frequency is a property of the atom or molecule in the laser; thus it is both stable and reproducible. In general, these lasers are simple to use and relatively inexpensive.

Neutral rare gas lasers were first studied from the early 1960s through the mid-1970s. The results of that work have been collected and evaluated by Bennett.[6] In the range 1.5–30 µm (frequency range 300–6000 cm^{-1}), there are 240 lines, of which 140 are from Xe and Ne. In most cases, there are unambiguous energy level assignments,[7] and thus the laser frequencies generally are known to an accuracy of about 0.01 cm^{-1}. In addition to the rare gases, there are many neutral atomic laser lines belonging to other atoms; these are of less interest, however, because in most cases the lasers are not amenable to stable, sealed off operation.

The absorption line frequencies and strengths of the principal molecules of atmospheric interest are found in the HITRAN database.[8,9] The lines in this compilation, a total of over 300,000, are computed from the known vibrational-rotational parameters of the 28 species included. The stronger line locations generally have an accuracy of better than 0.01 cm^{-1}, but weaker lines of common species, or even entire bands of rare species (such as H_2CO) may be missing or tabulated with less accuracy. Nevertheless, HITRAN has been found to generally be a reliable data source to use in determining the degree of interference expected at a given laser frequency.

To see if a given target species can be measured using a neutral rare gas laser source, its spectrum is compared with the tabulated laser lines and when a suitable coincidence is found, that frequency then is checked for interferences using the HITRAN tabulation. Since absorption lines at normal atmospheric pressure are collisionally broadened to approximately 0.1 cm^{-1} (full width), both the laser and interference frequencies usually are known with sufficient accuracy.

As noted above, we also require a reference laser line, i.e., one that is only weakly absorbed by the target species and that is close enough to the analytical line that factors such as optical losses in the system affect both lines equally. In many cases, a suitable line is available from the same laser (the methods of line selection are discussed later). Otherwise, the laser line may be Zeeman split to create a second line.

II. General Properties of Neutral Rare Gas Lasers

Most neutral rare gas lasers operate in the positive column of a glow discharge; RF excitation has also been used, especially in the earlier work.[10,11] Typical operating pressures are 0.01–10 torr. In the important case of the He-Ne laser, near-resonant excitation transfer from He to Ne is an important pumping mechanism,[12,13] while in other cases a gas mixture with He principal component is used to improve the electrical stability of the discharge. For the He-Ne laser, the variation of optical gain with pressure and current density has been studied in detail,[14] but few other neutral rare gas lasers have been studied in similar detail.

This work all refers to continuous wave (cw) laser operation; however, many lines that are weak or absent in cw operation become available with pulsed excitation.[15-19] In pulsed operation pumping effectiveness is enhanced because during the initial breakdown of the gas, a given current is carried by fewer, but more energetic, electrons;[20] this effect has been widely observed both in lasers[21] and in other gas discharge devices.[22-24] It follows from this picture of pulsed operation that the drive pulse should be actively terminated, i.e., once the plasma approaches equilibrium, the laser power output will be reduced or absent so that further drive current only increases power consumption, gas consumption, electrode degradation, etc.

A further advantage of pulsed operation is that the detector only needs to integrate the signal and noise while an IR pulse is present. The noise-equivalent energy of a measurement of the received signal is $(AT)^{1/2}/D^*$, where A is the detector area, T is the duration of the observing window, and D^* is the specific detectivity of the detector. The units of D^* are ordinarily given as cm $Hz^{1/2}$ W^{-1}, which is equivalent to cm $sec^{1/2}$ J^{-1}. Thus, when the laser transmits a series of short pulses, a smaller average power is needed to achieve the same signal-to-noise ratio. For example, the methane detection systems described by Eberle et al.,[3] Grant,[4] and Tai et al.[5] all achieved similar sensitivities. The systems of Grant (using two lasers)[4] and Tai et al.[5] used cw lasers with IR power output of around 2 mW. The authors do not specify primary power input, but this type of laser typically uses 50–100 W. The system of Eberle et al.[3] used pulsed lasers emitting average power below 100 μW; the primary power usage of that system was 6 W, an important factor in achieving a fully portable, battery-operated system. The smaller size and weight of the pulsed lasers (i.e., higher volumetric peak power output), compared to cw lasers, also contributed to portability; thus, the systems of Grant[4] and Tai et al.[5] had to be vehicle mounted, or used at a fixed location, whereas the system of Eberle et al.[3] could be carried in a backpack by a single operator.

III. Measurement of Atmospheric Methane

The most widespread use of neutral rare gas infrared lasers has been in the measurement of methane. This results from the fact that a strong absorption line (v_3P7) of methane coincides with the He-Ne laser line at 2947.9 cm^{-1}, while another strong laser line, at 2948.8 cm^{-1} is only weakly absorbed.[25] Moreover, methane (natural gas) is widely used as a fuel, and detection of methane leakage (at atmospheric concentrations in the >10 ppmv range) is important for its safe usage; the detected atmospheric level, though not hazardous in itself, is an indicator of higher concentrations close to the leak source. More recently, the role of methane as a greenhouse gas has been realized; in that context, the measurement of fluctuations of <10 ppbv in the 1.7 ppmv average ambient level is of interest.

The differential absorption between these two laser lines is approximately 8 (atm cm)$^{-1}$, at standard temperature and pressure.[4] One way to obtain these two lines is to use two separate lasers, one of which contains a methane cell to suppress the absorbed line, which otherwise is dominant; this is the method first demonstrated by Moore,[25] and has been used in several methane monitors.[3,4,26] Another approach is to use a single laser in which the spacing of the longitudinal modes is incommensurate to the spacing of the laser lines,[27] i.e., (n + 1/2) × mode spacing = (2947.9 – 2948.8) cm^{-1}. This allows the laser line to be selected by tuning the cavity length. This method has been used for a remote methane monitor,[5] and for one using an enclosed sample cell.[28] A complication of this technique is that the reference laser line has an optical gain that is a factor of approximately 2.7 smaller than for the analytical line.[29] In Tai et al.[5] the gains were equalized by adding a methane cell in the laser cavity, while in Kebabian[28] a short, low-gain plasma tube was used with optics optimized for the reference line.

Another approach to the generation of a reference line is to use the Zeeman effect to split the analytical line. Conventionally,[29,30] this is done with an axial magnetic field. However, that entails certain disadvantages: the magnetic field must be generated by a solenoid, which may require excessive power; and the two split components are circularly polarized with opposite handedness. To avoid these problems, a transverse magnetic field may be used.[31] Then, both split components are linearly polarized (perpendicular to B), and the magnetic field (up to several kilogauss [kG]) is readily obtained from permanent magnets. In a novel variation of this technique, we have used a laser in which a transverse magnetic field is applied to approximately 2/3 of the length of the plasma.[32,33] In this way, the plasma tube appears to have three, approximately equal gain lines: the analytical line and a reference line on either side. By suitable choice of the cavity node spacing, one can select among these three laser lines as done for the two laser lines just described. An exciting possibility of this approach is that since two differential absorptions can be measured, one can separately estimate methane concentration and the small interference from water vapor (actually HDO). Thus, a flux measurement can be made that is automatically corrected for water vapor flux. Figure 2 shows the calculated absorption spectra, at the splitting of ± 0.055 cm^{-1} used in our experiments. An instrument based on this laser, with an enclosed low pressure sample cell, was used in Alaska in the ABLE–3 program conducted by NASA in the summer of 1988. This instrument was used for approximately a month of nearly continuous measurement of methane production by the tundra ecosystem.[33a]

We have recently built another methane instrument that operates in a similar fashion.[33b] Since this instrument is designed to operate with a low pressure sample in an enclosed cell, the laser line is not fully split, but instead is simply broadened. By adjusting the length of the plasma in the magnetic field, as well as the field strength, we achieve tuning of ± 0.0085 cm^{-1} with virtually constant power output (a similar approach, but using an axial field along the entire plasma column, is described by Shimoda[34]). This methane monitor currently is installed in a mobile laboratory, where it is in use to study the relative importance of human and natural methane sources in the global methane budget. We have achieved noise levels of <5 ppbv root mean square (rms) at 1 sec averaging time.

IV. Measurement of Nitrous Oxide

Nitrous oxide, like methane, is a greenhouse gas. Recently,[35] there has been some concern that power plants may represent a significant source of atmospheric

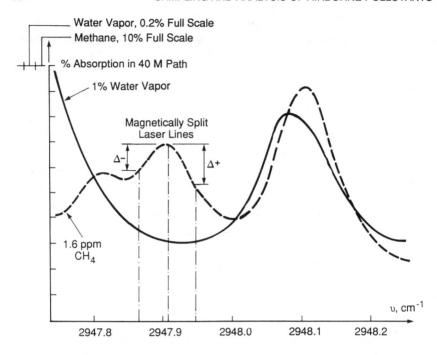

FIGURE 2.   Calculated absorption by methane and water vapor around the magnetically split He Ne laser line.

N_2O. Although it is now believed that the earlier measurements of high N_2O levels in combustion gases were the result of a sampling artifact,[36] there continues to be reason to believe there may be significant N_2O production by some types of combustion (e.g., fluidized-bed combustion).[37]

We recently studied the applicability of one of the neutral xenon laser lines for use in measuring N_2O produced by combustion sources.[38] The laser line is at 2567.37 cm^{-1} (3.895 μm), 0.06 cm^{-1} below the $2 v_1$ R4 line of N_2O. The pressure broadening coefficient of this line is 0.089 cm^{-1} atm^{-1}; thus although the frequency coincidence is not perfect, the absorption at the laser frequency of 0.74 (atm cm)$^{-1}$ is comparable to the line center absorption of 1.04 (atm cm)$^{-1}$.

This laser line, which is weak or absent in cw operation, is one of the stronger pulsed laser lines. Figure 3 shows the structure and driven circuitry of the laser used in these experiments, both of which are similar to those of the He-Ne laser used by Eberle et al.[3] The optical arrangement of the laser (shown in Figure 4) is conventional, with coupling of ~1% provided by the 0th order of the Littrow grating.[39] The grating had a nominal ghost level of 10^{-4}; the gain of the laser was high enough that it could oscillate on the ghost (as well as at the correct setting of the grating), which implies an optical gain of ~16%/cm. Within the spectral range studied, the only other lines that oscillated on the ghost were the well-known, high-gain lines at 2.026 μm and 3.508 μm.

PLASMA TUBE

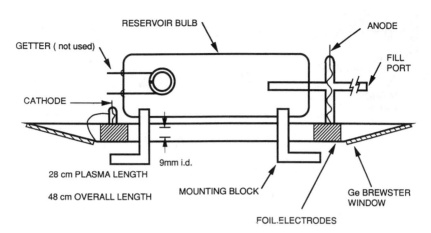

DRIVER CIRCUIT

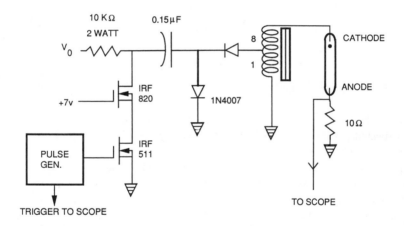

FIGURE 3.   Mechanical and electrical configuration of the laser, for low energy pulsed excitation.

Figure 5 shows the lines observed between 3.3 and 4.0 μm. In most cases, optimum conditions were: 5.0 torr He, 0.2 torr Xe, excited with a current pulse of 0.4-A peak amplitude and base width of 0.8 μsec. Note that this gives an average current density during the pulse that is similar to that used in cw lasers, and an order of magnitude less than used in other work.[16,19] The enhanced gain in pulsed operation is not simply the result of higher drive current.

The pulse energies shown in Figure 5 are those actually observed by Kebabian and Zahniser.[38] Subsequent experiments have shown that the grating used in those measurements had unexpectedly high losses (relative to the coupling into zero[th]

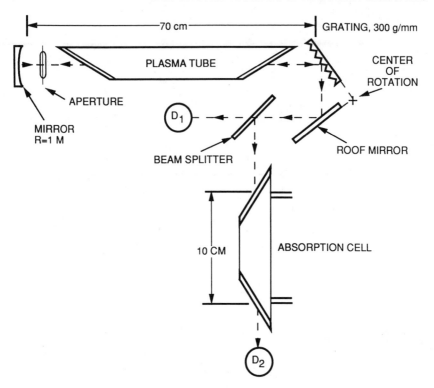

FIGURE 4.   Optical configuration of the laser.

order). With a more efficient means of line selection (e.g., a CaF$_2$ Littrow prism) the expected pulse energy would be ~10 times larger.

Figure 6 shows computed transmission spectra for two remote sensing scenarios: measurement of combustion gases of a coal-fired power plant in situ, over a 6-m path, such as through the gas stream in a stack or ductwork; and measurement of an extended path in the free atmosphere. In the case of combustion gases, the 3.895-μm laser line is used. The interfering spectral line is from H$_2$O vapor. If the laser line is Zeeman split into two components at ±0.06 cm^{-1} (a splitting comparable to that used in the He-Ne laser of Kebabian et al.[32] and McManus et al.[33]), then the interference corresponds to <1 ppmv N$_2$O, a level that would be acceptable in the context of combustion studies.

An extended path in the free air (perhaps closer to 1 km than the 10 km assumed in the figure) might be used to study the evolution of N$_2$O from biological systems, by observing concentration gradients with elevation. Since the expected gradients are small, interference from a highly variable species such as H$_2$O is undesirable; and we consider instead the use of the laser line at 2584.2 cm^{-1} (3.87 μm). In this case, the underlying interference is from methane, a much less variable atmospheric component. Assuming ±0.1 cm^{-1} Zeeman splitting, the interference from methane corresponds to 5 ppbv N$_2$O. This level of interference probably would be acceptable since gradients in methane are expected to be small

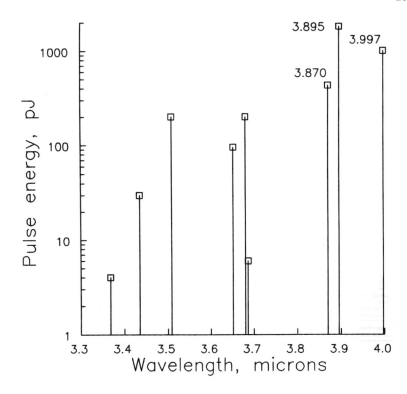

FIGURE 5.   Laser lines observed between 3.3 and 4.0 μm, using low-energy
             pulsed excitation of xenon.

and, in the contemplated experiment, methane probably would be measured at the
same time.

V. Measurement of Other Species

Many other molecular species exhibit line coincidences with neutral rare gas
laser lines. In this section, we shall briefly consider some of the more interesting
of these cases, especially those of environmental importance.

Formaldehyde (H_2CO) is a very minor component of the clean atmosphere, but
it is of some concern as an indoor pollutant,[40] and may become a serious outdoor
pollutant if methanol fueled vehicles come into wide use.[41] An exact coincidence
(within the molecule's Doppler width) exists with the high gain xenon laser line
at 3.508 μm.[42] Moreover, the absorption line shows a large enough Stark effect
to be of practical interest in an enclosed-cell measurement apparatus.[43]

Hydrogen fluoride (HF) is a by-product of certain industrial processes, such as
aluminum refining. The He-Ne laser line at 4173.98 cm^{-1} coincides with the R5
line of HF.[44] This, too, is an exact coincidence; and the frequency is about 2 cm^{-1}

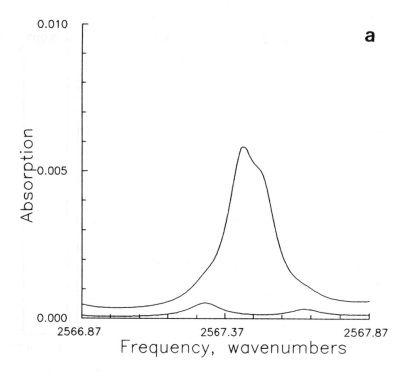

FIGURE 6.   Computed absorption spectra. (a) Stack gas: 8% H_2O, 12% CO_2, 500 K, 6-m path, 0- and 10-ppmv N_2O. (b) Ambient: 1.3% H_2O, 320-ppmv CO_2, 1.6-ppmv CH_4, 296 K, 10-km path, 0- and 300-ppbv N_2O.

away from the closest interference, a weak water vapor line. The laser operates under similar conditions to those in the 3.39 μm IR He-Ne laser, or the familiar visible He-Ne laser at 0.63 μm. Thus, a hydrogen fluoride monitor based on this laser could be a highly reliable and field-worthy device.

Methanol (CH_3OH) is a widely used solvent, and may become important as an automotive fuel in the near future. The Xe laser line at 1030.63 cm^{-1} coincides with a strong methanol absorption line, for which the Stark effect has been observed.[45,46] The chief potential interferent is ozone, which is not expected to be a problem in most situations.

Formic acid (HCOOH) is another potential product of methanol combustion in automobiles. It has an absorption line[47] at 1793.6 cm^{-1} that coincides with a very high-gain Xe laser line. The laser line has been extensively studied, both pulsed and cw, and has been used with Zeeman tuning.[48]

These line coincidences all appear to be exact, to within the accuracy of the spectroscopy of the molecules and laser lines. We shall now consider some cases of near coincidences, similar to that for N_2O discussed in the previous section.

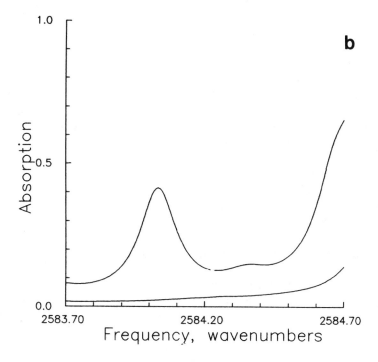

FIGURE 6 (continued).

Carbon monoxide (CO) is an important pollutant and also is a toxic hazard in many workplace situations. The Xe laser at 2202.99 cm^{-1} is 0.17 cm^{-1} away from a strong CO line, at 2203.16 cm^{-1}. Even at this relatively large spacing, the laser is absorbed strongly enough to be useful in pollution or hazard measurements. By using B~4 kG to Zeeman split the laser line, the absorption increases by about a factor of 10, making this an attractive candidate for long path monitoring of CO in the clean troposphere.

In this case, at 0.25 cm^{-1} below the laser frequency, there is a strong line of N_2O that could be observed using the other Zeeman split laser component. Thus, it may be possible to observe both of these atmospheric trace species using a single laser source.

Nitric oxide (NO) is a pollutant that participates in the chemistry of acid rain. The He-Ne laser at 1850.22 cm^{-1} is 0.045 cm^{-1} above a close pair of NO lines. This line falls in a spectral region that is relatively free of interference from H_2O: although there is significant absorption, it results from the wings of strong but remote H_2O lines, and thus differential absorption of the Zeeman split components of the laser line may be a good way to observe this important species. Absorption of this laser line by NO has been observed experimentally,[49] with a measured absorption coefficient in good agreement with theory.

Hydrogen cyanide (HCN) is a highly toxic gas that can occur as a workplace hazard where cyanides are used (e.g., electroplating, mining). The v_3P18 line of this molecule is at 3255.14 cm^{-1}, while a He-Ne laser line has frequency 3255.20 cm^{-1}. This laser line originates from the same upper state as that of the 2947.9 cm^{-1} line used for methane measurement.[50] It is obtained by suppressing that (stronger) line with a grating tuned cavity, as done with the Xe laser described above. It appears that, with a Zeeman split laser similar to that of Kebabian et al.,[32] sensitivity of 10 ppmv could be obtained in the presence of typical atmospheric water vapor levels. The only other species that interferes is acetylene (C_2H_2) which would not be present in significant quantities in most situations.

These species all have more or less well-resolved spectral structure in the vicinity of the laser lines. Heavier, more complex molecules may have spectra in which the individual lines are so numerous that they are not resolvable at atmospheric pressure. In that case laser absorption methods have less of a specificity advantage compared to other techniques, such as NDIR. The laser still has the advantage of producing a spatially coherent beam, however, and thus is more suitable in situations where an extended path is required, such as a perimeter alarm.

A case in point is phosgene[51] (carbonyl chloride, $COCl_2$). There are several neutral rare gas laser lines that fall in the strong bands around 850 and 1830 cm^{-1} of this molecule. Of particular interest are the Ne line at 843.1 cm^{-1} and the Xe lines at 1826.5 and 1793.6 cm^{-1}, where the latter could be used as a reference line. Given the toxicity of phosgene and its widespread use in the chemical industry, this application seems worth further study.

VI. SUMMARY AND CONCLUSIONS

The neutral rare gas lasers provide a large number of lines, many of which are in coincidence or near coincidence with spectral lines of trace atmospheric or pollutant species. These laser lines are a property of the gas used in the laser and thus are precisely, independently reproducible; in most cases, they are separated enough that a simple grating tuned laser cavity allows selection of the desired line. Most neutral rare gas laser lines can be Zeeman split, thus providing multiple, closely spaced infrared frequencies required for differential absorption measurements. The neutral rare gas lasers, especially in pulsed operation, consume little power; they operate at room temperature, with no significant temperature dependence in their operating properties.

In contrast, tunable Pb-salt diode lasers usually require cooling to 77°K or below. The frequency is determined by how the diode is manufactured (chemical composition, cavity length, etc.) and the exact temperature at which it is operated. It is not within the current state of the art to consistently produce diode lasers in the mid-IR that will reproducibly generate a given, single frequency. Thus, tunable diode lasers also require complicated and expensive frequency control and mode selection systems.

Another alternative light source, the molecular gas laser, suffers from the disadvantage that, in general, its lines do not show a usable Zeeman effect. Also, though these lines may be dense within specific regions, they are not uniformly available throughout the infrared.

Thus, we see that the neutral rare gas laser offers a number of unique advantages in this application, especially as a reliable trace species monitor for the real world outside the laboratory.

ACKNOWLEDGMENTS

We wish to note the collaboration of our colleagues at Aerodyne Research, Inc.: Drs. A. Freedman, J.B. McManus, and M.S. Zahniser. The work related to methane measurement was conducted with support from the National Science Foundation (ISI-8504143), and the National Aeronautics and Space Administration (NASW-4345). The nitrous oxide work was supported by NSF (ISI-8761052). The study of other target species has been pursued with Aerodyne internal research funding.

REFERENCES

1. Kogelnik, H., and T. Li. "Laser Beams and Resonators," Proc. IEEE 54:1312–1329 (1975).
2. Ku, R., E. Hinkley, and J. Sample. "Long Path Monitoring of Atmospheric Carbon Monoxide with a Tunable Diode Laser System," Appl. Opt. 14:854–861 (1975).
3. Eberle, A., J. Kruse, and P. Kebabian. "A Hand Held Remote Sensing Methane Detector for Leakage Survey," paper presented at American Gas Association Distribution/Transmission Conference, San Francisco, CA, April 1984.
4. Grant, W. "He-Ne and CO_2 Laser Long Path Systems for Gas Detection," Appl. Opt. 25:709–719 (1986).
5. Tai, H., H. Tanaka, and K. Uehara. "Quantitative Remote Sensing of Methane by Means of a Dual Wavelength He-Ne Laser," Kogaku 19:238–244 (1990) (In Japanese).
6. Bennett, W.R. Atomic Gas Laser Transition Data: A Critical Evaluation (New York: IFI/Plenum Publishing Corporation, 1979).
7. Moore, C.E. Atomic Energy Levels, NSRDS-NBS-35 U.S. Government Printing Office, (December 1971).
8. Rothman, L.S., R.R. Gamache, A. Goldman, L.R. Brown, R.A. Toth, H.M. Pickett, R.L. Poynter, J.-M. Flaud, A. Camy-Peyret, A. Barbe, N. Husson, C.P. Rinsland, and M.A.H. Smith. "The HITRAN Database: 1986 Edition," Appl. Opt. 26:4058–4097 (1987).
9. Park, J., L. Rothman, C. Rinsland, H. Pickett, D. Richardson, and J. Namkung. Atlas of Absorption Lines From 0 to 17,900 cm^{-1} NASA Reference Publication 1188, (1987). Available as NTIS number N87–28955.
10. Faust, W., R. McFarlane, C. Patel, and C. Garrett. "Noble Gas Optical Maser Lines at Wavelengths Between 2 and 35 μ," Phys. Rev. 133:A1476–A1486 (1964).
11. Patel, C., W. Faust, R. McFarlane, and C. Garrett. "Laser Action up to 57.355 μm in Gaseous Discharges (Ne, He-Ne)," Appl. Phys. Lett. 4:18–19 (1964).
12. White, A., and E. Gordon. "Excitation Mechanisms and Current Dependence of Population Inversion in He-Ne Lasers," Appl. Phys. Lett. 3:197–199 (1963).

13. Gordon E., and A. White. "Similarity Laws for the Effects of Pressure and Discharge Diameter on Gain of He-Ne Lasers," *Appl. Phys. Lett.* 3:199–201 (1963).
14. Moeller, G., and T. McCubbin. "Study of Helium-Neon Laser Amplification at 3.39 μm," *Appl. Opt.* 4:1412–1415 (1965).
15. Clark, P.O., R.A. Hubach, and J.Y. Wada. "Investigation of the DC Excited Xenon Laser," NASA CR-67298 (April 1965) (NTIS No. N65 – 35355).
16. Clark, P.O., "Pulsed Operation of the Neutral Xenon Laser," *Phys. Lett.* 17:190–192 (1965).
17. Linford, G., "High Gain Neutral Laser Lines in Pulsed Noble Gas Discharges," *IEEE J. Quantum Electron.* QE-8:477–482 (1972).
18. Linford, G., "New Pulsed Laser Lines in Krypton," *IEEE J. Quantum Electron.* QE-9:610–611 (1973).
19. Linford, G. "New Pulsed and CW Laser Lines in the Heavy Noble Gases," *IEEE J. Quantum Electron.* QE-9:611–612 (1973).
20. Targ, Russell, and Michael W. Sasnett. "High Repetition Rate Xenon Laser with Transverse Excitation," *IEEE J. Quantum Electron.* QE-8:166–169 (1972).
21. Courville, G.E., P.J. Walsh, and John H. Wasko. "Laser Action in Xe in Two Distinct Current Regions of AC and DC Discharges," *J. Appl. Phys.* 35:2547–2548 (1964).
22. Okamoto, Yukio. "Efficiency Enhancement in DC Pulsed Gas Discharge Memory Panel," *Jpn. J. Appl. Phys.* 22:L7–L9 (1983).
23. Pollack, S.A. "Short Duration Light Pulse During Electrical Breakdown in Gases," *J. Appl. Phys.* 36:3459–3465 (1965).
24. Barnes, B.T. "The Dynamic Characteristics of a Low Pressure Discharge," *Phys. Rev.* 86:351–358 (1952).
25. Moore, C. "Gas Laser Frequency Selection by Molecular Absorption," *Appl. Opt.* 4:252–253 (1965).
26. Gerritsen, H. "Methane Gas Detection Using a Laser," *Trans. Soc. Min. Eng.*, 20:428–432 (1966).
27. Johnston, T., and G. Wolga. "Difference-Frequency Resonance Measurement of the $3P_4$, $3P_2$ Neon Fine-Structure Splitting," *J. Opt. Soc. Am.* 59:190–194 (1969).
28. Kebabian, P., "Precision Measurement of Atmospheric Methane Concentration by Differential Absorption of Infrared 3.39 μm Laser Light," ERT, Inc. Technical Report 176-2812-100 (December 1980).
29. Menzies, R., A. Dienes, and N. George. "Axial Magnetic Field Effects on a Saturated He-Ne Laser Amplifier," *IEEE J. Quantum Electron.* QE-6:117–122 (1970).
30. Linford, G. "Experimental Studies of a Zeeman Tuned Xenon Laser Differential Absorption Apparatus," *Appl. Opt.* 12:1130–1139 (1973).
31. Fork, R., and C. Patel. "Magnetic Tuning of Gaseous Laser Oscillators," *Proc. IEEE* 52:208–209 (1964).
32. Kebabian, P., A. Freedman, and C. Kolb, "Magnetically Tuned He-Ne Laser Based Methane Flux Measurement Device," paper presented at 193rd National Meeting of American Chemical Society, Denver, CO, April 1987.
33. McManus, J., P. Kebabian, and C. Kolb. "Atmospheric Methane Measurement Instrument Using a Zeeman-Split He-Ne Laser," *Appl. Opt.* 28:5016–5022 (1989).
33a. Fan, S.M., S.C. Wofsy, P. Bakwin, D.J. Jacob, S.M. Anderson, P.L. Kebabian, J.B. McManus, C.E. Kolb, and D.R. Fitzjarrald. "Micrometeorological Measurements of CH_4 and CO_2 Exchange Between the Atmosphere and the Arctic Tundra.," *J. Geophys. Res.* In press (1992).

33b. "The Aerodyne Research Mobile Methane Monitor," McManus, J. B., P. L. Kebabian, and C. E. Kolb, in *Measurement of Atmospheric Gases, SPIE Proc. V1433* (January 1991).

34. Shimoda, K. "Absolute Frequency Stabilization of the 3.39-μm Laser on a CH_4 Line," *IEEE Trans. Instru. Meas.* IM-17:343–346 (1968).

35. Hao, W., S. Wofsy, M. McElroy, J. Beer, and M. Toquan. "Sources of Atmospheric Nitrous Oxide from Combustion," *J. Geophys. Res.* 92:3098–3104 (1987).

36. Muzio, L., and J. Kramlich "An Artifact in the Measurement of N_2O from Combustion Sources," *Geophys. Res. Lett.* 15:1369–1372 (1988).

37. Lyon, R., J. Kramlich, and J. Cole. "Nitrous Oxide: Sources, Sampling, and Science Policy," *Environ. Sci. Technol.* 23:392–393 (1989).

38. Kebabian, P., and M. Zahniser." Evaluation of the Neutral Xenon Laser for Infrared Measurement of Nitrous Oxide from Combustion Sources," Aerodyne Research, Inc. Report No. RR-672, October 1988.

39. Hard, T.M. "Laser Wavelength Selection and Coupling by a Grating," *Appl. Opt.* 9:1825–1830 (1970).

40. Pitts, J., H. Biermann, E. Tuazon, M. Green, W. Long, and A. Winer. "Time Resolved Identification and Measurement of Indoor Air Pollutants by Spectroscopic Techniques: Gaseous Nitrous Acid, Methanol, Formaldehyde, and Formic Acid," *J. Air Pollut. Control Assoc.* 39:1344–1347 (1989).

41. Williams, R., F. Lipari, and R. Potter. "Formaldehyde, Methanol, and Hydrocarbon Emissions from Methanol-Fueled Cars," *J. Air Waste Manage. Assoc.* 40:747–756 (1990).

42. Sakurai, K., and K. Shimoda. "High Resolution Spectroscopy of Formaldehyde by a Tunable Infrared Maser," *J. Phys. Soc. Jpn.* 21:1838 (1966).

43. Sakurai, K., K. Uehara, M. Takami, and K. Shimoda. "Stark Effect of Vibration-Rotation Lines of Formaldehyde Observed by a 3.5 μm Laser," *J. Phys. Soc. Jpn.* 23:103–109 (1967).

44. Eberhardt, J., and A. Pryor, "He-Ne Laser Frequencies Near 2.4 μm and Their Application to Hydrogen Fluoride Detection,"*IEEE J. Quantum. Electron.* QE-19:891–893 (1983).

45. Sattler, J., T. Worchesky, and W. Riessler. "Diode Laser Spectra of Gaseous Methyl Alcohol," *Infrared Phys.* 18:521–528 (1978).

46. Sattler, J., W. Riessler, and T. Worchesky. "Diode Laser Spectra of the C-O Stretch Band of Gaseous Methyl Alcohol," *Infrared Phys.* 19:217–224 (1979).

47. High Resolution FTIR Spectrum. Courtesy of J. Johns. Unpublished.

48. Schlossberg, H., and A. Javan. "Hyperfine Structure and Paramagnetic Properties of Xenon Studied with a Gas Laser," *Phys. Rev. Lett.* 17:1242–1244 (1966).

49. Abdullin, R., S. Boiko, S. Kotelnikov, and A. Popov. "Express Control of Some Industrial Air Pollutants (CO_2, C-H, HF, NO) Using Modified Commercial He-Ne Lasers," Fifteenth International Laser Radar Conference, Tomsk, U.S.S.R. (July 23–27, 1990).

49a. We wish to thank Dr. W. Grant of NASA-Langley for bringing this reference to our attention.

50. Brunet, H., and P. Laures. "New Infrared Gas Laser Transitions by the Removal of Dominance," *Phys. Lett.* 12:106–107 (1964).

51. Yamamoto, S., T. Nakanaga, H. Takeo, C. Matsumura, M. Nakata, and K. Kuchitsu. "Equilibrium Structure and Anharmonic Potential Function of Phosgene: Diode Laser Spectra of the v_1 and v_5 Bands," *J. Mol. Spec.* 106:376–387 (1984).

Data Interpretation Techniques

CHAPTER 17

Data Treatment Procedures in Ambient Air Monitoring

Robert A. McAllister

TABLE OF CONTENTS

0-87371-606-0/93/$0.00+$.50

I. INTRODUCTION

Data treatment procedures are designed to extract as much information as possible from a body of data. Monitoring data of any kind are expensive to obtain and usually commissioned with specific goals in mind. Data correlation and interpretation are also important and costly. In planning a monitoring program, safeguards should be designed into the experiment to ensure that the data are of sufficient (and known) quality to achieve the objectives for which the data were obtained. At every point in the planning, generation, and interpretation of a monitoring program, statistical techniques are available as tools to improve the probability of obtaining the most information for the least cost. Part of the purpose of this chapter is to show how some statistical techniques may be applied to data treatment in ambient air monitoring programs. Ambient air monitoring involves measurements of organic compounds in air at concentrations from zero to hundreds of parts per billion. Frequently the analyst is called on to estimate concentrations of specific compounds below well-defined detection limits.

The U.S. Environmental Protection Agency (EPA) has monitored ambient air concentrations of various organic compounds (and particulates) for many years. The Nonmethane Organic Compound (NMOC) Monitoring Program has provided data needed by states in developing revised ozone control strategies to meet the National Ambient Air Quality Standard for ozone. Integrated whole air samples have been collected at several urban centers from 6 to 9 A.M., Monday through Friday, June through September since 1984. Since 1987 in the Urban Air Toxics Monitoring Program (UATMP), 24-hr integrated whole air samples have been collected at urban centers on a 12-day schedule for a year and analyzed for 38 toxic organic compounds. Radian Corporation has participated in both these monitoring programs since their inception and in other regional and single-site monitoring programs. In all such monitoring programs, periodic samples are taken on a prearranged schedule, generally extending longer than a month. In each case, a well-defined quality assurance program is developed prior to the beginning of sampling to ensure that data of known and well-documented quality are obtained.

This chapter will review some of the data treatment procedures used in ambient air monitoring. The purpose of the data treatment procedures is to obtain as much useful and valid information from the monitoring data as possible. Some of the procedures that will be examined are calibration, data characterization, and quality monitoring. Treatment of calibration data near the detection limit can lead to erroneous results. An example is given of how this problem comes about, and a solution of the problem is offered.

Estimating the precision and accuracy of ambient air monitoring data is necessary to ensure that the data obtained are of known quality that is well documented. A number of procedures for treating precision are reviewed, giving specific examples from actual monitoring data. When the experimental design calls for performing replicate analyses on duplicate samples, the results of the analyses may be used to estimate significant differences between sampling variability and analytical variability. Examples are given, and conclusions are drawn.

Quality control charts are valuable tools for tracking precision for replicate analyses, duplicate samples, and calibration checks to ensure that the sampling, analysis, and calibration procedures are in statistical control. The use of control charts is discussed, and examples of their interpretation are suggested.

Ambient air monitoring data are characterized in terms of several statistics and graphical representations. Ambient air monitoring data use and interpretation have inherent special problems. For example, estimating annual average concentrations of a specific toxic compound causes a problem when several of the monitoring program samples have concentrations below the detection limit. The monitoring data are censored at the detection limit which can significantly affect the estimation of the true annual average concentration. Because of a great deal of recent interest in this topic, several approaches to uncensoring monitoring data have been proposed. Frequency distributions are a meaningful way to represent monitoring data graphically and numerically. Stem and leaf plots and box plots illustrate convenient ways that are available to display frequency distribution data. The use of the Pearson correlation matrix involving a number of organic compounds is shown to be a tool for "fingerprinting" a site, and offers possibilities for cataloging the types of emission sources in the vicinity of the site.

II. CALIBRATION DATA

The statistical theory of calibration was treated comprehensively by Scheffé.[1] For the nonstatistician, however, the paper is virtually incomprehensible. More practical approaches to the treatment of calibration data are presented by others.[2–5] An example problem involved in ambient air monitoring required that the calibration curve include: (1) four calibration standards of known concentrations from about 1.0 to 100 ppbv, (2) a measured zero concentration, (3) a correlation coefficient >.995, and (4) the requirement that the percent difference between the calculated value of nanoliters from the regression line and the calibration points be less than 30%. Table 1 shows the results for a particular compound in nanoliters vs area counts. Table 1 shows the calibration results for a particular compound in nanoliters in the calibration standards, x_i vs the flame ionization detection (FID) area counts. Columns 3 and 4 give the commonly used least squares results labeled "Linear Regression I" in the table. The parameter $\hat{x}_i$ in the table is the value of the independent variable calculated using the model (see Table 2), with b_0 and b_1,

Table 1. Calibration Data and Results

Calibration standard x_i, nL	FID area count y_i	Linear Regression I X_i^a	D^b	Linear Regression II X_i^a	%D	Linear Regression III X_i^a	%D	Constant Slope Method $\overline{RF}$ x 10^6	X_i^a	%D
0.00	1321	1.98	-100.00	—	—	—	—	—	—	—
1.48	10,526	3.29	-55.03	1.44	3.02	1.42	3.61	140.604	1.491	-0.77
3.69	28,223	5.81	-36.47	3.96	-6.92	3.89	-4.87	130.744	3.999	-7.73
167.96	1,125,237	161.84	3.78	160.65	4.55	162.19	3.56	149.226	159.43	5.35
263.60	1,764,079	252.70	4.31	251.89	4.65	255.74	3.07	149.426	249.95	5.46
344.34	2,487,852	355.64	-3.18	355.26	-3.07	362.28	-4.95	138.409	352.50	-2.32
Coefficient of regression		0.9987		0.9987		0.9998				

Average response factor, $\overline{RF}$ = 141.690 x 10^6

[a] $\hat{X}_i$ is the calculated or expected value of x_i, given y_i.

[b] Percent difference: $\%D = \dfrac{(X_i - \hat{X}_i)}{\hat{X}_i} * 100$

Table 2. Calibration Curve Equations

Linear Regression I — the method of least squares

Model:

$$y_i = b_o + b_i x_i$$
$$i = 1, 2, 3, \ldots, n$$

$$b_0 = \frac{n \sum x_i y_i - \sum x_i \sum y_i}{n \sum x_i^2 - \left(\sum x_i\right)^2}$$

$$b_1 = \sum y_i / n - b_0 \sum x_i / n$$

$$\hat{x}_i = \left(y_i - b_0\right)/ b_1$$

Linear Regression II — the method of least percent difference squared

Model:

$$y_i = b_0 + b_1 x_i$$

$$b_0 = \frac{\sum\left(1/x_i\right)^2 \sum\left(y_i^2 / x_i^2\right) - \left(\sum\left(y_i / x_i^2\right)\right)}{\sum\left(1/x_i^2\right)\sum\left(y_i / x_i\right) - \sum\left(1/x_i\right)\sum\left(\frac{y_i}{x_i^2}\right)}$$

$$b_i = \left(\sum\left(y_i / x_i\right) - b_0 \sum\left(1/x_i\right)\right) / \sum\left(1/x_i^2\right)$$

$$\hat{x}_i = \left(y_i - b_0\right)/ b_1$$

Linear Regression III — logarithmic regression

Model:

$$\ln y_i = \ln b_o + b_i \ln X_i$$
$$y_i = b_o X^{b_i}$$

$$\ln b_o = \frac{n \sum\left[\left(\ln y_i\right)\left(\ln X_i\right)\right] - \sum \ln y_i \sum \ln X_i}{n \sum\left(\ln X_i\right)^2 - \left(\sum \ln X_i\right)^2}$$

$$\hat{X}_i = \exp\left[\left(\ln y_i - \ln b_o\right)/ b_i\right]$$

Table 2 (continued).

Constant Slope Method

Model:[a]

$$\overline{RF} = \sum \left(RF_i \, / \, n \right)$$

$$\hat{X}_i = \overline{RF} \left(area \, counts \right)_i$$

[a] Response factor (a reciprocal slope), $RF_i = \chi_i/$(area counts).

estimated from Linear Regression I. As seen in Table 1, the coefficient of correlation for this regression is 0.9987, which is acceptable for the monitoring program, but the percent difference for the curve below $x_i = 3.69$ nL is greater than the acceptance criterion of 30%. This is not an atypical example for calibration results having such a large range of x_i, i.e., to have the percentage difference between the calculated (or expected) and measured values of x_i to become unacceptably large at low concentrations. Moreover, the expected value of x from the regression may become negative. Table 1 also shows "Linear Regression II," "Linear Regression III," and "Constant Slope Method" which treat the same calibration data. The equations and calculations used for all four treatments shown in Table 1 are given in Table 2.

The method of least squares shown in Linear Regression I may be found in most standard statistics books.[6,7] A number of assumptions are made when performing a simple linear regression using the method of least squares. The assumptions are: (1) the expected (or calculated) value of y, given x_1, $E(y/x_1)$, is a linear function of x, i.e., $E(y/x_1) = \hat{y}_i = b_0 + b_1 x_i$; (2) variance of y_i, given $x_i = \sigma^2$, where σ^2 is constant and independent of x_i; (3) x_i values are ndependently set and observed, or measured without error; (4) deviations $(y - \hat{y}_i)$ are independent of each other, (5) deviations $(y - \hat{y}_i)$ are normally distributed; (6) data are taken from the population about which inferences are drawn; and (7) there are no other independent variables which significantly affect the relationship between x_i and y_i. In practice, few if any of these assumptions are strictly adhered to in typical calibration procedures. Some consider the third assumption to be the most important. Fortunately, even moderate deviations from the first five assumptions are not likely to be of serious importance.[6] Especially in analytical work, however, assumptions 6 and 7 can be of overriding importance, because unsuspected effects of often unknown pollutants may significantly change the analytical results, in spite of the fact that the initial calibration was internally consistent and the correlation was good. The analyst needs always to keep in mind that matrix effects resulting from unknown interferents from the sample being analyzed may violate assumption 7 and seriously influence the results of the analysis.

Linear Regression II is a least squares method which minimizes the percent differences between the calculated and experimental differences of x, given y_i.[3,5]

As seen from Table 1, this procedure sharply reduces percent differences, especially at the lower concentrations of x_i. Because of the nature of the method, however, the absolute differences of x, given y_i, may actually increase somewhat when compared to the standard method of least squares (Linear Regression I). This disadvantage is usually not a serious one.

Another least squares method that may be used is given in Linear Regression III in which lognormal (ln) y_i is assumed to be a linear function of ln x_i. Table 2 shows this logarithmic regression to be nonlinear in x and y, but in most cases it improves the "fit" at lower concentrations because it is a form of weighting the lower concentrations of x more heavily than the higher concentrations. The logarithmic regression also enjoys the advantage of always returning positive values of x (or y), given actual (and positive) measured values of y_i (or x_i). Another interesting observation is that when the correlation coefficient for calibration data in the x and y mode (Linear Regression I) result in a value greater than 0.99, the correlation coefficient for Logarithmic Regression III is almost always greater for the same data set than the coefficient of correlation for Logarithmic Regression I.

Another treatment which provides an additional method to weight calibration data may be called the constant slope method.[2] Response factors, as shown in Table 2, are also reciprocal slopes which pass through the origin. Each calibration point, with the exception of one in which x_i and/or y_i is zero, may be used to calculate a response factor. An average response factor, $\overline{RF}$, which weights each response factor equally is shown in Table 1 and may be used to estimate the expected value of x_i, given y_i (see Table 2). The results shown in Table 1 distribute the percent error rather evenly among the calibration points.

Each treatment of the calibration data shown in Table 1 produces different results. For cases of interest in ambient air monitoring in which concentrations near the detection limit are involved, the standard method of least squares (Linear Regression I) generally gives the poorest results. The method of least squares weights larger values of x and y more than smaller values. Linear Regressions II and III give larger weights to the lower values of x_i, while the response factor, or Constant Slope Method equalize the weights among all calibration points. Many other weighting techniques for calibration data are possible. An excellent discussion of additional weighting techniques are given by Liteanu and Rîca.[2]

Table 2 summarizes the equations used to estimate the regression parameters for Linear Regression I, II, III, and the Constant Slope Method. Estimates of $\hat{x}_i$ from the calibration and regression parameters are also shown in Table 1.

III. PRECISION

Precision in ambient air monitoring is a measure of the repeatability of the measurement, for example, of concentration. Analytical precision can be determined by repeated analysis of aliquots from the same ambient sample in a canister, assuming that the sample concentration is not changing between analytical determinations. Overall precision of the measurement of ambient air concentration may also include a sampling precision, involving the reproducibility of: (1) collection of the sample into a container (adsorbent bed, liquid absorbent, plastic bag, or metal container), (2) transport of the sample to the laboratory for analysis, (3) removal of the sample from its container, and (4) introduction of the sample into the analytical device. In the examples presented in this study, duplicate ambient air samples were taken in SUMMA™-treated stainless steel canisters. Replicate analyses of each duplicate sample provided information to estimate analytical precision (variability), sometimes referred to as analytical "error." For ambient air samples collected in stainless steel canisters, experience suggested that repeated analyses be performed on successive days, giving at least 18 hr between removal of successive samples from the canisters. The time interval between removal of samples allowed adsorption equilibrium to be reestablished within the canister and enabled more representative sample aliquots to be taken. Replicate analyses of duplicate samples also provided the data necessary to separate sampling variability from analytical variability.

A. Replicate Analyses of Duplicate Samples

In Figure 1, two types of duplicate samples are illustrated. Type I duplicate samples are obtained in the U.S. EPA UATMP and NMOC monitoring programs. Type II duplicates are gathered in many other monitoring programs, generally involving sampling locations where electrical power is not easily available. Both types of duplicate samples are treated in the precision measurements discussed next.

Figure 2 sketches the result of two replicate, or repeated analyses, of duplicate samples. A Type I duplicate is shown, but the notation and treatment of data which follow are identical for both types of duplicates. All of the treatments of precision discussed later involve replicate analyses of duplicate samples.

Precision is a measure of the variability, or dispersion, of repeated measurements of the same quantity, e.g., concentration of a pollutant in an ambient air sample. To estimate precision, several measures may be used:

- difference, or range
- interquartile range (25th to 75th percentile of a cumulative distribution)
- standard deviation (SD)
- percent differences
- absolute percent differences
- percent coefficient of variation (% CV), also known as percent relative standard deviation (% RSD).

Figures 2 and 3 show how each of these measures of precision are calculated from the analytical results for both the replicate analyses and the duplicate samples. In Figure 3 the pooled standard deviations shown for analytical precision, s_A, require that each duplicate sample is analyzed twice and that the estimates of variance for analysis, s_A^2, be equal and independent of x_i. Likewise to pool the sampling and analytical variances by the equation shown in Figure 3, the variances must be for duplicate samples, equal, and independent of sample

Duplicate Sample Types

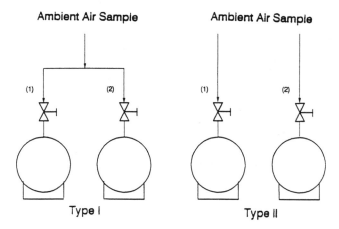

FIGURE 1.   Types of duplicate samples.

concentrations. The estimates of variance, s_i^2, must be tested to see if they are functions of concentration before a pooled standard deviation can appropriately be calculated.

Table 3 shows the results of the various measures of precision from the 1989 UATMP[6] for propylene. In six of the pairs of duplicated samples, propylene was detected in all four analyses for a total of 24 determinations of concentration in parts per billion by volume (ppbv). There were therefore six pairs of duplicate samples and 12 pairs of replicated pairs of analyses. Average percent differences, average absolute percent differences, and average % CVs are given under analytical precision and sampling and analytical precision. Analytical standard deviation, s_A, is the square root of the average variance (s_{ij}^2) of the replicate pairs (see Figure 2). The pooled standard deviation of the duplicate standard deviations is the square root of the average variance of the duplicate pairs. It is symbolized by $s_{S\&A}$, and is the measure of precision for sampling and analysis. Note that the s_A is almost identical with $s_{S\&A}$. Pooled % CV's are calculated from the pooled standard deviations and the overall mean concentration, 4.166 ppbv.

Pooled standard deviations for the UATMP propylene analytical precision and for the sampling and analytical precision are about equal at 0.43 ppbv. This result affirms the conclusion that the error or variability due to sampling for the Type I duplicate is very small or zero, at least for propylene. Pooled % CVs for analytical precision and for sampling and analytical precision with the UATMP propylene data are also estimated to be about equal at 10.42 and 10.35%, respectively.

Data measurements are inherently variable. In the measurement of ambient air concentrations by repeated analyses of duplicate samples taken simultaneously (see Figures 1 and 2), at least two sources of variability may be hypothesized: (1) the variability between the duplicate samples, and (2) the variability between

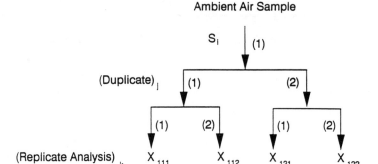

Replicate Pairs

$$S_{11} = |X_{112} - X_{111}|/\sqrt{2} \quad ; \quad S_{12} = |X_{122} - X_{121}|/\sqrt{2}$$

$$d_{11} = (X_{112} - X_{111}) \quad ; \quad d_{12} = (X_{122} - X_{121})$$

$$\overline{X}_{11} = (X_{112} + X_{111})/2 \quad ; \quad \overline{X}_{12} = (X_{122} + X_{121})/2$$

Duplicate Pairs

$$s_i = \left[\sum_{j}^{2} \sum_{k}^{2} (X_{ijk} - \overline{X}_i)^2/(4-1) \right]^{1/2}$$

$$\overline{X}_i = (\overline{X}_{11} + \overline{X}_{12})/2$$

FIGURE 2.   Schematic diagram of replicate analysis of duplicate samples.

the repeated analyses. The task of separating causative sources of variability from sets of related data (replicate analyses of duplicate samples) falls to a statistical procedure called the analysis of variance (ANOVA).

Separating the known sources of variabilities into their component parts allows determination whether one part is inordinately large and in need of spending some effort to reduce a given source of variability. In the case of analytical precision, for example, the ANOVA strips known or suspected sources of variability from the total variability give a better estimate of the analytical variability.

B. Analysis of Variance

To separate the sampling variability from the analytical variability, a nested ANOVA is used because of the structure of the sampling and analytical procedure described as a Type I duplicate in Figures 1 and 2.

Replicate Pairs for Each Duplicate

s_{11} ; $(\% \ CV)_{11}$ = $(s_{11} / X_{11}) \cdot 100$

s_{12} ; $(\% \ CV)_{12}$ = $(s_{12} / X_{12}) \cdot 100$

d_{11} ; $(\% \ Diff)_{11}$ = $(d_{11} / X_{11}) \cdot 100$; $(Abs \ \% \ Diff)_{11}$ = $| \ (\% \ Diff)_{11} \ |$

d_{12} ; $(\% \ Diff)_{12}$ = $(d_{12} / X_{12}) \cdot 100$; $(Abs \ \% \ Diff)_{12}$ = $| \ (\% \ Diff)_{12} \ |$

Duplicate Pairs for Each Sample

s_1 ; $(\% \ CV)_1$ = $(s_1 / X_1) \cdot 100$

d_1 ; $(\% \ Diff)_1$ = $(d_1 / X_1) \cdot 100$; $(Abs \ \% \ Diff)_1$ = $| \ (\% \ Diff)_1 \ |$

Analytical Precision

For n ambient samples, pooled $s_A = ((s_{11}^2 + s_{12}^2 + s_{21}^2 + s_{22}^2 + ... + s_{n1}^2 + s_{n2}^2) / 2n)^{1/2}$
Pooled % CV for analytical precision $= (s_A / ((\bar{X}_1 + \bar{X}_2 + \bar{X}_3 + ... + \bar{X}_n)/n)) \cdot 100$

Sampling and Analysis Precision

For n ambient samples, pooled $s_{S\&A} = ((s_1^2 + s_2^2 + ... + s_n^2) / n)^{1/2}$
Pooled % CV for overall precision $= (s_{S\&A} / ((\bar{X}_1 + \bar{X}_2 + \bar{X}_3 + ... + \bar{X}_n)/n)) \cdot 100$

FIGURE 3.   Calculation of selected measures of precision.

Equation 1 shows the model for the nested ANOVA required for the propylene data set from the 1989 UATMP.

$$X_{ijk} = \mu + S_i + D(1)_{j(1)} + D(2)_{j(2)} + D(3)_{j(3)} + D(4)_{j(4)} + D(5)_{j(5)} + D(6)_{j(6)} + \varepsilon_{k(ij)} \quad (1)$$

The experimental design used, i.e., replicate analyses of duplicate samples, is nested because the duplicate pairs are unique to the sample from which they were obtained. Duplicate Sample No. 1 from ambient Sample No. 1 is completely independent of Duplicate Sample No. 1 of ambient Sample No. 2. Likewise, the replicate analyses are nested within the six ambient samples and duplicate pairs. The model shows that each analysis, X_{ijk} equals an overall mean value μ, plus the sample effect S_i. Each duplicate effect $D(1)$, $D(2)$, $D(3)$, etc. is nested within the

Table 3. Propylene Precision: 1989 UATMP

Analytical precision		Sampling and analytical precision	
Replicate pairs	= 12	Duplicate pairs	= 6
Pooled s_A	= 0.4339 ppbv	Pooled $s_{S\&A}$	= 0.4313 ppbv
Average d_{ij}	= 0.1092 ppbv	Average d_i	= 0.1910 ppbv
Average Abs d_{ij}	= 0.4842 ppbv	Average Abs d_i	= 0.3108 ppbv
	Average x_{ijk} = 4.166 ppbv		
Average (% diff)$_{ij}$	= −1.748	Average (% Diff)$_i$	= 3.022
Average (abs % diff)$_{ij}$	= 11.599	Average (abs % diff)$_i$	= 6.694
Pooled s_A	= 0.4339	Pooled $s_{S\&A}$	= 0.4313
Pooled (% CV)$_A$	= 0.4339/4.166 x 100	Pooled (% CV)$_{S\&A}$	= 0.4313/4.166 x 100
	= 10.416		= 10.353

Table 4. ANOVA Table for Propylene Precision Results of Replicated Analyses of Duplicate Samples

Source	Sum of Squares	DF	Mean square	F ratio	p
S	0.396 E + 02	5	7.926798	0.4604 E + 02	0.000
D(1)	0.000900	1	0.000900	0.0052	0.944
D(2)	0.255025	1	0.255025	1.4817	0.246
D(3)	0.021025	1	0.021025	0.1221	0.733
D(4)	0.034225	1	0.034225	0.1988	0.664
D(5)	0.970225	1	0.970225	5.6350	0.035
D(6)	0.000225	1	0.000225	0.0013	0.972
Error[a]	2.066150	12	0.172179		

[a] For the application used in this discussion, "error" is taken as the analytical error, or precision.

sample from which it derived. $D(1)_{j(1)}$, for example, is nested within Sample 1. The final term in the model is $\varepsilon_{k(ij)}$ and is the residual error which is assumed to be the contribution of analytical and random variability factors.

Table 4 gives the ANOVA table resulting from using Equation 1 as a model with the propylene data. The first column in the table gives the source of variation. Sums of squares, degrees of freedom (DF) for each source, mean square, F ratio, and probability (p) complete remaining columns of the table. These ANOVA-table parameters are defined in statistics books dealing with analysis of variance.[7,8] The significance and use of some of these parameters are discussed later.

The final column, p, is the probability that the effect listed under Source is not significant. For example, for the row headed by S (the sample effect), $p = .000$, which indicates that the probability approaches zero that the sample effect is not significant. Stated positively, we may infer that each ambient sample has a significant effect on the propylene concentration or that the propylene concentration may be expected to be different in ambient air samples. In order to conclude that a source of variation has a significant effect, statisticians usually recommend that p be <.05. Probabilities <.01 are said to be highly significant. In the ANOVA

Table 5. Nested ANOVA Pooled for Propylene Data

Source	Sum of Squares	DF	Mean square	F ratio	P
S	0.396×10^2	5	7.926798	0.4604×10^2	0.000
D(1-6)	1.281625	6	0.213604	1.2405	0.201
Error	2.06615	12	0.172179		

Pooled s_A	=	$\sqrt{0.172179}$, *from which* s_A	=	0.4149 ppbv
and (% CV)$_A$	=	$(0.4149/4.166) \times 100$	=	9.96
cf. (% CV)$_A$		(see Table 2)	=	10.42
(% CV)$_{S\&A}$		(see Table 2)	=	10.35

given in Table 4, only the duplicate pair nested within Sample No. 5, D(5), is a significant factor. In Table 4, the column headed "Mean square" shows the variances of the duplicate effects. If the variances for the D(i) are equal, then the duplicate effects may be pooled to give the data shown in Table 5. By inspection, it is clear that the variances are not equal. There are also a number of statistical tests for equal variances, e.g., the Bartlett test.[7] In spite of this apparent contradiction, the ANOVAs for nested designs are commonly pooled to give data as in Table 5.[8] The model for the nested design in which all the duplicate effects are pooled is given by Equation 2.

$$X_{ijk} = \mu + S_i + D_j + \varepsilon_{k(ij)} \tag{2}$$

Table 5 shows $p = 0.201$ for duplicates which forms the basis for the conclusion that there is not a significant difference between duplicate sample pairs for this data set. Referring to Figure 1, and the fact that the propylene samples were Type I duplicates, it is not surprising that there is not a significant difference between duplicate samples. Table 5 goes on to use the error term in the ANOVA table to estimate the pooled analytical error. Note that the mean square for this term is the same in both Tables 4 and 5. The pooled analytical standard deviation (s_A) is therefore the square root of the mean square error of 0.172179, giving $s_A = 0.4149$ ppbv. The pooled $\%CV_A$ is thus 9.96, compared to 10.416 calculated earlier from the pooled s_A listed in Table 3.[6]

The ANOVA results for the pooled duplicate effects (Table 5) suggest the data do not support the position that the duplicate source of variation contributes significantly to the model; hence the variance for duplicates is estimated to be zero, causing the pooled % CV_A and % $CV_{S\&A}$ to be approximately equal. On the other hand, violation of the basic assumptions for the nested designs and the subsequent ANOVA weakens the conclusions an indeterminant amount. The use of the Bartlett test for homoscedasticity (equal variances) is especially sensitive to the assumption that the data (concentrations) come from a normal distribution. For this application, the Bartlett test may not be the optimum test for equal variances.

Table 6 gives an ANOVA from another monitoring program using Type II duplicate samples. The data set included eight ambient air samples collected in 16

Table 6. ANOVA Table for Sampling and Analysis Precision Determinations

Source	Sum of squares	DF	Mean square	F ratio	p
S	379.48965	7	54.21281	76.6906	0.344 E-10[a]
D(1)	8.32323	1	8.32323	11.7742	0.003[b]
D(2)	0.27040	1	0.27040	0.3825	0.545
D(3)	5.66440	1	5.66440	8.0130	0.012[c]
D(4)	6.94323	1	6.94323	9.8220	0.006[b]
D(5)	4.95063	1	4.95063	7.0033	0.018[c]
D(6)	5.97803	1	5.97803	8.4566	0.010[c]
D(7)	0.10890	1	0.10890	0.1541	0.700
D(8)	0.61623	1	0.61623	0.8717	0.345
Error	11.31045	16	0.70690		

[a] Source effect is very highly significant.
[b] Effects are said to be highly significant when $p < 0.01$.
[c] Effects (source in the table) cause a significant effect on dependent variable in model (Equation 1) when $P < 0.05$.

canisters, each of which was analyzed twice. The ANOVA table indicates that five of the duplicate pairs and the sample effect (S_i) showed significant differences in x_{ijk}, and are marked by asterisks. Again the sample factor (S) was highly significant, as expected. In addition, the duplicate variances appear to be approximately equal (their ratios must be more than 161 in order to be judged unequal) so that the duplicate effects may be pooled as shown in Table 7. The ANOVA table in this case shows the expected mean square (EMS) column. The table shows the duplicate effects to be highly significant. As before, the mean square error is the analytical variance, σ_ε^2. The mean square for duplicates is the analytical variance, σ_ε^2, plus twice the duplicate (or sampling) variance, $2\sigma_\delta^2$. Table 7 details how the sampling and analytical precision are estimated from the results tabulated in the ANOVA table when the effect of the Duplicate Source D(1-8) is significantly different from zero.

C. Conclusions About Precision

Pooled standard deviations for the UATMP propylene analytical precision and for sampling and analytical precision are about equal at 0.43 ppbv. Pooled % CVs for analytical precision are estimated to be about equal at 10.42 and 10.35%, respectively. Table 5 shows that for the nested ANOVA in which the duplicate mean squares are pooled, the s_A equals 0.41 and the analytical percent coefficient of variation is 9.96. Thus the several definitions of precision for this case are approximately equal. The most technically correct method is to use the ANOVAs for the individual duplicates given by Tables 4 or 6. It is common practice to use the pooled duplicate results shown in Tables 5 and 7, in spite of the fact that homoscedasticity does not obtain among the several mean squares for the separate duplicate effects.

Table 7. Pooled ANOVA Showing Separation of Sampling and Analysis Precision

Source	Sum of Squares	DF	Mean square	F ratio	p	EMS
S	379.48965	7	54.21281	76.6906	0.344	
D(1–8)	32.85505	8	4.10688	5.8098	0.002	$\sigma_\epsilon^2 + 2\sigma_\delta^2$
Error	11.31045	16	0.70690			σ_ϵ^2

$$\sigma_\epsilon^2 = 0.70690$$
$$\sigma_\epsilon^2 + 2\sigma_\delta^2 = 4.10688$$
$$\sigma_\delta^2 = 1.7000 \text{ ppbv}^2$$

Sampling precision	$= \sqrt{1.7000}$	$= 1.3038$ ppbv
Analytical precision	$= \sqrt{0.7069}$	$= 0.8408$ ppbv
Total measurement precision	$= \sqrt{1.7000 + 0.7069}$	$= 1.5514$ ppbv

The use of average values of absolute percent difference for replicates gives realistic precision results indicating the spread or dispersion of the data. The absolute percent difference statistic for duplicates gives a deceptively low indication of precision because the means of replicate analyses are used in its estimation.

Average percent difference values for replicate analyses are meaningful as a gauge of bias between the first and second analyses. As already discussed, time for canister equilibration of at least 18 hr between the first and second analysis is needed to reduce bias caused by transient phenomena. This equilibration time appears to be especially important when samples are collected at atmospheric (or below) pressure. Average percent differences for replicate analyses should always be nearly zero, and certainly the 95% confidence intervals for the mean percent differences should span zero. Average percent difference values for duplicate samples should also be near zero. Should average percent difference values for replicate analyses or duplicate samples differ significantly from zero, a systematic bias is to be expected and an investigation is made to attempt to determine the cause of the bias.

Average % CVs for duplicates and replicates are useful and valid measures of precision, but should not be used to separate analytical from sampling variance.

IV. CONTROL CHARTS

Analytical precision may be monitored by plotting percent difference of replicate analyses vs analysis number on a quality control chart.[9] Percent difference is expected to be a random variable with a mean, $\mu = 0$, and a variance of σ^2. If each determination of a pair of replicate analyses is plotted against a consecutive analysis number, a series of randomly variable ups and downs results. Should the pattern of random behavior change then the result may often be seen more quickly on a control chart. To maximize the efficiency of control chart use, the parameter to be monitored (e.g., percent difference of replicate analyses) should be measured frequently and regularly; and numerical results should be plotted as soon as possible (at least the same day if possible) after measurement.

Sampling and analysis precision may be monitored by plotting the standard deviation of the duplicate sample (see Table 3) vs the consecutive duplicate sample pair. In a similar manner, calibration checks for selected compounds may also be used in control charts to monitor analytical performance. Duncan[9] shows how to determine when the process or procedure is in statistical control.

When a process or procedure is in statistical control, a steady state prevails. Random behavior continues, but the overall mean value and the variance do not change. When a process is no longer "in control," something in the process or procedure has changed. The control chart does not identify the cause of the change. The analyst must determine the cause of the change.

V. MONITORING DATA CHARACTERIZATION

Ambient air monitoring data, whether from a single sampling site or from several sites, produce a number of discrete data points (concentrations) and a range of concentrations. For example, the 1988 NMOC monitoring data[10] included 3069 data points ranging from 0.016 to 5.890 parts per million carbon (ppmC) by volume.

Part of data characterization involves being able to describe a large number of data points with as few words as possible or with as few statistics (numbers or symbols that are well defined) or parameters as possible. Frequency distributions are useful for this purpose because experience has shown that ambient air monitoring data generally approximate a two-parameter lognormal frequency distribution. A mean and standard deviation completely describes a lognormal frequency distribution and permits a great deal of information to be estimated from just two parameters. Graphical frequency distribution displays are useful for visually determining some of the character of the frequency distribution.

A. Graphical Displays

Velleman and Hoaglin[11] describe stem and leaf plots and box plots, which are useful ways to display monitoring data concentrations graphically. Figure 4 shows the stem and leaf plot of the 1988 NMOC data, and Figure 5 shows box plots of NMOC concentrations for each hour of the day for an automated diurnal monitoring program conducted in Raleigh, NC during the summer of 1989. Several personal computer and mainframe computer software packages[12,13] will generate stem and leaf plots and box plots.

Both stem and leaf diagrams and box plots show not only the range of data (maxima and minima), but also the median and quartiles. The median divides the sorted concentration data into halves, i.e., it locates the 50th percentile. One-half of the sorted data have concentrations smaller than the median, and one-half have

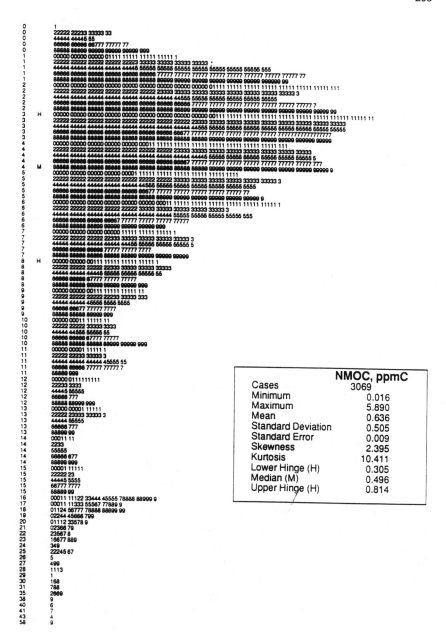

FIGURE 4. Stem and leaf plot of the 1988 morning NMOC data.

concentrations which are larger. The quartiles divide each sorted half into quarters. The lower quartile locates the 25th percentile while the upper quartile marks the 75th percentile.

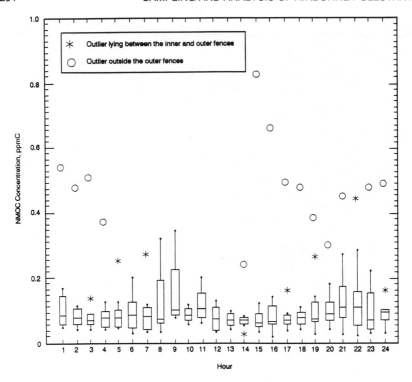

FIGURE 5.   Box plots of NMOC concentrations for each hour of the day.

B. Stem and Leaf Plots

Stem and leaf diagrams plot the numerical value of each datum, truncated to two or three digits. Stems are on the left of the vertical open channel and leaves are on the right. The minimum value, 0.016 ppmC, is shown as "0 1" at the top of the diagram in Figure 4; and the maximum value, 5.890 ppmC, is shown as the last digit in the leaf, "58 9", at the bottom of the figure. The M in the open channel between the stems and the leaves locates the stem of the median, 0.496 ppmC. The Hs identify the stems for the lower and upper quartiles, 0.305 ppmC and 0.814 ppmC, respectively. The stem and leaf plots show details of the frequency distribution. For example, in Figure 4 there appears to be a small node beginning at about 1.6 ppmC. Careful examination of the stem and leaf plot shows that beginning with 1.6 ppmC (16 on the stem in the plot) the scale changes, which accounts for the apparent node beginning at that point. Another observation needs to be made regarding Figure 4. Although the NMOC monitoring data are approximately lognormally distributed, the presence of the very long tail representing larger values of NMOC concentrations violates the expectation of the lognormal distribution. The very large concentration values unduly bias determination of means, standard deviations, and other statistics that are generally used to characterize monitoring data. To obviate these difficulties, Sager et al.[19] and McAllister et al.[20] recommend the use of ranks instead of concentration values to compare one distribution with another distribution. Ranks number the sorted concentration data from the largest to the smallest. The use of ranks have a number of advantages over the use of the raw concentration data. Many statistical procedures using nonparametric statistics with ranks do not require the assumption of normal distribution to characterize or compare data distributions.[21]

C. Box Plot

Hourly NMOC concentration data are characterized using box plots in Figure 5. For each hour of the day, the box plot shows the range and skewness of the concentration data. Each rectangle stretches between the 25th and 75th percentile and is called the interquartile or midrange of the data. The horizontal line in the rectangle marks the locus of the 50th percentile, or the median. The whiskers reach out of the ends of the rectangles to span the remainder of the data, except for outliers. The maximum length of a whisker is the distance between the 25th (or 75th) percentile and the inner fences (lower and upper), as follows:

$$\text{upper inner fence} = \text{75th percentile} + (1.5 \times \text{midrange}) \qquad (3)$$
$$\text{lower inner fence} = \text{25th percentile} - (1.5 \times \text{midrange}) \qquad (4)$$

Outer fences are defined in Equations 4 and 5:

$$\text{upper outer fence} = \text{75th percentile} + (3.0 \times \text{midrange}) \qquad (5)$$
$$\text{lower outer fence} = \text{25th percentile} - (3.0 \times \text{midrange}) \qquad (6)$$

Data between the inner and outer fences are shown as asterisks in Figure 5. Data outside the outer fences are shown as circles. The outliers shown in the figure may be erroneous, but in the absence of a specific case-by-case explanation why a datum is erroneous, outliers should be viewed simply as data lying outside the probability distribution described by the box plot.

D. Statistics

Other statistics used to characterize monitoring data are mean, standard deviation, skewness, and kurtosis. The standard deviation is a measure of how "spread out" the data are and is the second moment about the mean. Skewness is the third moment about the mean, and kurtosis is the fourth moment about the mean. Skewness, as the name implies, quantifies the symmetry of the frequency distribution, and kurtosis quantifies how pointed or peaked a distribution is. A Gaussian normal distribution has a skewness of zero and a kurtosis of 3.0. Distributions more pointed than a normal distribution have a kurtosis greater than 3.0. The values of the statistics described here are shown on Figure 4 for the 1988 NMOC data.

E. Pearson Correlation Matrices

In studying UATMP site monitoring results, seven organic compounds were found in each sample analyzed — 1,1,1-trichloroethane, carbon tetrachloride,

ppbv1	1,1,1-Trichloroethane
ppbv2	Carbon tetrachloride
ppbv3	Toluene
ppbv4	Styrene/*o*-xylene
ppbv5	*m/p*-Xylene
ppbv6	Ethylbenzene
ppbv7	Benzene

	ppbv1	ppbv2	ppbv3	ppbv4	ppbv5	ppbv6	ppbv7
ppbv1	1.000						
ppbv2	0.603	1.000					
ppbv3	0.473	0.652	1.000				
ppbv4	0.329	0.605	0.957	1.000			
ppbv5	0.454	0.654	0.995	0.967	1.000		
ppbv6	0.423	0.634	0.987	0.981	0.992	1.000	
ppbv7	0.522	0.730	0.962	0.925	0.965	0.959	1.000

Number of observations: 31

FIGURE 6.   Pearson correlation matrix for W1KS.

toluene, styrene/*o*-xylene, ethylbenzene, and benzene. Graphs of these concentration data vs Julian date collected illustrated an interdependence or correlation among the aromatic compounds. This interdependence led to the study of Pearson correlation matrices at each site to ascertain if there were some pattern among the site correlation matrices which would help to characterize and/or explain the monitoring data. In the Pearson correlation matrix, coefficients of correlation among several organic compounds present in each ambient sample are arranged in a triangular matrix. For example, a correlation matrix for one location (W1KS) in Wichita, KS is shown in Figure 6. The data are taken from the 1989 UATMP and show correlations among seven air toxic target compounds at W1KS. The data in Figure 6 represent 31 ambient air samples taken from January 6, 1989, through January 23, 1990. For example, the correlation coefficient for toluene concentrations vs ethylbenzene concentrations in 31 ambient air samples was 0.987. Note that the correlation coefficients among all the aromatic hydrocarbons in the list are greater than 0.925, while the correlations among the chlorohydrocarbons are all below 0.730. This means that the concentrations of chlorocarbons cannot be predicted well by knowing the concentrations of the aromatic hydrocarbons. Why this is true is unknown at this time. Figure 7 shows a Pearson correlation matrix for W2KS, a second monitoring site in Wichita, covering the same time period as that in Figure 6. The correlation among the aromatic hydrocarbons is similar to the correlation at site W1KS with the exception of the coefficients involving styrene/ *o*-xylene, which are significantly lower for site W2KS. The coefficients involving the chlorocarbons in this case are much lower, but are all positive.

Figures 8 and 9 show Pearson correlation matrices for two District of Columbia monitoring sites, W1DC and W2DC. Again the correlations among the aromatic hydrocarbons are seen to be very strong, all greater than 0.900. The correlation coefficients involving 1,1,1-trichloroethane and carbon tetrachloride show similar

ppbv1	1,1,1-Trichloroethane
ppbv2	Carbon tetrachloride
ppbv3	Toluene
ppbv4	Styrene/o-xylene
ppbv5	m/p-Xylene
ppbv6	Ethylbenzene
ppbv7	Benzene

	ppbv1	ppbv2	ppbv3	ppbv4	ppbv5	ppbv6	ppbv7
ppbv1	1.000						
ppbv2	0.270	1.000					
ppbv3	0.137	0.144	1.000				
ppbv4	0.228	0.052	0.804	1.000			
ppbv5	0.162	0.162	0.972	0.854	1.000		
ppbv6	0.120	0.137	0.963	0.839	0.993	1.000	
ppbv7	0.158	0.263	0.947	0.787	0.965	0.958	1.000

Number of observations: 31

FIGURE 7.   Pearson correlation matrix for W2KS.

ppbv1	1,1,1-Trichloroethane
ppbv2	Carbon tetrachloride
ppbv3	Toluene
ppbv4	Styrene/o-xylene
ppbv5	m/p-Xylene
ppbv6	Ethylbenzene
ppbv7	Benzene

	ppbv1	ppbv2	ppbv3	ppbv4	ppbv5	ppbv6	ppbv7
ppbv1	1.000						
ppbv2	0.317	1.000					
ppbv3	0.667	-0.217	1.000				
ppbv4	0.532	-0.344	0.944	1.000			
ppbv5	0.599	-0.284	0.972	0.985	1.000		
ppbv6	0.584	-0.311	0.944	0.988	0.984	1.000	
ppbv7	0.684	-0.145	0.971	0.906	0.944	0.902	1.000

Number of observations: 27

FIGURE 8.   Pearson correlation matrix for W1DC.

patterns between the two Washington sites. A striking feature about the Washington site matrices is that all the correlation coefficients involving carbon tetrachloride are negative. The reasons for the negative coefficients are not known at this time. While the Washington site matrices are similar, each displays a unique fingerprint.

Figures 10 and 11 show Pearson correlation matrices for Houston, TX (H1TX), and Baton Rouge, LA (BRLA). The correlation coefficients among the aromatic hydrocarbon pairs are noticeably smaller, with the possible exceptions of the

ppbv1	1,1,1-Trichloroethane
ppbv2	Carbon tetrachloride
ppbv3	Toluene
ppbv4	Styrene/o-xylene
ppbv5	m/p-Xylene
ppbv6	Ethylbenzene
ppbv7	Benzene

	ppbv1	ppbv2	ppbv3	ppbv4	ppbv5	ppbv6	ppbv7
ppbv1	1.000						
ppbv2	-0.038	1.000					
ppbv3	0.299	-0.278	1.000				
ppbv4	0.333	-0.278	0.936	1.000			
ppbv5	0.319	-0.268	0.960	0.986	1.000		
ppbv6	0.342	-0.300	0.952	0.984	0.990	1.000	
ppbv7	0.237	-0.228	0.961	0.929	0.952	0.925	1.000

Number of observations: 27

FIGURE 9.   Pearson correlation matrix for W2DC.

ppbv1	1,1,1-Trichloroethane
ppbv2	Carbon tetrachloride
ppbv3	Toluene
ppbv4	Styrene/o-xylene
ppbv5	m/p-Xylene
ppbv6	Ethylbenzene
ppbv7	Benzene

	ppbv1	ppbv2	ppbv3	ppbv4	ppbv5	ppbv6	ppbv7
ppbv1	1.000						
ppbv2	0.274	1.000					
ppbv3	0.400	0.101	1.000				
ppbv4	0.656	0.221	0.779	1.000			
ppbv5	0.509	0.051	0.647	0.878	1.000		
ppbv6	0.413	0.159	0.552	0.792	0.672	1.000	
ppbv7	0.390	0.124	0.942	0.780	0.683	0.610	1.000

Number of observations: 34

FIGURE 10. Pearson correlation matrix for H1TX.

benzene-toluene coefficient for H1TX, and the styrene/o-xylene vs m/p-xylene coefficient at BRLA. An explanation for this pattern of behavior that immediately comes to mind is that Houston and Baton Rouge are the locations of petrochemical processing facilities with independent hydrocarbon emissions that reduce the correlations among the aromatic hydrocarbons. Neither Washington nor Wichita has high concentrations of petrochemical processing facilities.

An hypothesis to explain similarities between the Wichita sites and the Washington sites might relate to a major source of air pollution common to both sites,

ppbv1	1,1,1-Trichloroethane					
ppbv2	Carbon tetrachloride					
ppbv3	Toluene					
ppbv4	Styrene/o-xylene					
ppbv5	m/p-Xylene					
ppbv6	Ethylbenzene					
ppbv7	Benzene					

	ppbv1	ppbv2	ppbv3	ppbv4	ppbv5	ppbv6	ppbv7
ppbv1	1.000						
ppbv2	-0.005	1.000					
ppbv3	0.558	0.199	1.000				
ppbv4	0.153	-0.124	0.337	1.000			
ppbv5	0.232	-0.112	0.339	0.928	1.000		
ppbv6	0.127	-0.027	0.273	0.665	0.791	1.000	
ppbv7	0.432	0.418	0.819	0.152	0.188	0.262	1.000

Number of observations: 31

FIGURE 11. Pearson correlation matrix for BRLA.

emissions from fossil-fueled engines. These emissions combined with photochemical reactions occurring at both sites might satisfactorily account for the similar matrix patterns, among the aromatic hydrocarbons, at least.

Much more research is needed to understand and explain the patterns displayed by the Pearson correlation matrices for monitoring sites; however, it is clear that each site pattern is unique, and site patterns in the same city or perhaps the same air shed are similar. Pearson correlation matrices offer possibilities of characterizing and comparing air monitoring data in meaningful ways when using the combinations of organic compound concentrations appropriate for the site(s). For the Pearson correlation matrix to be useful, periodic and regular monitoring data must be available. For a compound to make a useful comparison, each compound must be present in the sample each day and paired with each other compound analyzed on the same day. The Pearson correlation matrices given here extend over a 1-year monitoring period. It may be desirable to compare Pearson correlation matrices spanning 3 months at a time (quarterly) by the calendar to study possible seasonal effects on the pattern of the matrix.

VI. CENSORED DATA

In risk analysis, the annual average concentration of a pollutant is important. In calculating annual averages, the question frequently arises how to treat concentrations below the detection limit, i.e., the nondetected concentrations. A number of ways are recommended.[14-17] Often zero is used for "nondetects," or one-half the analytical detection limit. Either approach biases the annual average. Research is ongoing to be able to generate an unbiased estimate of an average, using censored data. Newman et al.,[14] Helsel,[17] and Rao et al.[18] are among the researchers who are using more sophisticated methods for estimating the mean and variance for environmental samples with below detection limit observations.

VII. CONCLUSIONS

Some of the conclusions that result from the data treatment procedures discussed here are:

1. A linear calibration curve that minimizes the sum of squares of percent difference between the data point and the regression line is superior to the usual least squares regression when dealing with concentrations near (or below) the detection limit.

2. The mean-slope method is useful for calculating concentrations near the detection limit.

3. Pooled % CV is a useful statistic to measure analytical precision, or sampling and analytical precision.

4. A nested analysis of variance can provide the statistics to separate analytical precision from sampling precision when the two sources of variation are significantly different.

5. For propylene collected as a Type I duplicate, the sampling precision was zero. Thus, there was not a significant difference between duplicate samples for propylene. This conclusion is not to imply that there may never be a significant difference between Type I duplicates.

6. For an ambient air sample collected as a Type II duplicate, there was a significant difference between duplicate samples collected simultaneously. This is not to imply that there may always be a difference between Type II duplicate samples.

7. NMOC ambient air monitoring data display features of a lognormal distribution when plotted as stem and leaf plots and as box plots.

8. Quality control charts are useful in monitoring programs to ensure that sampling and analytical procedures are not changing significantly.

9. Pearson correlation matrices are useful in characterizing ambient monitoring data in which specific compounds exist in each sample. The correlation matrices show unique patterns for each site and also unique patterns for a given air shed.

10. Several statistical techniques are available to estimate ambient air monitoring average concentrations, and standard deviations for data sets in which some samples have concentration levels below the detection limit.

11. The use of ranks is a useful technique for comparing distributions of concentrations of air monitoring data.

ACKNOWLEDGMENTS

Support for the UATMP and the NMOC monitoring programs was provided by the U.S. EPA Office of Air Quality Planning and Standards, Research Triangle Park, NC. Mr. Neil J. Berg, Jr. is the current Project Officer for the U.S. EPA. Radian Corporation, Austin, TX, supported part of the work and supplied additional data. The assistance of Phyllis L.O'Hara and Dr. Joan T. Bursey at the Research Triangle Park laboratory of Radian Corporation is gratefully acknowledged. Several discussions with Dr. M.W. Hemphill of the Texas Air Control Board and Dr. Thomas Sager of the University of Texas at Austin were especially helpful in improving the accuracy of statistical assertions. Should any statistical assertions be wrong or inaccurate, the author takes full responsibility for them.

REFERENCES

1. Scheffé, H. *Ann. Stat.* 1:1–37 (1973).
2. Liteanu, L., and I. Rîca. *Statistical Theory and Methodology of Trace Analysis* (New York: John Wiley & Sons, Inc., 1980).
3. Plesch, R. *Z. Ann. Chem.* 275: 269–274 (1975).
4. Bocek, P., and J. Novák *J. Chromatog.* 51: 375–383 (1970).
5. Smith, E. D., and D. M. Mathews. *J. Chem. Ed.* 44: 757–759 (1967).
6. McAllister, R. A., W. H. Moore, Joann Rice, E. Bowles, D-P. Dayton, R. F. Jongleux, R. G. Merrill, Jr., and J. Bursey. "1989 Urban Air Toxics Monitoring Program," Radian Corporation, DCN No. 90-262-045-93, Research Triangle Park, NC (June 1990).
7. Neter, J., W. Wasserman, and M. H. Kutner. *Applied Linear Statistical Models* (Homewood, IL: Richard D. Irwin, 1989).
8. Milliken, G. A., and D. E. Johnson. *Analysis of Messy Data. Volume I: Designed Experiments* (Belmont, CA: Lifetime Learning Publications, 1984).
9. Duncan, A. J. *Quality Control and Industrial Statistics* (Homewood, IL: Richard D. Irwin, 1965).
10. McAllister, R. A., P. L. O'Hara, W. H. Moore, D-P. Dayton, J. Rice, R. F. Jongleux, R. G. Merrill, Jr., and J. T. Bursey. "1989 Nonmethane Organic Compound Monitoring Program," Final Report, Vol. 1 "Nonmethane Organic Compounds," Office of Air Quality Planning and Standards, U.S. EPA Report EPA-450/4-89-003 (1988).
11. Velleman, P. F., and D. C. Hoaglin. *Applications, Basics, and Computing of Exploratory Data Analysis* (Belmont, CA: Wadsworth, 1981).
12. Wilkinson, L. *SYSTAT: The System for Statistics* (Evanston, IL: SYSTAT, 1988).
13. *SAS® Procedures Guide for Personal Computers*, Version 6 ed. (Cary, NC: SAS Institute, 1985).
14. Newman, M. C., P. M. Dixon, B. B. Looney, and J. E. Pinder, III. *Water Resour. Bull.* 24:905–915 (1989).
15. Gleit, A. "Estimation for Small Normal Data Sets with Detection Limits," *Environ. Sci. Technol.* 19:1201–1206 (1985).

16. Travis, C. C., and M. L. Land. "Estimating the Mean Value of Data Sets with Nondetectable Values," *Environ. Sci. Technol.* 24(7):961–962 (1990).

17. Helsel, D. R. "Less than Obvious. Statistical Treatment of Data below the Detection Limit," *Environ. Sci. Technol.* 24(12): 1766–1774 (1990).

18. Rao, S., Trivikrama, Jia-Yeong Ku, and K. Shankar Rao. "Analysis of Toxic Air Contaminant Data Containing Concentrations Below the Limit of Detection," *J. Air Waste Manage. Assoc.* 41:442–448 (1991).

19. Sager, T. W., Vaquiax, and M. W. Hemphill. "Statistical Assumptions Matter in Data Analysis for Texas Ozone Nonattainment Sites," *J. Air Waste Manage. Assoc.* 40(2):199–205 (1990).

20. McAllister, R. A., T. W. Sager, J. P. Gise, and M. W. Hemphill. "Importance of NO_x Control Suggested by NMOC-NO_x-Ozone Data Analysis," *Proceedings of the 84th Annual Meeting of the Air & Waste Management Association*, paper 91-75.1, Vancouver, B.C., June 17–21, 1991.

21. Hollander, M., and D. A. Wolfe. *Nonparametric Statistical Methods* (New York: John Wiley & Sons, Inc., 1973).

CHAPTER 18

Assessing the Visibility Impairment Associated with Various Sulfate Reduction Scenarios at Shenandoah National Park

James F. Sisler and William C. Malm

TABLE OF CONTENTS

0-87371-606-0/93/$0.00 + $.50

© 1993 by Lewis Publishers

I. INTRODUCTION

A major challenge facing scientists, decision makers, and lay persons in assessing the effect of emission control programs on visibility is to link aerosol chemical and physical properties to the ability of the atmosphere to transmit visual stimuli. It is possible, with varying degrees of accuracy, to model the effect that a particular mix of aerosols and meteorologic conditions have on various optical parameters such as visual range, contrast, or extinction coefficient. However, the goals of many emission control programs are frequently expressed as reductions of annual emissions or aerosol loadings for large regions. How then, does one translate such reductions in emissions/loadings to expected changes in atmospheric optical properties in a way that is meaningful?

One such method is to compare the probability of occurrence of various levels and types of aerosol loadings and meteorologic conditions for present and projected conditions, and to model the expected levels and changes in extinction. Given this information, computer-generated images of scenic vistas may be utilized to demonstrate the impact of emission control strategies on visibility by comparing "before and after" images. The necessity of computer imaging the effect of aerosol loading for a particular scene is due to the fact that atmospheric extinction affects many aesthetic qualities that are difficult to quantify, but are easily integrated by the interaction of the eye/brain system with visual stimuli.[1,2]

A historical database of atmospheric aerosol loadings is prerequisite for such an analysis. The National Park Service (NPS) has operated aerosol sampling devices for several years at three eastern National Parks (NP): Shenandoah NP (since 1982), Great Smoky Mountains NP (since 1984), and Acadia NP (since 1985). Using the historical record at Shenandoah NP, state-of-the-art thermodynamic/MIE scattering models, and advanced image-processing techniques, three sulfate reduction scenarios will be described. The scenarios will compare current aerosol/visibility conditions to projected conditions assuming 20, 40, and 60% reductions in sulfate aerosol loadings.

II. AEROSOL DATA

Baseline aerosol conditions for 1985 will be derived from aerosol sampling devices operated by the University of California at Davis.[3] Various techniques and assumptions are employed to characterize ambient aerosol concentrations using raw sampler data. Table 1 summarizes the analysis techniques and assumptions for the various components of ambient aerosols used in this study, and are discussed in detail as follows.

Until December 1987, the aerosol sampling devices employed were stacked filter units (SFU). The SFU is a two-stage particulate sampler consisting of an 8-μm pore size Nuclepore membrane filter in series with a 100% efficient Teflon® filter of reduced area. The inlet tube is designed to pass particles as large as 15 μm

Table 1. Analysis Techniques Used to Convert SFU Filter Samples to Raw Data and Assumptions Used to Convert the Raw Data to Aerosol Species Concentrations

Species	Analysis method	Composite/derived variable equation
Coarse mode	Gravimetric	—
Fine mass	Gravimetric	—
Fine mass elements	PIXE[a]	—
Ammonium sulfate	PIXE	4.125*S
Fine soil	PIXE	Al*2.2 + Si*2.49 + Ca*1.63 + Fe*2.42 + Ti*1.94
Light absorbing carbon	LIPM[b]	$b_{abs}/\varepsilon, \varepsilon = 10 m^2/g$
Ammonium nitrate	PIXE	0.04*(4.125*S)
Organics	Gravimetric, PIXE, LIPM	Fine mass − soil − sulfate − nitrate − carbon

[a] PIXE — Proton induced X-ray emission
[b] LIPM — Laser Integrating Plate Method

in diameter; the capture efficiency at the 15-μm cut point is 50%.[3] Coarse material (2.5 to 15 μm in diameter) is deposited on the first-stage filter of the SFU sampler, and the remaining fine mass is captured on the second-stage filter. At the 2.5-μm cut point, the capture efficiency of the two stages is 50%, causing some of the fine mass to be deposited on the coarse filter and vice versa.[3] Fortunately, the bulk of the optically active aerosol exists as fine mass with a diameter range of 0.3-1.0 μm and is not affected by the softness of 2.5-μm cut point.

The mass concentrations for the coarse and fine particles are determined from the difference between gravimetric measurements of the filters made before and after the sample collection.

Because the coarse fraction is in a size range that is not efficient for the scattering of light and because the coarse fraction is presumed to primarily consist of airborne soil particles, only the mass of the coarse fraction will be used for this analysis.

The fine fraction, which accounts for the bulk of atmospheric extinction, is presumed to consist of ammonium sulfate, ammonium nitrate, organics, light-absorbing carbon, and fine soil. Since the various components of the fine fraction have different chemical and optical properties, individual treatment is required to effectively model visibility impairment associated with the fine fraction.

Proton induced X-ray emission (PIXE)[3] was used to determine the bulk concentrations of atomic elements that compose the aerosol particles trapped by the fine stage filter, and are reported in units of nanograms per cubic meter. Assumptions about sources of the various elements must be made to estimate the actual aerosol composition. For this study, all elemental sulfur is presumed to be from

Table 2. Descriptive Statistics[a] for Summertime Coarse Aerosol Mass and Components of Fine Aerosol Mass at Shenandoah NP

Variable	Minimum	Maximum	Mean	SD
Coarse	836	25,500	8,475	3,579
Fine	905	76,941	16,323	9,524
Ammonium sulfate	45	28,128	8,933	4,726
Ammonium nitrate	1	1,040	330	174
Soil	10	2,266	581	438
Soot	20	5,327	532	532
Organics	157	47,858	5,946	5,497

[a] ng/m^3

ammonium sulfate; therefore elemental sulfur mass is multiplied by 4.125 to obtain the mass of ammonium sulfate. Fine soil is typically comprised of oxides of aluminum, silicon, iron, titanium, calcium, and potassium, thus, masses of these elements are appropriately scaled to account for their oxides and summed to estimate fine soil.

Light-absorbing carbon is determined optically by the laser-integrating plate method (LIPM),[3] and is reported in units of optical absorption ($b_{abs} = 10^{-8}/m^3$). If an absorption efficiency of 10 m^2/g is assumed, then the absorption coefficient is equal to a concentration in nanograms per cubic meter.

Ammonium nitrate was not measured by the SFU sampler. However, current aerosol monitoring programs that have been in place since the winter of 1988, show that the summertime ratio of ammonium nitrate to ammonium sulfate is about 0.04.[1] Thus, ammonium nitrate concentrations are estimated by multiplying ammonium sulfate concentrations by 0.04.

Finally, the remaining fine mass not accounted for by ammonium sulfate, ammonium nitrate, light-absorbing carbon, and fine soil is presumed to be organic. The mass of organic aerosol is then estimated by subtracting from fine mass the sum of the masses of ammonium sulfate, ammonium nitrate, fine soil, and light-absorbing carbon. Table 2 has descriptive statistics for summertime coarse aerosol mass and components of fine aerosol mass at Shenandoah NP.

III. AEROSOL GROWTH AND MIE SCATTERING MODELS

Models are used to calculate the scattering and absorption properties of the aerosol as a function of size, index of refraction, relative humidity, and other physical properties such as the relative mix of different chemicals within one particle. The model used here, developed by Tsay and Stephens,[4] allows for externally mixed spherical aerosols as well as multicomponent aerosols that consist of a spherical core with shells of various thickness and physicochemical characteristics.

Figure 1 is a schematic for the modeling process employed here to generate optical parameters used for the NPS image-processing system.[1] Were it not for the

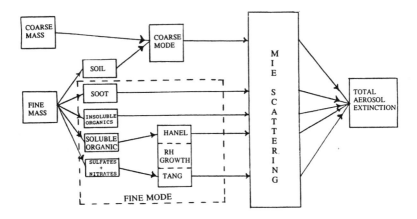

FIGURE 1.   Flow diagram showing the procedure used to calculate the extinction, scattering, SVR, and the amount of water due to RH for a given
aerosol.

input requirements of the image-writing procedure, bulk optical properties such as
extinction, contrast, and visual range could be adequately calculated with empirical and theoretical models which require fewer detailed assumptions and inputs
than are employed here. What follows is a brief description of the model, the
inputs, and the assumptions.

For modeling purposes, the various components of an aerosol are lumped into
two size distribution modes — coarse and fine — and are presumed to be
lognormally distributed. A coarse mode with a mass median radius of 5.55 μm,
a geometric standard deviation of 2.5, and a density of 2.55 g/cm³ was used, while
the values for the fine mode were 0.2 μm, 2.0, and 1.75 g/cm³, respectively. Since
the coarse aerosol fraction is largely comprised of soil particles, fine soil was
assumed to be on the small side of size distribution of the coarse fraction.
Therefore, fine soil mass was added to the coarse aerosol fraction mass to obtain
the total mass of the coarse mode. The four remaining components of the fine
fraction comprise the fine mode. Light-absorbing carbon is associated with the
soot component of the fine mode; while ammonium sulfate, ammonium nitrate,
and one-half the organic mass are assumed to form one water soluble component.
The remaining organic mass forms the insoluble nonsoot component of the fine
mode.

Aerosol growth as a function of relative humidity (RH) is calculated following
an approach outlined by Hanel, and was used to calculate the amount of water and
the effect of water on particle size associated with each water soluble species
belonging to the fine mode.[4] Using thermodynamic theory and assuming an
atmosphere in equilibrium, the growth of a soluble particles radius for a particular
RH can be given as:

$$r(a_w) = r_d[1 + p^*M_w(a_w)/M_d]^{1/3}$$ (1)

where

r_d = radius of dry aerosol
p   = density of aerosol relative to that of liquid water
$M_w(a_w)/M_d$ = mass ratio of condensed water to dry aerosol[6]

The water activity parameter (a_w) is defined by:

$$a_w(r) = RH*exp[-0.314688/(T*r(a_w))] \qquad (2)$$

where T is the absolute temperature (K), and is valid for temperature range between 245 and 300K. Due to the form of Equations 1 and 2, there is no analytic expression for aerosol size as function of RH; and an iterative procedure must be used until convergence.[6]

Two water activity functions were used to simulate growth of the soluble portion of the fine mode due to RH. One water activity function was parameterized to reproduce growth curves published by Tang;[7] ammonium sulfate and ammonium nitrate mass were associated with this growth curve. Soluble organics, presumed to be less hygroscopic than ammonium sulfate and ammonium nitrate, were associated with a water activity function parameterized to data published by Hanel.[5]

MIE theory and the assumption of externally mixed spherical particles were used to model the interaction of light with aerosols in a scattering/absorbing atmosphere at a wavelength of 0.63, 0.55, and 0.45 μm (red, green, blue). The index of refraction used for the coarse mode was 1.7-0.008 i.[4] Four indices of refraction were assumed for the fine mode. An index of 1.7-0.008 i was assigned to the insoluble portion; an index of 1.53–0.006 i, for the soluble portion; an index of 1.9–0.6 i was used for the soot portion; and the index was taken to be $1.33–2^{-9}i$ for liquid water.[4] Once the radii of wet aerosol particles are found using Equations 1 and 2, the effective index of refraction (m) for a soluble aerosol particle, which is homogeneously mixed with liquid water, is simply the volume weighted average of:

$$m = m_w + (m_h - m_w)[r_d/r(a_w)]^3 \qquad (3)$$

where

m_h = complex refractive index of soluble aerosol
m_w = complex refractive index of liquid water

After growing the soluble components of the aerosol and calculating their effective indices of refraction, the extinction and scattering for each component of the fine aerosol is calculated. Using the principal of superposition, the total non-Rayleigh extinction for the aerosol is the sum of extinction due to the coarse mode and each component of the fine mode. By adding Rayleigh scattering to aerosol extinction, the total atmospheric extinction is obtained, and the visual range for the aerosol can be calculated. Visual range is defined as the maximum distance that a large black object on the horizon can be seen.[8] By using a value of 0.01

km^{-1} for Rayleigh scattering, which is the atmospheric scattering at an altitude of about 1000+ m and is chosen to represent a standard atmosphere, the standard visual range (SVR) can be calculated.[8] On a "perfect" day with no aerosol loading, the extinction would be due only to Rayleigh scattering and the maximum SVR of 391 km would be obtained as given by:

$$SVR = 3.912/b_{ext} \qquad (4)$$

where b_{ext} denotes the sum of aerosol extinction plus Rayleigh scattering.

IV. VISIBILITY ASSESSMENT PROCEDURE

Using the historical database of aerosol species concentrations at Shenandoah NP, a frequency of occurrence analysis was used to characterize ambient aerosol conditions. Past attempts to do this have focused on total fine mass as the frequency analysis variable which enables a ranking varying between clean (good visibility) and dirty (poor visibility).[1] However, the chemical fractional composition of an ambient aerosol can vary independently of total fine mass. For example, some of the cleanest days were days where sulfate composed the largest as well as the smallest fraction of the ambient fine aerosol; the same holds for the dirtiest days. Using the sulfate fraction as an analysis variable, the expected impairment of visibility due to sulfate can be characterized by the fractional composition of the aerosol rather than the total fine mass.

Using sulfate fraction as a frequency analysis variable, the data were partitioned into three subgroups, low, median, and high. The low subgroup consists of cases where the cumulative frequency of occurrence is between 0 and 20%. The median subgroup consists of cases with a cumulative frequency of occurrence between 40 and 60%, and the high subgroup was between 80 and 100%. A characteristic aerosol for a subgroup was derived by using the mean attributes of the subgroups data. To further characterize high relative sulfate conditions, a single-data record from the high subgroup was selected to constitute a fourth extreme high sulfate condition for subsequent assessment.

The water and extinction for any water soluble aerosol requires explicit consideration of RH. Figure 2 shows the distribution of summertime relative humidity values at Shenandoah NP. Due to the nonsymmetric distribution of RH and the nonlinear nature of aerosol growth, using the average relative humidity would seriously underestimate the average water and average extinction associated with a representative aerosol. Therefore, the water and extinction associated with a particular subgroup characteristic aerosol was weighted to the frequency of occurrence of RH assuming the distribution of RH is independent of the distribution of sulfate fraction, while an RH of 95% is used for the fourth extreme high sulfate condition.

By estimating extinction of an aerosol (and other optical properties) for the three primary colors (red, green, and blue), radiative transfer models[9,10,11] coupled with image processing techniques[1,2] can simulate the haze for a scenic vista or

RH

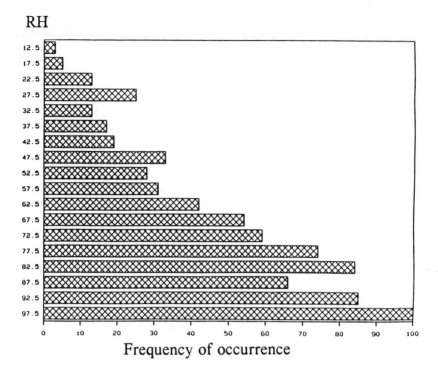

Frequency of occurrence

FIGURE 2.   Distribution of summertime relative humidities at Shenandoah NP.

view. To simulate haze requires a photograph of a view when the atmosphere is extremely clean or is as close to Rayleigh conditions as possible. By digitizing the image and knowing the distance to each feature from the camera, haze can be added for any level of extinction or atmospheric loading. Figure 3 is a computer-imaged photograph of a typical Appalachian scene at near Rayleigh conditions. This photograph will be used to demonstrate the visibility assessment by illustrating the difference in visibility between a distribution of ambient summer season aerosol conditions and three levels of sulfate reduction. For a more complete description of the image-processing techniques and input requirements employed here, the reader is referred to Malm et al.[1]

V. RESULTS

Table 3 has the mass budgets of the four aerosols used to characterize ambient conditions and the extreme sulfate case at Shenandoah NP. Shown is the mass of each component of the fine fraction, the total mass of the components, the fraction of the total for each component, and the water associated with the soluble components due to RH. Table 4 details the extinction budget at a wavelength of 0.55 μm for the four aerosols assuming one-half the organic mass is soluble.

FIGURE 3. Typical Appalachian vista at near Rayleigh conditions.

Shown for each aerosol is the average extinction due to the coarse mode; and shown for each component are the fine mode, the total non-Rayleigh extinction, and the fractional contribution to total extinction with and without Rayleigh scattering added.

Table 5 details the extinction budget for the high subgroup at current conditions and the three levels of sulfate reduction. Table 6 shows the change in total extinction (Rayleigh included), SVR, and percent change in extinction and SVR between current conditions and the three sulfate reduction scenarios for the three sulfate fraction subgroups, high, median, low, and the extreme high sulfate case.

Using relative humidity weighted extinctions, Figures 4 through 7 photographically show the effect of reducing atmospheric sulfate loadings for the high sulfate fraction group (not the extreme sulfate case). Figure 4 shows the current visibility condition; Figures 5 through 7 show the effect of 20, 40, and 60% reductions of sulfate loading, respectively, by contrasting the left half of the picture, which is unchanged from current conditions (Figure 4), with the right half, which shows the improvement in visibility.

VI. CONCLUSIONS

The analysis of historical aerosol databases, coupled with state-of-the-art optical and thermodynamic equilibrium models, can effectively characterize the expected improvement in visibility resulting from reductions of aerosol sulfate loading. The optical and thermodynamic models estimate scattering and absorption effects of multicomponent aerosols, and the effects of relative humidity on growth of various aerosol species. Radiative transfer models and state-of-the-art image processing techniques can integrate changes of visibility into an easily

SAMPLING AND ANALYSIS OF AIRBORNE POLLUTANTS

Table 3. Mass Budgets of the Four Aerosols Used to Characterize Ambient Average and Extreme High Sulfate Conditions at Shenandoah NP

	Mass type	Current mass budget	Current average mass (ng/m³)	Associated water (ng/m³)
Low	Sulfate	0.37	7,440	22,600
	Nitrate	0.01	275	835
	Soil	0.04	711	0
	Soot	0.05	987	0
	Organics	0.53	10,718	11,001
		1.00	20,131	34,436
Median	Sulfate	0.56	7,999	24,298
	Nitrate	0.04	552	1,676
	Soil	0.03	363	0
	Soot	0.03	451	0
	Organics	0.34	4,888	5,017
		1.00	14,253	30,991
High	Sulfate	0.73	10,541	32,020
	Nitrate	0.02	306	929
	Soil	0.04	563	0
	Soot	0.03	364	0
	Organics	0.18	2,611	2,680
		1.00	14,385	35,629
High sulfate	Sulfate	0.86	4,909	43,406
	Nitrate	0.02	142	1,255
	Soil	0.03	157	0
	Soot	0.03	181	0
	Organics	0.06	310	1,368
		1.00	5,699	46,029

understandable pictorial format. The National Park Service visibility image processing system was used to demonstrate the difference between current visibility conditions and the improvement associated with various sulfate reduction scenarios for a typical Appalachian view. The effect on visibility impairment caused by different aerosol loadings can readily be seen by comparing photographs depicting different levels of haziness for the same scene.

DISCLAIMER

The assumptions, findings, conclusions, judgments, and views presented herein are those of the authors and should not be interpreted as necessarily representing official National Park Service policies.

Table 4. Extinction Budgets of the Four Aerosols Used to Characterize Ambient Average and Extreme High Sulfate Conditions at Shenandoah NP

	Mass type	Extinction budget non-Rayleigh	Extinction budget with Rayleigh	Non-Rayleigh extinction
Low	Sulfate	0.55	0.519	0.123
	Nitrate	0.02	0.019	0.004
	Soot	0.03	0.025	0.006
	Organics	0.40	0.376	0.089
	Coarse	0.00	0.007	0.001
	Case	1.00	0.946	0.223
Median	Sulfate	0.71	0.668	0.133
	Nitrate	0.05	0.046	0.009
	Soot	0.02	0.014	0.002
	Organics	0.22	0.205	0.040
	Coarse	0.00	0.006	0.001
		1.00	0.939	0.185
High	Sulfate	0.85	0.804	0.175
	Nitrate	0.02	0.023	0.005
	Soot	0.01	0.010	0.002
	Organics	0.11	0.100	0.021
	Coarse	0.01	0.005	0.001
		1.00	0.942	0.204
High sulfate case	Sulfate	0.94	0.889	0.196
	Nitrate	0.03	0.025	0.005
	Soot	0.01	0.005	0.001
	Organics	0.02	0.019	0.004
	Coarse	0.00	0.005	0.001
		1.00	0.943	0.207

**Table 5. Extinction Budget for the High Sulfate Fraction Subgroup at
Current Conditions and the Three Levels of Sulfate Reduction**

	Mass type	Extinction budget non-Rayleigh	Extinction budget with Rayleigh	Non-Rayleigh extinction km^{-1}
Current	Sulfate	0.85	0.807	0.175
	Nitrate	0.02	0.023	0.005
	Soot	0.01	0.007	0.001
	Organics	0.11	0.100	0.021
	Coarse	0.01	0.005	0.001
		1.00	0.942	0.203
20% Reduction	Sulfate	0.82	0.770	0.140
	Nitrate	0.03	0.027	0.005
	Soot	0.01	0.009	0.001
	Organics	0.13	0.120	0.021
	Coarse	0.01	0.006	0.001
		1.00	0.932	0.168
40% Reduction	Sulfate	0.78	0.715	0.105
	Nitrate	0.04	0.034	0.005
	Soot	0.01	0.011	0.001
	Organics	0.16	0.148	0.021
	Coarse	0.01	0.008	0.001
		1.00	0.916	0.133
60% Reduction	Sulfate	0.70	0.626	0.070
	Nitrate	0.05	0.045	0.005
	Soot	0.02	0.014	0.001
	Organics	0.22	0.195	0.021
	Coarse	0.01	0.011	0.001
		1.00	0.891	0.098

Table 6. Summary of the Total Extinction (Rayleigh Scattering Included), SVR, and Percent Change in Extinction and SVR Between Current Conditions and the Three Sulfate Reduction Scenarios for the Three Sulfate Fraction Subgroups, and the Extreme High Sulfate Case

	Sulfate reduction (%)	Extinction km^{-1}	% Change in extinction	Standard visual range (km)	% Change in visual range
Low	0	0.24	0	16	0
	20	0.21	−10	18	11
	40	0.19	−20	21	26
	60	0.16	−31	24	45
Median	0	0.20	0	19	0
	20	0.17	−13	22	15
	40	0.15	−26	27	37
	60	0.12	−40	33	68
High	0	0.21	0	18	0
	20	0.18	−16	21	19
	40	0.15	−32	26	48
	60	0.11	−48	35	95
High sulfate case	0	0.22	0	17	0
	20	0.18	−17	21	21
	40	0.14	−35	27	56
	60	0.10	−53	38	116

FIGURE 4.   Appalachian vista showing the average visibility conditions for the high sulfate fraction subgroup.

FIGURE 5.   Appalachian vista showing the average visibility conditions for a sulfate loading reduction of 20%. Current visibility is shown on the left, while the improvement is shown on the right.

FIGURE 6. Appalachian vista showing the average visibility conditions for a sulfate loading reduction of 40%. Current visibility is shown on the left, while the improvement is shown on the right.

FIGURE 7. Appalachian vista showing the average visibility conditions for a sulfate loading reduction of 60%. Current visibility is shown on the left, while the improvement is shown on the right.

REFERENCES

1. Malm, W. C., J. F. Sisler, and L. L. Mauch. "Assessing the Improvement in Visibility of Various Sulfate Reduction Scenarios," in *Proceedings of the 83rd Annual Meeting of the Air and Waste Management Association*, (June 24–29, 1990).
2. Williams, M. D., E. Treiman, and M. Wecksung. "Plume Blight Visibility Modeling with a Simulated Photographic Technique," *Air Pollut. Control Assoc.* 30: 131, 1980.
3. Cahill, T. A., R. A. Eldred, and P. J. Feeney. "Particulate Monitoring and Data Analysis for the National Park Service 1982–1985," NPS Contract No. USDICX-0001-3-0056, University of California, Davis (1986).
4. Tsay, S., and G. L. Stephens. "A Physical/Optical Model for Atmospheric Aerosols with Application to Visibility Problems," Final Report, Cooperative Institute for Research in the Atmosphere, Colorado State University, Fort Collins, CO (1990).
5. Hanel, G. "The Properties of Atmospheric Aerosol Particles as Functions of the Relative Humidity at Thermodynamic Equilibrium with the Surrounding Moist Air," *Adv. Geophys.* 19:73–188.
6. Shettle, E. P., and R. W. Fenn. "Models for the Aerosols of the Lower Atmosphere and the Effects of Humidity on Their Optical Properties," Report No. AFGL-TR-0214, Hanscom AFB, MA (1979).
7. Tang, I. N., W. T. Wong, and H. R. Munkelwitz. "The Relative Importance of Atmospheric Sulfates and Nitrates in Visibility Reduction," *Atmos. Environ.* 15(12):2463 (1981).
8. Malm, W. C. "Considerations in the Measurement of Visibility," *Air Pollut. Control Assoc.* 29(10): (1975).
9. Lenoble, J., Ed. *Radiative Transfer in Scattering and Absorbing Atmospheres: Standard Computational Procedures*, (Deepak Publishing: Hampton, VA (1985), p. 300.
10. Greenwald, T. J., and G. L. Stephens. "Application of an Adding Doubling Radiation Model to Visibility Problems," Final Report, Cooperative Institute for Research in the Atmosphere, Colorado State University, Fort Collins, CO (1987).
11. Tsay, S., J. M. Davis, G. L. Stephens, S. K. Cox, and T. B. McKee. "Backward Monte Carlo Computations of Radiation Propagating in Horizontally Inhomogeneous Media," Part I. Description of Codes, Final report, Cooperative Institute for Research in the Atmosphere, Colorado State University, Fort Collins, CO (1987).

CHAPTER 19

Evaluating Exposure to Environmental Tobacco Smoke

S. Katharine Hammond

TABLE OF CONTENTS

0-87371-606-0/93/$0.00 + $.50

I. INTRODUCTION

In the past decade there has been an increasing interest in the adverse health effects which may be associated with passive smoking. This has led to the necessity of evaluating exposure to environmental tobacco smoke (ETS). Any appraisal of the effectiveness of controls that may be used to reduce exposure, e.g., increased ventilation or segregation of smokers from nonsmokers, requires some means of measuring the concentrations of ETS. Possible regulations other than outright bans on smoking might mandate measurements of ETS levels. Finally, evaluations of ETS exposure are critical to epidemiological studies which examine whether ETS does indeed present a health risk to the nonsmoking public.

II. TOOLS FOR ASSESSING EXPOSURE TO ENVIRONMENTAL TOBACCO SMOKE

Three general types of tools have been used to assess exposure to ETS: questionnaires, measurement of air contaminants, and measurement of biological markers. The measurement of air contaminants is more versatile, because it describes both the environment and the individual's exposure; and hence it is useful both in epidemiological studies and in evaluating controls intended to reduce exposure. Questionnaires have been used extensively in epidemiological studies, while the application of biological markers is usually restricted to more intensive studies. This chapter will focus on the measurement of air contaminants, but the usefulness of questionnaires and biological markers of exposure will be discussed briefly.

A. Questionnaires

Several methods of assessing ETS exposure have been used in various epidemiological studies. One common method is to classify spouses of smokers as exposed to ETS and spouses of nonsmokers as unexposed. Clearly this leads to some misclassification of exposure, because those married to nonsmokers may be exposed at work or in public places, and that level of exposure may exceed the exposure in homes with smokers.[1] A second method utilizes questionnaires, which commonly ask people how many hours they are exposed to ETS. Because people have varying tolerances of and sensitivities to ETS and each actual exposure may be to very different levels of ETS, these estimates may be highly inaccurate or biased. More recently, a questionnaire which also asks the number of smokers and their proximity in each location was developed and validated with 53 nonsmoking volunteers who also wore a passive nicotine monitor for a week and completed a 7-day diary.[2] An exposure index incorporating all these variables predicted the actual exposures the following week much better than hours of exposure only.

B. Biological Markers of ETS Exposure

Biological markers of ETS exposure integrate all exposures. This is an advantage for an epidemiological study, which should address the total exposure, but a disadvantage in evaluating the sources of the exposure, e.g., determining the contribution of workplace exposure to the total. A further disadvantage is the difficulty in collecting biological samples, which may be as easy as expired air or as difficult as blood or tissue samples.

Cotinine, a metabolite of nicotine, can be measured in the blood, serum, saliva, and urine of nonsmokers exposed to environmental tobacco smoke. These levels are generally two to three orders of magnitude below the levels found in smokers. Because the half-life of cotinine is less than 24 hr for most people,[3] these levels typically reflect only very recent exposures. For young children, whose primary exposure is in the home and is probably stable, urinary cotinine measurements collected on a midweek morning have been found to reflect home exposures (as measured by air concentrations of nicotine or by number of cigarettes smoked).[4] However, the exposure of adults is probably more variable due to the multitude of possible exposures in diverse settings.

The emissions of carbon monoxide in cigarette smoke lead to high levels of carboxyhemoglobin (COHb) in smokers. However, COHb also has a short half-life, 3–4 hr; thus levels are elevated only a short time after exposure. Other common sources of carbon monoxide such as automobile exhaust and gas stoves may contribute as much or more as ETS to nonsmokers' COHb. Some of the compounds in environmental tobacco smoke form adducts with DNA or hemoglobin. Hemoglobin adducts to 4-aminobiphenyl and tobacco specific nitrosamines have been measured in smokers and nonsmokers.[5-8] These adducts may be appropriate as possible markers of ETS exposure. Thus, for example, an association was found between the levels of 4-aminobiphenyl hemoglobin adducts and the exposures to environmental tobacco smoke over a 1-week period.[8] Collection of the blood or tissue samples is complicated, and the analysis is expensive and time-consuming. The repair mechanisms of DNA adducts further complicate the interpretation of their measurements, but hemoglobin adducts are not subject to repair mechanisms and therefore are more stable.

C. The Use of Airborne Markers

Environmental tobacco smoke (ETS) is a combination of sidestream emissions from cigarettes and exhaled mainstream smoke (that drawn through the cigarette directly into the lungs). ETS is a complex mixture of thousands of compounds which are distributed in the gaseous, vapor, and particulate phases.[9-11] Dozens of carcinogens have been identified in ETS, but the causative agents for lung cancer and other cancers associated with cigarette smoking have not been identified. These facts present a challenge to investigators who are attempting to measure exposure to ETS: what compound(s) should be measured in environments contaminated with ETS?

One approach is to choose a marker to serve as a surrogate for environmental tobacco smoke. The marker may be a specific compound, such as nicotine, or a more integrative measure, such as respirable suspended particulate mass. The use of markers to measure exposures to complex mixtures has been discussed in detail elsewhere.[12]

Markers are particularly valuable when the active agent, or the agent of interest, is not known, but the complex mixture is suspected of causing adverse health effects. For instance, markers of exposure have been used successfully in epidemiological studies of the relationship between cigarette smoking and lung cancer. Although thousands of compounds have been identified in cigarette smoke, those responsible for lung cancer have yet to be positively identified. The marker used for the complex mixture of mainstream tobacco smoke has been the number of cigarettes smoked. Epidemiological studies have found very good correlations between this marker and the risk of lung cancer among smokers.[13]

Even if an agent has been identified as either being the cause or the suspected cause of disease, a marker may be useful. The concentration of the agent may be too low to measure in real environments, or the analytical costs may be too high for large numbers of samples.

There are several attributes of a good marker. Clearly the concentration of the marker must increase proportionately as the source strength increases, e.g., as the number of cigarettes smoked increases in an environment which is otherwise held constant. Ideally a marker would be specific, or at least selective, for ETS. That is, ETS would be the unique environmental source of the marker. If the marker is a major component of the emissions and sensitive analytical methods exist, the limit of detection may be quite low, an important condition for environmental samples. The sampling and analysis of the marker should be sufficiently simple and inexpensive that large numbers of samples can be collected and analyzed to characterize exposures in diverse settings, to study possible adverse health effects epidemiologically, and to conduct other studies. The samples should be easy to collect in the field, and the methods lend themselves to personal sampling. Where possible, the ratio of emissions of the marker compound to other compounds of interest should be established, so that exposure to these other compounds can be calculated from measurements of exposure to the marker compound. Finally, different brands of cigarettes should emit similar amounts of the marker.

D. Markers for Environmental Tobacco Smoke

Several markers have been proposed for ETS. Since the smoke is so visible, one obvious choice has been the concentration of particles, which are submicron in size. Furthermore, many of the carcinogens in ETS are in the particulate phase. Many techniques have been developed to measure the concentration of particles in the air: air may be drawn through a preweighed filter which is weighed after sampling, or various direct reading instruments using piezoelectric or light-scattering principles can measure the instantaneous concentration. These methods may be very easy or relatively inexpensive. Measurements made in a wide variety of locations clearly show increases of particle concentration related to the amount of smoking occurring. Thus, the average concentration of particles in homes without smokers is typically about 20 μg/m^3, while homes with smokers have levels two to three times higher.[14-16] Similarly, public places such as restaurants typically have much higher particle concentrations when cigarette smoking is occurring. One survey found a mean of 43 μg/m^3 in restaurants with no smoking, in contrast to a mean of 161 μg/m^3 in those with smoking.[17]

The principle disadvantage to the use of particles as a marker for ETS is that there are many other environmental sources of particles. ETS particles are submicron, so the use of size selective sampling eliminates larger, and hence heavier, particles (the cut points most often used are 10, 3.5, and 2.5 μm); smaller particles, especially from other combustion processes, will continue to contribute mass to the sample. When the concentration of ETS is high, for example, greater than 100 μg/m^3 particles, this interference is generally of minor importance. However, at lower concentrations careful measurement of background levels or more specific measures of ETS concentration are needed.

Many compounds have been proposed or used as markers for ETS. Carbon monoxide is easy to measure with sufficient sensitivity, but it is also nonspecific to ETS. Polycyclic aromatic hydrocarbons have been used, but they are also nonspecific. Furthermore, they are relatively low in concentration and thus more difficult to measure.

Several nitrosamines have been identified and measured in ETS. Some of these, e.g., 4-(methylnitrosoamino)-1-(3 pyridyl)-1-butanone (NNK) and N'-nitrosonornicotine (NNN), are unique to tobacco and have been shown to be carcinogenic in rats, mice, and hamsters.[6] These characteristics would make the tobacco specific nitrosamines good candidates for marker compounds, but their concentrations are too low for detection in most environmental settings. Furthermore, the analysis costs are high because a gas chromatograph equipped with a thermal energy analyzer is required.[6,18]

3-Ethenylpyridine has been suggested as a marker because ETS is reportedly the only source and relatively large amounts are emitted in ETS.[19] Presumably 3-ethenylpyridine decays more slowly than nicotine.[19-20] Much less 3-ethenylpyridine than nicotine is emitted in sidestream smoke, and the methods for the analysis are not as sensitive. This means that the limit of detection is over an order of magnitude greater than for nicotine. However, a more important problem with the use of 3-ethenylpyridine is that its production does not appear to be linear with the number of cigarettes burned, according to one study in which both respirable particles (RSP) and nicotine concentrations were proportional to the source strength (Figure 1). Clearly the first prerequisite for a marker is that its concentration should reflect the concentration of the complex mixture, here ETS. 3-Ethenylpyridine fails this most fundamental test for a marker.

E. Nicotine as a Marker for ETS

Nicotine has been used in more studies than any other compound. ETS is the only source of nicotine in most environments (nicotine is used as a pesticide). Nicotine is one of the major constituents of cigarette emissions. Recently, methods

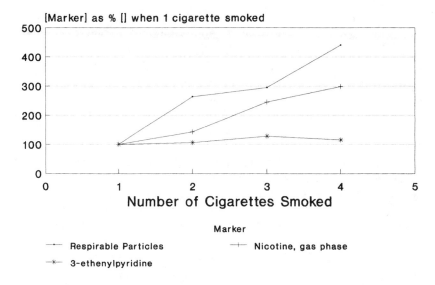

FIGURE 1. Concentrations of three proposed markers for environmental tobacco smoke as a function of the number of cigarettes smoked in a
Teflon® chamber. Marker concentration is plotted as a percentage of
the concentration when one cigarette was smoked. (Calculated and
plotted from data in Eatough, D. J. et al. *Environ. Sci. Technol.*
23:679–687 [1989].)

have been developed to measure nicotine simply and accurately.[21-25] These methods are quite sensitive, so that low limits of detection (<0.1 μg/m³) can be
achieved with personal sampling. Furthermore, the methods are simple to use in
the field and relatively inexpensive, so that they are appropriate for large studies,
e.g., epidemiological or exposure evaluations.

Nicotine concentrations do vary with source strength, i.e., number of cigarettes
smoked in controlled environments (Figure 1). Field studies in 200 homes have
shown good correlation between the number of cigarettes smoked and the weekly
average nicotine concentrations.[15,16,26,27] Furthermore, excellent correlation has
been found between personal samples of nicotine collected over 1 week and a
concurrent daily diary of ETS exposures.[2]

Emissions of nicotine and various compounds by cigarettes have been reported, and further work is currently underway. The emissions of nicotine in
sidestream smoke and environmental tobacco smoke show little variability with
the brand of cigarettes, despite the variation in the yield of nicotine in mainstream
smoke, which is achieved primarily through ventilation.[15,28]

One suggested disadvantage is that nicotine may have a different decay rate
than other components of tobacco smoke.[29] The authors state that field data
exhibit a wide variability in the ratio of particles to nicotine. However, the authors
fail to recognize that at low concentrations of environmental tobacco smoke the
particles from sources other than ETS will dominate the particle concentration,

and thus the ratio would not be expected to be consistent. A more appropriate analysis would examine the correlation between nicotine and RSP. In a plot of the concentration of particles as a function of nicotine concentration, the intercept is an estimate of the background particle level, often found to be about 20–30 μg/m^3 in homes, and the slope is the ratio of ETS associated RSP to nicotine. For example, in a study of 96 homes in New York State, such a plot yielded a slope of 10.8 and an intercept of 18, r^2 = 0.71.[15] Although this model did not include the contribution of other known combustion sources of particles which were present, including kerosene heaters and wood burning stoves or fireplaces, the correlation between nicotine and particle concentrations was excellent, the intercept was similar to background levels found in other studies, and the slope was in agreement with other studies.

Considering all the advantages and the above analyses of the purported disadvantages, most investigators consider nicotine to be a good marker for ETS. As such, nicotine has been used is several studies both to serve as a marker of personal exposure and to trace exposures in various locations, e.g., homes, workplaces, and public spaces.

F. Measuring Exposure to Nicotine

Although predominantly in the particle phase in mainstream smoke, nicotine is predominantly in the gas phase in sidestream smoke. Over 95% of the nicotine in environmental tobacco smoke is generally found in the gas phase.[21,25] Nicotine in the particle phase has been used as a marker for ETS,[29–30] but such methods are intrinsically less sensitive since there is much less nicotine emitted in that phase. Furthermore, there is more variability in replicate measurements, again due to the low concentrations.[25] The particles are collected by active sampling on either Teflon®-coated glass fiber filters or Teflon® filters. The nicotine is desorbed from the filters in either dichloromethane[30] or water[31] and is analyzed by gas chromatography or ion chromatography.

Both active and passive sampling methods for gas-phase nicotine have been reported in the literature.[21-25,32-35] In the most commonly used active sampling methods, air is drawn at 1–4 Lpm through a filter which has been impregnated with a weak acid, such as sodium bisulfate, to trap nicotine on the filter.[21,25] Another general approach is to use a solid sorbent, such as XAD resin or Tenax, to collect the nicotine, usually with air flows of 0.2–1 liters per minute (Lpm).[25,32-34] Each of these methods, with a treated filter or a solid sorbent, can be used either for area sampling or for personal sampling with battery-operated pumps. A third technique is to collect the gas-phase nicotine on annular denuders coated with a weak acid, such as citric acid or benzenesulfonic acid.[23,24] The annular denuders may operate at 20 Lpm[23] or, for personal sampling, at 2 Lpm.[24] After appropriate desorption from the sampling media, the nicotine is analyzed by gas chromatography, frequently with nitrogen-selective detection to enhance the sensitivity.

Three types of passive samplers for nicotine have been described.[22,25,35] In the first, a filter impregnated with sodium bisulfate is held in a modified 37-mm diameter polystyrene cassette with a windscreen so that diffusion alone carries nicotine to the filter. This sampler is inexpensive, lightweight (16 g), and has a sampling rate of 0.024 Lpm.[22] The other two methods used a commercial stainless steel sampler (Scientific Instrumentation Specialists, Moscow, ID) containing either XAD-IV sorbent[35] or a glass fiber filter treated with benzenesulfonic acid.[23] The analyses were the same as for the active samplers with the same collection media.

In our laboratory, the analytical limit of detection is 0.01-µg nicotine collected, but occasionally field blanks have this level of nicotine. Therefore, we use an operational limit of detection of 0.02-µg nicotine collected. The limit of detection in ambient environments depends on the sampling rate and duration. Thus, the active sampler operated at 4 Lpm can accurately measure exposures as low as 0.1 µg/m³ in an hour of sampling. Longer sampling times would yield lower limits of detection. The passive monitor is intended for longer sampling times, but would have a limit of detection of 2 µg/m³ for 8 hr of sampling, 0.6 µg/m³ for 24 hr of sampling, and 0.1 µg/m³ for a full week of sampling. These are time-weighted average exposures; clearly the profile of exposure to cigarette smoke varies greatly with time.

Recently, an intercomparison study of the sampling techniques used to measure nicotine was reported.[25] Six experiments were conducted in a controlled chamber with cigarette smokers, at very high (particle concentration ca. 950 µg/m³) or moderately high (ca. 190 µg/m³ particles) environmental tobacco smoke levels and for 1- to 6-hr duration. The active methods which used acid-treated filters (sodium bisulfate, citric acid, or benzenesulfonic acid), both annular denuders (the area sampler at 20 Lpm and the personal sampler at 2 Lpm), and two of the passive samplers gave similar results. The active system using XAD resin to adsorb the nicotine gave low results. The stainless steel passive sampler with the treated filter also gave low results; it differed from the other stainless steel passive sampler (with XAD resin) in that it had not been passified by the manufacturer. Subsequent experiments on the passified stainless steel sampler indicated that it, too, had problems.[35] Therefore, the only passive sampler currently recommended is the one contained in a modified polystyrene cassette, which also is the least expensive and lightest to wear.[22]

III. RANGE OF LEVELS OF NICOTINE SEEN IN REAL ENVIRONMENTS

The choice of an appropriate marker enables researchers to quantify exposures to environmental tobacco smoke in various settings. Only respirable particles and nicotine have been used extensively as markers of ETS in a wide variety of environments, because these two markers have been most accepted among researchers as both appropriate and feasible. For coherence only one marker, gas-phase nicotine, will be discussed in the final sections of this report. Similar assessments may be made of other markers.

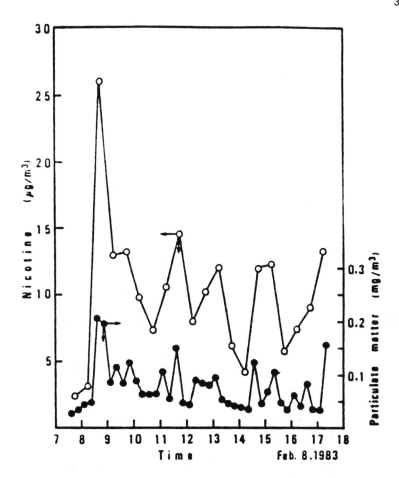

FIGURE 2. Typical profile of changes in concentrations of ambient nicotine, open circles, and particulate matter, closed circles. The time-weighted average through the full day is approximately 10 μg/m³ nicotine. (From Muramatsu, M. et al. *Environ. Res.* 35:218–227 [1984]. With permission.)

Because cigarettes are not continuous sources, but rather point sources emitting intermittently (as cigarettes are smoked or not), the concentration of ETS and hence the markers of ETS will vary greatly over time. For instance, sequential 30-min samples taken in one office over 10 hr indicated that nicotine concentrations rose from 2–3 to 26 μg/m³ early in the morning; then they varied between 5 and 15 μg/m³ through most of the rest of the day (Figure 2).[33] Longer sampling times will produce time-weighted average exposures, which will tend to be lower than short-term samples collected in "smoky" environments. Most of the results given below are from longer term sampling, from full work shifts of 1 day to 1 week. When samples are collected over 24 hr to 1 week, clearly

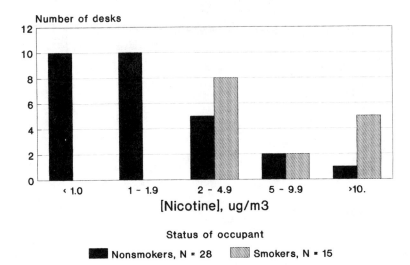

FIGURE 3.   Distribution of office exposures to environmental tobacco smoke.
Data collected in two office buildings with area passive nicotine
monitors placed at workers' desks from Monday morning to Friday
evening. Averaging time was 45 hr. (From data in Vaughan, W. M.,
and S. K. Hammond. *J. Air Waste Manage. Assoc.* 40:1012–1017
[1990]; Hammond, S. K. Unpublished results [1987].)

much of the sampling time occurs when there is no exposure at all, e.g., during
sleep. Therefore, comparisons of concentrations should always account for the
sampling times.

Most occupational exposures to environmental tobacco smoke have been made
in offices. Ten personal samples collected in two Japanese offices had a range of
ca. 10–30 $\mu g/m^3$, with an average nicotine concentration of about 20 $\mu g/m^3$.[33]
Most measurements taken in U.S. offices tend to have lower concentrations,
although many offices have policies restricting smoking. Among office workers
for one railroad company, 44 full shift samples had an average of 8 $\mu g/m^3$ and a
median of 7 $\mu g/m^3$ of nicotine.[36] Two office buildings were studied before a
policy restricting smoking was implemented.[37,38] In each building approximately
one-third of the samples were collected at smokers' desks and two-thirds at
nonsmokers' desks. Passive samplers collected nicotine through a 5-day work
week, and an 8-hr active sample was collected on Wednesday. The daily samples
were generally consistent with the weekly average nicotine concentrations. Although the two office buildings were in different parts of the country and of
different design, the distribution of exposures was similar in both; the combined
distribution is presented in Figure 3. Not surprisingly, there tended to be more
nicotine at smokers' desks than at those of nonsmokers. The nicotine concentrations at the nonsmokers desks ranged from 0.2 to 21 $\mu g/m^3$, while those at the
smokers' desks ranged from 3 to 33 $\mu g/m^3$. Nearly all the exposures fell to less
than 1 $\mu g/m^3$ 2 weeks after smoking was restricted, and most were under 0.5 $\mu g/m^3$.

Table 1. Occupational Exposure to Environmental Tobacco Smoke Flight
Attendants and Railroad Workers

| Job | N | [Nicotine], $\mu g/m^3$ | | |
		Mean	($\pm$ SE)	Median
Flight attendants[a]	16	4.7	($\pm$1.0)	4
Railroad workers[b]				
Clerks	44	8.2	($\pm$1.2)	7
Carmen	34	11.6	($\pm$2.4)	7
Engineers	14	13.0	($\pm$4.4)	5
Brakers	30	5.6	($\pm$1.8)	1
Shop workers	158	3.4	($\pm$1.8)	0.2

[a] Source: Mattson et al.[40]
[b] Source: Schenker et al.[36]

Other studies which have selected just a few office locations to sample have found results consistent with these, with one office with a heavy smoker having a very high nicotine concentration of 48 $\mu g/m^3$.[39]

In a study of railroad workers 280 personal samples were taken of occupational exposure to nicotine.[36] The full shift average nicotine concentrations varied with the job, and are presented in Table 1. The concentrations found for nonsmoking nonoffice workers ranged from less than 0.1–25 $\mu g/m^3$. Nonsmoking flight attendants on transcontinental flights (before restrictions on smoking) also had a wide range of exposures, from 0.1 to 10 $\mu g/m^3$, with mean exposures of 5 $\mu g/m^3$.[40]

Several studies have been conducted to examine the weekly average nicotine exposures in homes. Nicotine concentrations in homes without smokers in residence tend to average less than 0.2 $\mu g/m^3$.[15,26,27] Homes with smokers tend to have higher weekly nicotine concentrations. A study of 96 homes in New York State found 47 homes with nicotine >0.1 $\mu g/m^3$, and these had a weekly average of 2.2 $\mu g/m^3$ nicotine;[15] 13 North Carolina homes with smokers averaged 3.4 $\mu g/m^3$ nicotine over approximately 14 hr from 5 P.M. to 7 A.M.;[16] and 25 Minnesota homes with smokers averaged 5.6 $\mu g/m^3$ nicotine over a week.[27]

In addition to occupational and home exposures, people are exposed to environmental tobacco smoke in many other locations. Before smoking was banned on domestic flights in the United States, two studies measured levels of nicotine on several flights.[34,40] The distribution of nicotine exposures in the smoking and the nonsmoking sections is given in Figure 4. While most samples were between 1 and 20 $\mu g/m^3$, a few samples collected in the nonsmoking sections were over 30 $\mu g/m^3$, and one was over 70 $\mu g/m^3$. A Japanese study found an average of 15 (range 6–29) $\mu g/m^3$ of nicotine on seven airplanes, consistent with the U.S. aircraft exposures.[33]

Exposures in restaurants would be expected to be highly variable. A study of 33 restaurants reported a wide range, <0.1–35 $\mu g/m^3$, with a mean of 10.5 $\mu g/m^3$.[41] Other studies in the United States have reported between 2 and 70 $\mu g/m^3$.[2,29,39] Seven Japanese tea-rooms had average exposures of 33 $\mu g/m^3$, while the exposures in five restaurants averaged 15 $\mu g/m^3$.[33] Six French cafes had

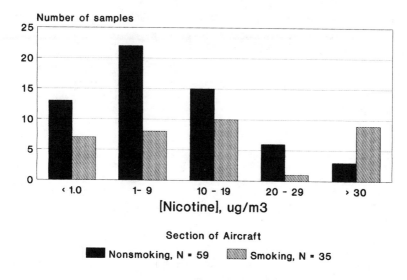

FIGURE 4. Distribution of exposures to environmental tobacco smoke on commercial aircraft flights in the United States. Data collected on multiple flights from active nicotine samplers either worn by passengers or placed in seats in passengers' breathing zone. Averaging time was length of flight, from 45 min to 5 hr. (From data in Oldaker, G. B., III, and F. W. Conrad, Jr. *Environ. Sci. Technol.* 21:994–999 [1987]; Mattson, M. E. et al. *J. Am. Med. Assoc.* 261:867–872 [1989].)

25–52 $\mu g/m^3$ of nicotine.[42] As would be expected, bars, taverns, and discos tend to have higher ETS levels, with nicotine concentrations between 6 and 115 $\mu g/m^3$ reported.[2,29,39,43] Other places in which the public is exposed to ETS include buses and trains (8–50 $\mu g/m^3$ nicotine);[33,42] hospitals and clinics (1–17 $\mu g/m^3$);[33,39,42] bowling alleys (13–20 $\mu g/m^3$);[2] and bingo parlors and casinos (3–85 $\mu g/m^3$).[44,45]

All these results point to the wide range of ETS exposures that exist in the home, in the workplace, and in public places. Table 2 summarizes some of the findings of the studies just cited. The next challenge becomes placing these values in context.

IV. INTERPRETING EXPOSURES TO NICOTINE

Researchers have often tried to interpret measurements of nicotine exposure by converting these to "cigarette equivalents."[33,34] For instance, they may say that being in a particularly smoky environment is the equivalent of smoking x number of cigarettes. This is calculated by comparing the ratios of nicotine in mainstream and in sidestream cigarette smoke. Such a calculation yields the number of cigarettes one would have to smoke to receive the same amount of nicotine as breathing the air in the given environment for a particular period of time.

Table 2. Range of [Nicotine] in Various Indoor Environments

Environment	Averaging time	[Nicotine] $\mu g/m^3$	Ref.
Homes	1 week	<0.1–30	15, 26, 27
Offices	hours	0.2–33	33, 36, 37
Hospitals and clinics	hours	2–5	33, 37, 39, 42
Trains and buses	hours	8–50	33, 42
Restaurants	hours	<0.1–70	2, 29, 33, 37, 39, 41, 42
Bars, pubs	hours	6–82	2, 43
Airlines	hours	<0.1–100	34, 40

However, the ratio of sidestream to mainstream emissions is quite specific to individual compounds, and is highly variable among these compounds. Nicotine is usually not the suspected agent or compound of ultimate interest. If "cigarette equivalents" were calculated on the basis of many of the carcinogens in environmental tobacco smoke, the cigarette equivalents would be much higher than if calculated on the basis of nicotine. This seeming anomaly is due to the differences between mainstream and sidestream smoke.

Environmental tobacco smoke is composed of a mixture of sidestream smoke, which is emitted from the end of smoldering cigarettes between puffs; and the mainstream smoke, which is exhaled by a smoker after taking a puff. Approximately 80% of ETS is from sidestream and 20% from exhaled mainstream smoke, so that ETS resembles sidestream smoke more than mainstream smoke or that which a smoker inhales directly through the cigarette. The physicochemical conditions of combustion are different as each of these are formed (Table 3). Sidestream smoke is formed at cooler temperatures and in a reducing environment, while mainstream smoke is slightly more acidic than sidestream smoke.[9] As a result, mainstream smoke is enriched with some compounds (such as HCN) whereas sidestream smoke is enriched with others (such as ammonia). Table 4 presents the relative emissions of various compounds in mainstream and in sidestream smoke. Clearly the amount of a compound inhaled in passive smoking compared to the amount inhaled by a smoker while smoking a cigarette will depend not only on the concentration of an ETS marker, but also on the relative amounts of that compound in sidestream and in mainstream smoke.

The number of "cigarette equivalents" that a passive smoker has inhaled may be calculated according to Equation 1:

$$\text{cigarette equivalents} = \frac{\text{amount from passive exposure}}{\text{amount from smoking 1 cigarette}} \qquad (1)$$

The goal, therefore, is to determine the amount of a compound inhaled during an exposure to ETS. This may be estimated with the knowledge of the sidestream and mainstream emissions of the compound of interest, the sidestream and mainstream emissions of a marker compound, and the concentration of the marker compound. Thus, if nicotine is the marker compound, the mass of a compound of interest may

Table 3. Physicochemical Characteristics of Mainstream and Sidestream Cigarette Smoke

	Mainstream smoke	Sidestream smoke
Peak temperature of combustion	900°C	600°C
pH of smoke	5.8–6.2	6.4–8.5
Percent oxygen	12–16	1.5–2

Sources: National Research Council[9] and International Agency for Research on Cancer.[11]

Table 4. Emissions of Selected Compounds in Mainstream and Sidestream Cigarette Smoke

	Emissions, ng/cigarette	
Compound	Mainstream smoke	Sidestream smoke
Nicotine[a]	1500.0 µg	4140.0 µg
HCN[b]	450.0 µg	50.0 µg
Benzene[a]	48.0 µg	453.0 µg
Benzo(a)pyrene[a]	17.8 ng	45.7 ng
N-nitrosodimethylamine[a] (NDMA)	4.3 ng	597.0 ng
N'-nitrosonornicotine[a] (NNN)	448.0 ng	307.0 ng
4-(Methylnitrosamino)-1-(3-pyridyl)-1-butanone[a] (NNK)	180.0 ng	752.0 ng
4-aminobiphenyl[b]	2.4–4.6 ng	143.0 ng

[a] For a filter cigarette, with medium tar delivery. Source: Adams, et al.[18]
[b] Source: IARC.[11]

be estimated by Equation 2:

$$Y = \frac{[\text{nicotine}] \times V \times Y(SS)}{\text{nicotine}(SS)} \qquad (2)$$

where

Y = mass of compound Y inhaled by passive smoker
V = volume of air breathed by passive smoker
= minute volume × minutes exposed
$Y(SS)$ = mass of compound Y in SS emissions of one cigarette
nicotine(SS) = mass of nicotine in SS emissions of one cigarette

Equations 1 and 2 can be combined to calculate the compound-specific number of cigarette equivalents smoked by a passive smoker in a given exposure:

$$\text{cigarette equivalent compound } Y = \frac{Y}{Y(MS)}$$

$$= \frac{[\text{nicotine}] \times V \times Y(SS)}{\text{nicotine}(SS) \times Y(MS)}$$

where Y(MS) = mass of Y in mainstream emissions of one cigarette.

As an example, assume a person is exposed to ETS for 8 hr and the average nicotine concentration is 10 $\mu g/m^3$. The volume of air breathed is:

$$V = 0.075 \text{ m}^3/\text{min} \times 8 \text{ hr} \times 60 \text{ min/hr} = 3.6 \text{ m}^3$$

The cigarette equivalent is calculated using data from Table 4 as follows:

$$\text{cigarette equivalent} = \frac{[\text{nicotine}] \times V \times Y(SS)}{\text{nicotine(SS) x Y(MS)}}$$

$$\text{nicotine cigarette equivalent} = \frac{(10 \text{ } \mu g/m^3)(3.6 \text{ m}^3)(4140 \text{ } \mu g)}{(4140 \text{ } \mu g)(1500 \text{ } \mu g)}$$

$$= 0.024 \text{ cigarettes}$$

$$\frac{\text{4-aminobiphenyl}}{\text{cigarette equivalent}} = \frac{[\text{nicotine}] \times V \times \text{4-ABP(SS)}}{\text{nicotine(SS)} \times \text{4-ABP(MS)}}$$

$$= \frac{(10 \text{ } \mu g/m^3)(3.6 \text{ m}^3)(143 \text{ ng/cigarette in SS})}{(4140 \text{ } \mu g) \text{ } (2.4 \text{ ng/cigarette in MS})}$$

$$= 0.52 \text{ cigarettes}$$

Thus, one might describe the same exposure as being the equivalent of smoking only 2% of one cigarette or of smoking half a cigarette, depending on the compound chosen for comparison. That is to say, a person working in the environment described here will breathe as much nicotine as someone smoking 0.02 of a cigarette, but as much 4-aminobiphenyl as someone smoking half a cigarette. Table 5 presents some calculations of "cigarette equivalents" for several compounds of possible interest. Three hypothetical exposure situations are assumed: one at levels found at nonsmokers desks in the workplace, one similar to levels found in restaurants with smoking, and one at the very high levels found in some bars and taverns. Each exposure is assumed to last for 2 hr, and a resting ventilation rate of 7.5 Lpm is assumed. Longer exposures will lead to proportionately higher "cigarette equivalents," as will higher breathing rates due to work or dancing. However, for any given set of assumptions, the calculated "cigarette equivalent" exposure ranges over three orders of magnitude, and the value depends on the compound selected.

Clearly the number of "equivalent cigarettes smoked" is an ambiguous method of describing exposure. Because nicotine is commonly used as a marker for ETS, it is often chosen for calculations of "equivalent cigarettes." However, unless nicotine is definitely the compound of interest, such a choice is inappropriate. Generally, nicotine is not the agent of interest. As can be seen in Table 5, a description of exposure in nicotine "cigarette equivalents" yields a substantial

Table 5. Compound Specific Cigarette Equivalents for 2-hr Exposures to Environmental Tobacco Smoke

	Environmental tobacco smoke concentration		
Contaminant	Intermediate[a]	High[b]	Very high[c]
[Nicotine]: $\mu g/m^3$	3	11	55
Nicotine	0.001	0.007	0.033
Benzene[d]	0.006	0.023	0.113
Hydrogen cyanide	<0.001	<0.001	0.001
4-Aminobiphenyl[d]	0.020	0.074	0.371
4-Nitrosodimethylamine[e]	0.091	0.332	1.660

[a] For example, levels found at nonsmokers desks in large offices with smokers, or in nonsmoking sections of cafeterias with smoking sections.
[b] For example, high levels found in restaurants.
[c] For example, levels found in smoky taverns or bars.
[d] Human carcinogen.
[e] Animal carcinogen.

underestimate of the exposure to some of the known carcinogens; this underestimate may be orders of magnitude. If the compound of interest is known, the "cigarette equivalents" for that compound can be calculated from the concentrations of nicotine measured and the relative emissions of nicotine and that compound. If the compound of interest is not known, which is most commonly the case, then any such calculations should be used with caution.

V. SUMMARY

Environmental tobacco smoke is a complex mixture of thousands of compounds, many of which are carcinogens. Marker compounds, such as nicotine, are quite useful in evaluating exposure to ETS; and sensitive and simple methods exist for measuring area concentrations of, and personal exposures to, nicotine. ETS exposure is extremely variable in time and space: in one location the exposure may change over an order of magnitude in an hour, and several orders of magnitude of differences have been found in different locations. Therefore, sampling times should be carefully chosen to match the goals of a particular study. If the aim is to characterize a particular environment, such as a restaurant, a workplace, or an airline flight, then the sample should be taken over a few hours. Longer sampling times, several days to a full week, are recommended to characterize an individual's integrated exposure. Small differences in characteristics of marker or agent compounds, such as emissions from different brands of cigarettes or deposition and decay rates of various compounds, are dwarfed in importance by the wide range of ETS exposures that have been observed.

Nicotine measurements enable quantitations of exposures and comparisons of exposures in different environments. However, the data must not be overinterpreted. Uncertainties in the causative agents or agents of interest must be acknowledged.

Where these agents are known, experiments should be conducted to establish the relationships between these agents and the marker compounds. Then, concentrations of the active agents can be calculated from measurements made of the marker compounds. In the interim, marker compounds such as nicotine are useful to distinguish the wide range of environmental tobacco smoke exposures.

REFERENCES

1. Hammond, S.K., J. Coghlin, and B.P. Leaderer. "Field Study of Passive Smoking Exposure with Passive Sampler," in *Indoor Air '87, Proceedings of the 4th International Conference on Indoor Air Quality and Climate,* Vol. 2; Environmental Tobacco Smoke, Multicomponent Studies, Radon, Sick Buildings, Odours and Irritants, Hyperreactivities and Allergies, Berlin, West Germany (1987) 131–136.

2. Coghlin, J., S.K. Hammond, and P. Gann. "Development of Epidemiologic Tools for Measuring Environmental Tobacco Smoke Exposure," *Am. J. Epidemiol.* 130:696–704 (1989).

3. Benowitz, N.L., F. Kuyt, P. Jacob, R.T. Jones, and A. Osman. "Cotinine Disposition and Effects," *Clin. Pharmacol. Ther.* 34:604–611 (1983).

4. Henderson, R.W., H.F. Reid, R. Morris, Ou-Li Wang, P.C. Hu, R.W. Helms, L. Forehand, J. Mumford, J. Lewtas, N.J. Haley, and S.K. Hammond. "Home Air Nicotine Levels and Urinary Cotinine Excretion in Young Children," *Am. Rev. Respir. Dis.* 140:197–201 (1989).

5. Bryant, M.S., P.L. Skipper, S.R. Tannenbaum, and M. Maclure "Hemoglobin Adducts of 4-Aminobiphenyl in Smokers and Nonsmokers," *Cancer Res.* 47:602–608 (1987).

6. Hecht, S.S., and D. Hoffmann. "Tobacco-Specific Nitrosamines, an Important Group of Carcinogens in Tobacco and Tobacco Smoke," *Carcinogenesis* 9:875–884 (1988).

7. Carmella, S.G., S.S. Kagan, M. Kagan, P.G. Foiles, G. Palladino, A.M. Quart, E. Quart, and S.S. Hecht. "Mass Spectrometric Analysis of Tobacco-Specific Nitrosamine Hemoglobin Adducts in Snuff Dippers, Smokers, and Nonsmokers," *Cancer Res.* 50:5438–5445 (1990).

8. Hammond, S.K., P.H. Gann, J. Coghlin, S.R. Tannenbaum, and P.L. Skipper. "Tobacco Smoke Exposure and Carcinogen-Hemoglobin Adducts," in *Indoor Air '90, Proceedings of the 5th International Conference on Indoor Air Quality and Climate,* Vol. 2; Characteristics of Indoor Air, (Ottawa, Ontario, Canada: Canada Mortgage and Housing Corporation, 1990), 157–162.

9. National Research Council. *Environmental Tobacco Smoke: Measuring Exposures and Assessing Health Effects* (Washington, DC: National Academy Press, 1986).

10. *The Health Consequences of Involuntary Smoking: A Report of the Surgeon General* (Washington, DC: U.S. Department of Health and Human Services, 1986).

11. International Agency for Research on Cancer. *IARC Monographs on the Evaluation of the Carcinogenic Risk of Chemicals to Humans,* Vol. 38 (Lyon, France: IARC, 1987).

12. Hammond, S. K. "The Uses of Markers to Measure Exposures to Complex Mixtures," in *Exposure Assessment for Epidemiology and Hazard Control,* S. Rappaport and T.J. Smith, Eds. (Chelsea, MI: Lewis Publishers, Inc., In press).

13. Doll, R., and A.B. Hill. "Mortality in Relation to Smoking: Ten Years' Observations of British Doctors," *Br. Med. J.* 1:1399 (1964).

14. Spengler, J.D., R.D. Treitman, T.D. Tosteson, D.T. Mage, and M.L. Soczek. "Personal Exposures to Respirable Particulates and Implications for Air Pollution Epidemiology," *Environ. Sci. Technol.* 19:700–707 (1985).

15. Leaderer, B.P., and S.K. Hammond. "An Evaluation of Vapor Phase Nicotine and Respirable Suspended Particle Mass as Markers for Environmental Tobacco Smoke," *Environ. Sci. Technol.* In press.

16. Mumford, J.L., J. Lewtas, R.M. Burton, F.W. Henderson, L. Forehand, J.C. Allison, and S.K. Hammond. "Assessing Environmental Tobacco Smoke Exposure of Preschool Children in Homes by Monitoring Air Particles, Mutagenicity, and Nicotine," in *Measurement of Toxic and Related Air Pollutants,* Environmental Protection Agency/Air and Waste Management Association International Symposium (Pittsburgh, PA: Air and Waste Management Association, 1989), 606–610.

17. Repace, J.L., and A.H. Lowrey. "Indoor Air Pollution, Tobacco Smoke, and Public Health," *Science* 298:464–472 (1980).

18. Adams, J.D., K.J. O'Mara-Adams, and D. Hoffmann, "Toxic and Carcinogenic Agents in Undiluted Mainstream Smoke and Sidestream Smoke of Different Types of Cigarettes," *Carcinogenesis* 8:729–731 (1987).

19. Eatough, D.J., C.L. Benner, J. M. Bayona, F.M. Caka, G. Richards, J.D. Lamb, E.A. Lewis, and L.D. Hansen, "The Chemical Composition of Environmental Tobacco Smoke. I. Gas Phase Acids and Bases," *Environ. Sci. Technol.* 23:679–687 (1989).

20. Eatough, D.J., F.M. Caka, J. Crawford, S. Braithwaite, Ld. Hansen, and E.A. Lewis "Environmental Tobacco Smoke in Commercial Aircraft," in *Indoor Air '90, Proceedings of the 5th International Conference on Indoor Air Quality and Climate,* Vol. 2; Characteristics of Indoor Air, (Ottawa, Ontario, Canada: Canada Mortgage and Housing Corporation, 1990), 311–316.

21. Hammond, S.K., B.P. Leaderer, A.C. Roche, and M. Schenker. "Collection and Analysis of Nicotine as a Marker for Environmental Tobacco Smoke," *Atmos. Environ.* 21:457–461 (1987).

22. Hammond, S.K., and B.P. Leaderer. "A Diffusion Monitor to Measure Exposure to Passive Smoking," *Environ. Sci. Technol.* 21:494–497 (1987).

23. Eatough, D.J., C.L. Benner, J.M. Bayona, F.M. Caka, H. Tang, H. L. Lewis, J.D. Lamb, M.L. Lee, E.A. Lewis, and L.D. Hansen. in *Measurement of Toxic and Related Air Pollutants* (Pittsburgh, PA: Air Pollution Control Association, 1987) 132–139.

24. Koutrakis, P. A.M. Fasano, J.L. Slater, J.D. Spengler, J.F. McCarthy, and B.P. Leaderer. *Atmos. Environ.* 23:2767–2773 (1989).

25. Caka, F.M., D.J. Eatough, E.A. Lewis, H. Tang, K. Gunther, J. Crawford, P. Koutrakis, J.D. Spengler, J. McCarthy, M.W. Ogden, S.K. Hammond, B.P. Leaderer, and J. Lewtas. "An Intercomparison of Sampling Techniques for Nicotine in Indoor Environment," *Environ. Sci. Technol.* 24:1196–1203, (1990).

26. Hammond, S. K., J. Lewtas, J. Mumford, and F.W. Henderson. "Exposures to Environmental Tobacco Smoke in Homes," in *Measurement of Toxic and Related Air Pollutants,* Environmental Protection Agency/Air and Waste Management Association International Symposium (Pittsburgh, PA: Air and Waste Management Association, 1989), 590–595.

27. Marbury, M.C., S.K. Hammond, and N.J. Haley, "Assessing Exposure to Environmental Tobacco Smoke in Epidemiological Studies of Acute Health Effects," in *Indoor Air '90, Proceedings of the 5th International Conference on Indoor Air Quality and Climate,* Vol. 2, Characteristics of Indoor Air, (Ottawa, Ontario, Canada: Canada Mortgage and Housing Corporation, 1990), 189–194.

28. Rickert, W.S., J.C. Robinson, and N. Collinshaw. "Yields of Tar, Nicotine, and Carbon Monoxide in the Sidestream Smoke for 15 Brands of Canadian Cigarettes," *Am. J. Public Health* 74:228–231 (1984).

29. Eatough, D.J., C.L. Benner, H. Tang, V. Landon, G. Richards, F.M. Caka, J. Crawford, E.A. Lewis, L.D. Hansen, and N. L Eatough. "The Chemical Composition of Environmental Tobacco Smoke. III. Identification of Conservative Tracers of Environmental Tobacco Smoke," *Environ. Int.* 15:19–28 (1989).

30. Hammond, S.K., T.J. Smith, S.R. Woskie, B.P. Leaderer, and N. Bettinger. "Markers of Exposure to Diesel Exhaust and Cigarette Smoke in Railroad Workers," *Am. Ind. Hygiene Assoc. J.* 49:516–522 (1988).

31. Benner, C.L., J. M. Bayona, F.M. Caka, H. Tang, L. Lewis, J. Crawford, J. D. Lamb, M.L. Lee, E.A. Lewis, L.D. Hansen, and D.J. Eatough. "The Chemical Composition of Environmental Tobacco Smoke. II. Particle Phase," *Environ. Sci. Technol.* 23:688–699 (1989).

32. "Nicotine," *NIOSH Manual of Analytical Methods*, Vol. 3, 2nd ed., U.S. Dept. of Health Education and Welfare, Publication No. 77-157-C. (1977).

33. Muramatsu, M., S. Umemura, T. Okada, and H. Tomita. "Estimation of Personal Exposure to Tobacco Smoke with a Newly Developed Nicotine Personal Monitor," *Environ. Res.* 35: 218–227 (1984).

34. Oldaker, G.B., III, and F.W. Conrad, Jr. "Estimation of the Effect of Environmental Tobacco Smoke (ETS) on Air Quality within Aircraft Cabins of Commercial Aircraft," *Environ. Sci. Technol.* 21:994–999 (1987).

35. Ogden, M.W., C.W. Nystrom, G.B. Oldaker, III, and F. W. Conrad, Jr. "Evaluation of a Personal Passive Sampling Device for Determining Exposure to Nicotine in Environmental Tobacco Smoke," in *Measurement of Toxic and Related Air Pollutants,* Environmental Protection Agency/Air and Waste Management Association International Symposium (Pittsburgh, PA: Air and Waste Management Association, 1989), 552–558.

36. Schenker, M.B., S.J. Samuels, N.Y. Kado, S.K. Hammond, T.J. Smith, and S.R. Woskie. *Markers of Exposure to Diesel Exhaust* (Cambridge, MA: Health Effects Institute) Research Publication Number 33 (1990).

37. Vaughan, W. M., and S.K. Hammond. "Impact of "Designated Smoking Area" Policy on Nicotine Vapor and Particle Concentrations in a Modern Office Building," *J. Air Waste Manage. Assoc.* 40:1012–1017 (1990).

38. Hammond, S.K. Unpublished results (1987).

39. Miesner, E.A., S.N. Rudnick, F-C. Hu, J.D. Spengler, H. Ozkaynak, L. Preller, and W. Nelson. "Particulate and Nicotine Sampling in Public Facilities and Offices," *J. Air Pollut. Control Assoc.* 39:1577–1582 (1989).

40. Mattson, M.E., G. Boyd, D. Byar, C. Brown, J.F. Callahan, J.W. Cullen, J. Greenblatt, N. Haley, N., S.K. Hammond, J. Lewtas, and W. Reeves. "Passive Smoking on Commercial Airline Flights," *J. Am. Med. Assoc.* 261: 867–872 (1989).

41. Oldaker, G.B., M.W. Ogden, K.C. Maiolo, J.M. Conner, F.W. Conrad, and P.O. DeLuca. "Results from Surveys of Environmental Tobacco Smoke in Restaurants in Winston-Salem, North Carolina," in *Indoor Air '90, Proceedings of the 5th International Conference on Indoor Air Quality and Climate*, Vol. 2. Characteristics of Indoor Air, (Ottawa, Ontario, Canada: Canada Mortgage and Housing Corporation, 1990), 281–285.

42. Badre, R., R. Guillerm, N. Abran, M. Bourdin, and C. Dumas, "Pollution Atmospherique par la Fumee de Tabac," *Ann. Pharm. Fr.* 36:443–452 (1978).

43. Lofroth, G., R. Burton, L. Forehand, S.K. Hammond, R. Seila, R. Zweidinger, and J. Lewtas, "Characterization of Environmental Tobacco Smoke," *Environ. Sci. Technol.* 23: 610–614 (1989).

44. McCurdy, S.A., N.Y. Kado, M.B. Schenker, S.K. Hammond, and N.L. Benowitz. "Measurement of Personal Exposure to Mutagens in Environmental Tobacco Smoke," in *Indoor Air '87, Proceedings of the 4th International Conference on Indoor Air Quality and Climate,* Vol. 2, Environmental Tobacco Smoke, Multicomponent Studies, Radon, Sick Buildings, Odours and Irritants, Hyperreactivities and Allergies, Berlin, West Germany (1987) 91–96.

45. Kado, N.Y., S. McCurdy, S.J. Tesluk, S.K., Hammond, D.P.H. Hsieh, J. Jones, and M.B. Schenker. *Measuring Personal Exposure to the Mutagens in Environmental Tobacco Smoke.* Submitted for publication.

CHAPTER 20

Bioavailability of Benzene Through Inhalation and Skin Absorption

Ronald C. Wester, Howard I. Maibach, Larry Gruenke, and John Craig

TABLE OF CONTENTS

0-87371-606-0/93/$0.00+$.50

© 1993 by Lewis Publishers

I. INTRODUCTION

Benzene, a widely used chemical with an annual production of about 10 billion pounds,[1] is a small compound (mol wt 78.11); it is a clear, colorless, highly flammable liquid that is soluble in 1430 parts of water and also is miscible with alcohol, chloroform, ether, carbon disulfide, carbon tetrachloride, glacial acetic acid, acetone, and oils. Because of its low boiling point (80°C), benzene exists both as liquid and vapor; the industrial worker and the public are exposed to both forms. Human toxicity manifests itself as irritation of mucous membranes, restlessness, convulsions, excitement, and depression. Chronic exposure causes bone marrow depression and aplastic leukemia. Wahlberg and Bomar[2] exposed guinea pigs to benzene topically and saw no mortality, but there was an effect on weight gain, suggesting percutaneous absorption and subsequent toxicity.

Concerns about the toxicity and carcinogenicity of benzene have led to continuing pressure to lower the levels of allowable occupational exposures and have raised the issue of possible health risks to the general population from atmospheric benzene pollution.[1-4] If realistic decisions about minimizing risk are to be made as allowable levels become even lower, questions about the relative importance of various routes of exposure and the role of distribution and metabolism must be reconsidered. This chapter discusses the analytical procedures and bioavailability results related to benzene exposure through the air and through the skin.

II. EXPERIMENTAL

A. Air Samples

1. Materials

Benzene-1,3,5-d, (98% D, KOR Isotopes) was used as the internal standard. The water used was purified using the Milli-Q system (Millipore Corporation). A block heater (Reacti-therm heating module, Pierce Chemical Company) was used as a temperature bath. Headspace samples were taken with a 1.0-mL gas syringe with a push button valve (Precision Sampling Pressure Lok series A-2) and a side port needle.

Breath samples were obtained in 5-L gas sample bags made of a five-layer plastic/metal laminate (Varian, No. 00-996893-02). Bags are repeatedly flushed with nitrogen followed by silica-purified nitrogen before use. Hoses and mouthpieces were supplied by Chesebrough Ponds Inc. (part of the Triflo II deep breathing exerciser No. 5-7173).

Ambient air samples were taken with a battery-operated pump which delivers flows from 5 to 2000 mL/min (Supelco, PAS-3000). Flows were measured with a flow meter (Brooks Instrument Company, size R-2-65-10) which was calibrated by measuring the time required to empty a sample bag containing a known amount of air (delivered with a 3-L calibration syringe from Biotrine Corporation, model 1100) at a constant measured flow.

Ambient air or breath samples were trapped using traps containing 520 mg of silica in the upstream section and 260 mg in the downstream (SKC Inc., No. 26-15 or equivalent).

2. Preparation of Standards

Standard solutions were made in water to avoid contamination or interference from organic solvents. Benzene was added either as a neat liquid or as a solution in ethanol. If added neat, solutions were sonicated for 15 min to ensure dissolution. Containers having a minimum of deadspace volume were used and solutions were handled at 4°C to minimize losses to the headspace volume. Transfers were made when possible through Mininert valves using Hamilton syringes. Solutions of the internal standard contained 10.0 μL of benzene-d$_3$ in 767 mL of water. This was divided into 20-mL sealed ampoules for long-term storage. Working aliquots were kept in 20-mL vials capped with Mininert valves. Calibration samples were prepared from the standard solutions and analyzed with the study samples for every gas chromatography/mass spectrometry (GC/MS) run.

3. Sample Storage

Silica samples from traps were placed in 2.0-mL screw-capped vials (Supelco), sealed with Teflon®-silicone disks (8-mm diameter, 10-mil [0.010-in.] Teflon® layer, Pierce Chemical Company, No. 12708), and stored over activated charcoal.

4. Sample Collection and Preparation

Ambient air samples were obtained either directly, using the sample pump at a calibrated flow rate to draw a sample through a silica trap, or indirectly, using the sampling pump to fill a gas sample bag. Flow rates as high as 500 mL/min through the silica traps were used without sample breakthrough. In air samples trapped by the direct method, upstream and backup sections of the trap were saved in separate 2-mL vials. An aliquot of the benzene-d$_3$ standard solution (generally 150 ng of benzene-d$_3$ in 25 μL of water) was added to each vial. If the air sample was collected by the indirect method, an aliquot of the benzene-d$_3$ solution was added by injecting it into the bag. After allowing at least 1 hr for equilibrium, the contents of the bag were withdrawn at a measured flow rate (generally between 200 and 500 mL/min) through a silica trap and the time required to empty the bag was recorded. Only the upstream section of the silica trap was saved for analysis.

Breath samples were collected by asking the subject to exhale normally through a mouthpiece and hose into a gas sample bag. Breath samples were analyzed in the same manner used for the indirect method of air analysis.

5. Calibration

Calibration samples for the analysis of breath or ambient air were prepared by adding aliquots of the internal standard solution and of the standard benzene solution to empty 2-mL vials. This method of calibration was shown to give results identical to those obtained by spiking sample bags containing silica purified nitrogen, provided that samples were not stored in bags for more than a few hours.

6. GC/MS Analyses

Selected ion records were obtained using a model 2468/2600 gas chromatograph (Infotronics) coupled to a Kratos/AEI MS-12 mass spectrometer which had been modified for selected ion monitoring.[5] Records were obtained at m/z 81 for the internal standard (M+ of benzene-d$_3$) and for the analyte at m/z 78 for natural benzene.

Sample vials were placed in a 70°C heated block for at least 1 hr before analysis. Headspace samples of about 1 mL were generally taken for direct injection onto the GC column.

Several GC column materials were tested during this study. Initially, a 1m × 2 mm glass U-tube column packed with 60/80 mesh Tenax GC (Applied Science) was used. Typically, the column was operated at 105°C for 2 min; then the temperature was increased at 2°/min for 8 min with a helium flow of 27 mL/min. The injector was operated at 110°C and the detector at 250°C. Under these conditions, the retention time for benzene was about 5 min. It was subsequently found that by using an 80/100 mesh Tenax GC column at 115°C and a lower helium flow rate (20 mL/min), significantly sharper peaks could be obtained. However, two difficulties were noted with Tenax columns. A broad interfering peak was often seen just ahead of the benzene peak when samples which had been stored in the gas sample bags were analyzed. Furthermore, sample elution and background levels from the column sometimes fluctuated unpredictably. Switching to a Chemipack C18 column (Alltech Associates, Inc.) eliminated both problems, although extensive conditioning of the column at elevated temperatures was generally required between sample injections to maintain low background ion current at m/z 78. Typically, a 1 m x 2 mm U-tube column operated at 90°C with a helium flow of 25 mL/min gave a retention time of 3 min for benzene. Further discussions of this methodology are contained in Gruenke et al.[7]

7. Percutaneous Absorption

Percutaneous absorption in vivo is usually determined by the indirect method of measuring radioactivity in excreta following topical application of the labeled compound. In human studies, plasma levels of the compounds are generally extremely low following topical application (often below assay detection level) so that it is necessary to use tracer methodology. The labeled compound, usually containing carbon-14 or tritium, is applied to the skin. The total amount of radioactivity excreted in urine (or urine plus feces) is then determined. The amount of radioactivity retained in the body or excreted by some route not assayed (CO_2, sweat) is corrected by determining the amount of radioactivity excreted following parenteral administration. The final amount of radioactivity is then expressed as the percent of the applied dose that was absorbed.[8,9]

In vitro percutaneous absorption utilizes a section of fresh human skin from surgical reduction, firmly held above a continuous flowing water reservoir. The ^{14}C-labeled chemical is placed on the outer surface of skin. The ^{14}C-labeled chemical on the surface water is in contact with the skin for a set time period; then the surface of skin is washed to removed nonabsorbed chemical. That chemical that diffused through the human skin into the underlying reservoir would be considered the portion that was percutaneously absorbed and would become systematically available in humans in vivo. The skin can be cellophane tape stripped twice to determine how much material was surface bound to the stratum corneum (outer layer of skin). The skin is fragile after the in vitro procedure, and two tape strippings pull the stratum corneum away from the rest of the skin. The remaining portion of the inner skin (epidermis and dermis) is digested (Soluene 350 for 5 h at 45°C) and its portion of ^{14}C-labeled chemical is also determined. That amount of radioactivity removed in the surface wash and that which was residual on the apparatus are also determined for total accountability of applied chemical.[10]

8. Human Stratum Corneum Binding

This is an in vitro model that utilizes the partition coefficient of the chemical contaminant in water with that of powdered human stratum corneum. Adult foot calluses are ground with dry ice and freeze dried to form a powder. That portion of the powder that passed through a 40-mesh but not 80-mesh sieve is used. The ^{14}C-labeled chemical as a solution in 1.5-mL water is mixed with 1.5-mg powdered human stratum corneum, and the mixture is allowed to set for 30 min. The mixture is centrifuged, and the proportions of chemical bound to human stratum corneum and that remaining in water are determined by scintillation counting.[11]

III. RESULTS

A. Benzene Bioavailability Through Air

Wester et al.[12] determined benzene levels in human breath and in ambient air, comparing the urban area of San Francisco (SF) and a more remote coastal pristine setting of Stinson Beach, CA (SB) (Tables 1, 2, and 3). Benzene analysis was done by GC/MS. Ambient benzene levels were seven fold higher in SF (2.6 ± 1.3 ppbv, n = 25) than in SB (0.38 ± 0.39 ppbv, n = 21). In SF, benzene in smokers' breath (6.8 ± 3.0 ppbv) was greater than in nonsmokers' breath (2.5 ± 0.8 ppbv) and in smokers' ambient air (3.3 ± 0.8 ppbv). In SB, the same pattern was observed: benzene in smokers' breath was higher than in nonsmokers' breath and ambient air. Benzine in SF nonsmokers' breath was greater than in SB nonsmokers' breath.

Table 1. Summary of Atmospheric Benzene Levels

San Francisco, CA

Average 2.6 ± 1.3[a] ppbv benzene (n = 25)[b,c]
Range 0.8 to 5.2 ppbv
Dates of measurement between November 4 and December 6, 1984, at six different urban locations

Stinson Beach, CA		
Day	n[b]	Atmospheric benzene (ppbv)
1[d]	5	1.02 ± 0.09
2[d]	6	0.23 ± 0.18
3	6	0.13 ± 0.11
4	4	0.16 ± 0.12
Total	21	0.38 ± 0.39[c]

[a]   Standard deviation (SD).
[b]   Number of samples, n.
[c]   Significantly different, $p < 0.001$ (Student's t-test), between benzene in San Francisco and Stinson Beach.
[d]   Days when breath measurements were taken.

Table 2. Benzene Concentrations in Breath of Smokers and Nonsmokers in Respective Ambient Air

Environment concentration	n[a]	Benzene (ppbv)[b]
San Francisco		
Smokers' breath	15	6.8 ± 3.0
Smokers' ambient air	5	3.3 ± 0.8
Nonsmokers' breath	15	2.5 ± 0.8
Nonsmokers' ambient air	5	1.4 ± 0.1
Stinson Beach (day 1)		
Smokers' breath	8	12.1 ± 9.6
Nonsmokers' breath	6	1.8 ± 0.2
Marijuana-only smokers' breath	3	2.5 ± 1.3
Ambient air	5	1.0 ± 0.1
Stinson Beach (day 2)		
Smokers' breath	6	4.8 ± 2.3
Nonsmokers' breath	3	1.3 ± 0.3
Ambient air	6	0.23 ± 0.18

[a]   n = Number of samples.
[b]   Mean ± SD.

Marijuana-only smokers had benzene breath levels between that of smokers and nonsmokers. There was little correlation between benzene in breath and number of cigarettes smoked, or with other benzene exposure such as diet. Of special interest was the finding that benzene in breath of SF nonsmokers (2.5 ± 0.8 ppbv) was

Table 3. Statistical Comparison of Benzene Concentrations

Environment	Comparison A	B	Statistic
San Francisco			
	Smokers' breath	> Nonsmokers' breath	$p = 0.001$[a]
	Smokers' breath	> Ambient air	$p = 0.02$[a]
	Nonsmokers' breath	> Ambient air	$p = 0.02$[a]
Stinson Beach			
	Smokers' breath	> Nonsmokers' breath	$p = 0.02$[a]
	Smoker's breath	> Ambient air	$p = 0.03$[a]
	Nonsmokers' breath	> Ambient air	$p < 0.001$[a]
	Smoker's breath	≯ Marijuana-only smokers' breath	$p = 0.13$[a] (NS)
	Nonsmokers' breath	≯ Marijuana-only smokers' breath	$p = 0.18$[a] (NS)
	Marijuana-only	> Ambient air smokers' breath	$p = 0.03$[a]
Stinson Beach (day 2)	Smokers' breath	> Nonsmokers' breath	$p = 0.04$[a]
	Smokers' breath	>Ambient air	$p < 0.001$[a]
	Nonsmokers' breath	>Ambient air	$p < 0.001$[a]

Note: (NS) not statistically different; > greater than; < less than; ≯ not greater than.

[a] Statistically significant difference (ANOVA and Student's *t*-test).

greater than in nonsmokers' ambient air (1.4 ± 0.1 ppbv). The same was true in SB where benzene in nonsmokers' breath was greater than in ambient air (1.8 ± 0.2 ppbv vs 1.0 ± 0.1 ppbv on day 1, and 1.3 ± 0.3 ppbv vs 0.23 ± 0.18 ppbv on day 2). A person walking the streets can expect a 100-fold range in exposure to benzene (0.1–10 ppbv) depending on environment and personal habit (i.e., smoking).

B. Benzene Bioavailability Through Skin

Interactions of benzene and human skin were examined with an in vitro model that utilizes the partition coefficient of the chemical between water and powdered human stratum corneum. [^{14}C]Benzene in water solution (217 μg/mL) exposed to powdered stratum corneum in an open system for 30 min had $16.6 \pm 1.4\%$ of the dose partition into the skin.[13] This value is compared to those for PCBs and p-nitroaniline under similar conditions (Figure 1).

In vitro percutaneous absorption of benzene was determined by placing [^{14}C]benzene (217 μg/mL water) over human cadaver skin in an open diffusion system for 30 min. Only 0.15% applied dose was absorbed (systemic and skin) and only 2.7% could be explained.[13] Evaporation into the atmosphere was the major route of benzene dispersion (Table 4 and Figure 2).

Jakobsen et al.[14] determined the concentration of benzene in blood during and after percutaneous exposure to anesthetized guinea pigs. The absorption of benzene

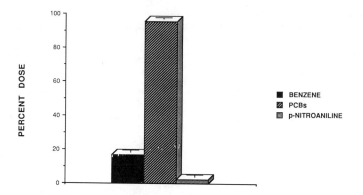

FIGURE 1.   The percent benzene dose which partitions from water into pow-
            dered human stratum corneum. Value for benzene compared to
            PCBs and p-nitroaniline.

produced a maximum concentration in blood at 0.5 hr (1.6 μg/mL), and this
remained elevated at 6 hr (1.3 μg/mL). This suggests that the skin absorption was
rapid and that absorption and elimination were constant during the exposure
period.

Maibach and Anjo[15] determined the percutaneous absorption of benzene through
the skin of the rhesus monkey using [14C]benzene and quantitating the labeled
material and its metabolite in urine (Table 5). They showed that percutaneous
absorption of benzene is approximately 0.2% of applied dose. The application of
benzene in multiple doses increases absorption. Wester et al.[13] also showed that
the absorption of benzene increases with multiple applications (Figure 3).

Maibach[16] used benzene containing [14C]benzene to study the absorption in
man. Total radioactivity was estimated in urine. The results indicate that only a
small amount of benzene was absorbed through human skin. The mean value was
0.065% of the applied dose (range was 0.035–0.1%). Because of the low level of
skin penetration, many of the counts were near background level. The amount of
benzene in urine was quantified as the percentage of applied dose. It must be
assumed that the majority of the dose evaporated and that this represents the
amount of material that actually entered the body.

Franz[17] studied the percutaneous absorption of benzene in the monkey, minipig,
and man using both in vitro and in vivo techniques. In vitro studies were conducted using special diffusion chambers in which the skin was mounted as a
barrier between two half cells. [14C]Benzene was applied to the epidermis and its
rate of movement through the skin was assessed by serial sampling of the solution
bathing the dermis. Percutaneous absorption in vivo was measured by applying
[14C]benzene to the skin and monitoring the secretion of radioactivity in the urine.

Percutaneous absorption in vivo following the application of a thin layer (5μL/
cm^2) of benzene was found to average less than 0.2% in all species studied. Under
the conditions of these experiments, the remainder of the applied material quickly
volatilized (less than 30 sec) and was lost into the atmosphere. Total absorption

Table 4. In Vitro Percutaneous Absorption and Skin Distribution of Benzene in Water Solution for 30-min Exposure

Parameter	% Dose
Percutaneous absorption (systemic)	0.045 ± 0.037
Surface bound/stratum corneum	0.036 ± 0.005
Epidermis and dermis	0.065 ± 0.057
Total (skin/systemic)	0.15
Skin wash/residual	2.51 ± 0.94
Apparatus wash	0.006 ± 0.005
Total (accountability)	2.67

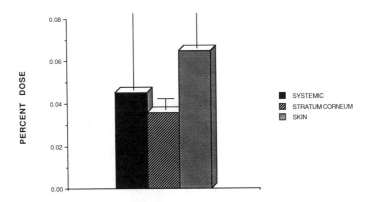

FIGURE 2.   The percent dose of benzene in stratum corneum (upper skin layer), skin (epidermis and dermis) and systematically absorbed (receptor fluid) from water after 30-min exposure.

Table 5. Skin Absorption of Benzene in Rhesus Monkey

Exposure	% Dose absorbed
Benzene (0.36%) in solvent	0.08 ± 0.03
Cutaneous administration of benzene	0.17 ± 0.14
Multiple exposures to full-strength benzene	0.85 ± 0.81
Multiple exposures to benzene (0.35% in rubber solvent)	0.43 ± 0.26
Damaged skin exposed to full-strength benzene	0.91 ± 0.63
Palmar exposure to benzene in rubber solvent	0.65 ± 0.48

was 0.14% in the monkey, 0.09% in the minipig, and 0.7% in man. Peak excretion of radioactivity occurred in the first 2 hr and decreased rapidly thereafter. Figure 4 and Table 6 present the urinary recovery of [14C]benzene and rate of urinary excretion following topical application in monkeys and man. In vitro studies demonstrated rapid penetration of benzene through the monkey, minipig, and human skin. When the same dose used in vivo (5 μL/cm²) was used in vitro, the peak rate of absorption occurred at 15–20 min and total absorption was similar to

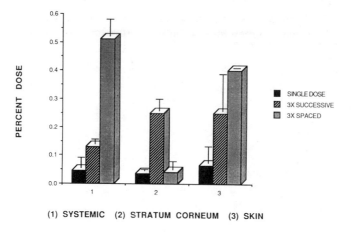

(1) SYSTEMIC   (2) STRATUM CORNEUM   (3) SKIN

FIGURE 3.   Benzene skin absorption increases with multiple dose application.

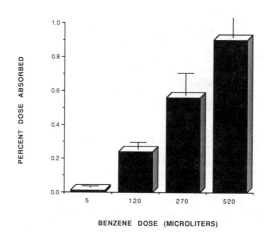

BENZENE DOSE (MICROLITERS)

FIGURE 4.   Benzene skin absorption increases as the dose on skin is increased.

that measured in vivo. Absorption was 0.19% in the monkey, 0.23% in the minipig, and 0.2% in man.

Benzene absorption was found to be a function of its contact time with the skin. Application of progressively larger doses that persisted on the skin for up to 3 hr resulted in 10 to 100 times greater absorption. Total absorption was directly related to the length of time benzene remained on the skin. A summary of in vivo and in vitro data of benzene absorption is shown in Table 6. The human data of Franz confirms the earlier work of Maibach. Franz concluded that a major factor controlling percutaneous absorption of benzene is its contact time with the skin. Under ideal conditions in which the contact time is short due to its inherent volatility, less than 0.2% of the applied dose will be absorbed, or ca. 0.1 $\mu L/cm^2$.

Table 6. Benzene Absorption (% of Dose)

	In vivo Man			In vivo Animal		In vitro	
	Maibach	Franz		Maibach	Franz	Franz	
Palm	0.13	0.14	Monkey	0.2	0.14	Monkey	0.19
Arm	0.7	0.5	Pig		0.9	Pig	0.23
						Man	0.10

It must be noted that the radioactivity assay is a direct measurement of the material that is absorbed through the skin, becomes systematically available, and is excreted in the urine. This is, therefore, a measure of total radioactivity and does not differentiate between metabolized benzene absorbed through the skin and the metabolites that are formed subsequently.

IV. DISCUSSION

Although the prevailing flow of air from the ocean gives the San Francisco Bay Area a unique climate, we found that atmospheric benzene levels in the city were typical of those reported[5] for other urban areas (1–10 ppbv). The lowest levels were observed near the coast, and the highest levels were observed inland near sources of heavy auto traffic. Atmospheric benzene may thus determine some of the amount of benzene that gets into the body. This was consistent with our findings that benzene levels in the breath of nonsmokers was higher in the urban environment than in the pristine environment.

Benzene in the breath of nonsmokers for both urban and pristine environments was higher than the respective ambient air concentrations suggesting an additional source of benzene exposure other than environmental air. Other possible sources of benzene exposure include cigarette smoke, solvents (including gasoline), and diet (especially the number of eggs eaten). The pharmacokinetics of benzene have been described as a three-compartment model. It may be that benzene is accumulated in the body in the home or workplace, and the higher concentration is seen in body elimination of this concentrated benzene. However, no correlation could be found between breath levels and exposure to any of these sources as indicated on questionnaires filled out by each subject. Benzene in indoor air may be an important source of exposure. However, it is not known whether the magnitude of exposure to benzene from indoor air is sufficient to explain the elevation in breath levels we have observed in our subjects. Thus, we also suggest the possibility of an in vivo source of benzene production.

The volatility of benzene demonstrates that ambient air and smoking are primary routes of exposure. However, skin is a route of exposure, and benzene is absorbed through skin. Exposure can be long term, despite the volatility of benzene. Any type of clothing or "protection" glove will act as a transdermal delivery system. This will provide a continual delivery of benzene through skin and into the systemic circulation.[9]

V. SUMMARY

Benzene exists in the ambient air that we breathe, the exposure depending on environmental location and personal habits such as smoking. In the industrial setting where benzene is used, the amount in the vapor phase would be greatly increased and the body burden of benzene would increase appropriately. Benzene as a liquid when contacting skin would be absorbed and become systemic. The literature indications are that skin absorption would be rapid, and dependent on dose and time of exposure. Benzene is a toxic substance and is able to enter the human body from a variety of routes (ambient air, skin, smoking, food). All of these routes of exposure must therefore be considered when health hazard assessments are made.[18]

REFERENCES

1. Lee, S. D., K. M. Dourson, D. Murkerjee, J. F. Stars, and J. Kawecki. "Assessment of Benzene Health Effects in Ambient Water," in *Advances in Modern Environmental Toxicology, Vol. 4, Carcinogenicity and Toxicity of Benzene,* M. A. Mohlman, Ed. (Princeton, NJ: Scientific Publishers, 1983), pp. 91–125.
2. Wahlberg, J. E., and A. Boman. "Comparative Percutaneous Toxicity of Ten Industrial Solvents in the Guinea Pig," *Scand. J. Work Environ. Health* 5:345–351 (1979).
3. Snyder, R. "The Benzene Problem in Historical Perspective," *Fundam. Appl. Toxicol.* 4:692–699 (1984).
4. Zenz, C. "Benzene-Attempts to Establish a Lower Exposure Standard in the United States," *Scand. J. Work, Environ. Health* 4:103–113 (1978).
5. Brief, R. S., J. Lynch, T. I. Bernath, and R. A. Scala. "Benzene in the Workplace," *Am. Ind. Hyg. Assoc. J.* 41:616–623 (1980).
6. Mehlman, M. A. *Advances in Modern Environmental Toxicology, Vol. 4, Carcinogenicity and Toxicity of Benzene,* (Princeton, NJ: Princeton Scientific Publishers, Inc., 1983), pp. 111–116.
7. Gruenke, L. D., J. C. Craig, R. C. Wester, and H. I. Maibach. "Quantitative Analysis of Benzene by Selected Ion Monitoring/Gas Chromatography/Mass Spectrometry," *J. Anal. Toxicol.* 10:225–232 (1986).
8. Wester, R. C., and H. I. Maibach. "In Vivo Methods for Percutaneous Absorption Measurements," in *Percutaneous Absorption,* R. Bronaugh and H. Maibach, Eds. (New York: Marcel Dekker, 1989), pp. 215–220.
9. Wester, R. C., and H. I. Maibach. "Cutaneous Pharmacokinetics: 10 Steps to Percutaneous Absorption," *Drug Metab. Rev.* 14:169–205 (1983).
10. Wester, R. C., and H. I. Maibach. "In vitro Testing of Topical Pharmaceutical Formulations," in *Percutaneous Absorption,* 2nd Ed., R. Bronaugh and H.I. Maibach, Eds. (New York: Marcel Dekker, 1989), pp. 653–659.
11. Wester, R. C., and H. I. Maibach. "Dermatopharmacokinetics in Clinical Dermatology," *Semin. Dermatol.* 2:81–84 (1983).
12. Wester, R. C., H. I. Maibach, L. D. Gruenke, and J. C. Craig. "Benzene Levels in Ambient Air and Breath of Smokers and Nonsmokers in Urban and Pristine Environments," *J. Toxicol. Environ. Health* 18:567–587 (1986).

13. Wester, R. C., M. Mobayen, and H. I. Maibach. "In vivo and in vitro Absorption and Binding to Powdered Human Stratum Corneum as Methods to Evaluate Skin Absorption of Environmental Chemical Contaminants from Ground and Surface Water," *J. Toxicol. Environ. Health* 21:367–374 (1987).

14. Jakobson, I., J. E. Wahlberg, B. Holmberg, and G. Johansson. "Uptake via the Blood and Elimination of Ten Organic Solvents Following Epicutaneous Exposure in the Anesthetized Guinea Pig," *Toxicol. Appl. Pharmacol.* 63:181–187 (1981).

15. Maibach, H. I., and D. M. Anjo. "Percutaneous Penetration of Benzene and Benzene Contained in Solvents Used in the Rubber Industry," *Arch. Environ. Health* 36:256–260 (1981).

16. Maibach, H. I. Unpublished data (1980).

17. Franz, T. J. "Percutaneous Absorption of Benzene," in *Advances in Modern Environmental Toxicology, Vol. 6, Applied Toxicology of Petroleum Hydrocarbons,* H. N. MacFarland, Ed. (Princeton, NJ: Princeton Publishers, 1984), pp. 61–70.

18. Wester, R. C., and H. I. Maibach. "Benzene Bioavailability Through Air and Skin," in *Occupational and Industrial Dermatology,* 2nd ed., H. I. Maibach and G. Gellin, Eds. (Chicago: Year Book Medical Publishers, 1987), pp. 258–264.

Index

INDEX